Environmental Sustainability Using Green Technologies

Environmental Sustainability Using Green Technologies

Edited by V. Sivasubramanian

CRC Press
Taylor & Francis Group
Boca Raton London New York

CRC Press is an imprint of the
Taylor & Francis Group, an **Informa** business

CRC Press
Taylor & Francis Group
6000 Broken Sound Parkway NW, Suite 300
Boca Raton, FL 33487-2742

First issued in paperback 2020

ISBN 13: 978-0-367-57457-4 (pbk)
ISBN 13: 978-1-4987-5305-0 (hbk)

This book contains information obtained from authentic and highly regarded sources. Reasonable efforts have been made to publish reliable data and information, but the author and publisher cannot assume responsibility for the validity of all materials or the consequences of their use. The authors and publishers have attempted to trace the copyright holders of all material reproduced in this publication and apologize to copyright holders if permission to publish in this form has not been obtained. If any copyright material has not been acknowledged please write and let us know so we may rectify in any future reprint.

Visit the Taylor & Francis Web site at
http://www.taylorandfrancis.com

and the CRC Press Web site at
http://www.crcpress.com

Contents

List of Figures

List of Tables

Preface

Sustainability is balancing environmental protection and social responsibility with a healthy economy over time. Green engineering and chemistry play an important role in creating the options that enable sustainability by developing chemicals, processes, value-added products, and systems that are environmentally preferable, more energy and resource-efficient, and often more cost-effective. Achieving sustainable economic growth will require changes in industrial processes, in the type and amount of resources used, and in the products that are manufactured. The world must move to a more energy-efficient society by using resources more responsibly and organizing industrial processes in ways that minimize and reuse wastes. Sustainable chemical processes are dealing with the problems of waste, inefficiencies in processes, and adopting the life-cycle approach to process and product development. Sustainability is an action-oriented variant of sustainable development. It is needed to develop the ability to make a choice that respects the relationship between the three E's: economy, ecology, and equality. If all the three E's were incorporated in the national goals of countries, then it would be possible to develop a sustainable society.

There are tremendous challenges when man moves toward sustainable development but fails to use the necessary tools to overcome the challenges. It is our duty to adapt the nature and not vice versa. The flora and fauna of ecosystems are being seriously threatened by the impact of human activities. Environmental sustainability is about making responsible decisions that will reduce our negative business impact on the environment. It is not simply about reducing the amount of waste produced or using less energy, but is concerned with developing processes that will lead to businesses becoming completely sustainable in the future. Hopefully, it would reduce the burden on the earth.

This book, *Environmental Sustainability Using Green Technologies*, aims to provide comprehensive coverage on emerging dimensions on environmental sustainability. It contains 16 articles by eminent experts on different aspects of the subject. Articles on wastewater treatment technologies, nanomaterials in environmental applications, green synthesis of ecofriendly nanoparticles, the role of phytoremediation in maintaining environmental sustainability, algal biosorption of heavy metals, mass production of microalgae for industrial applications, an integrated biological system for the treatment of sulfate rich wastewater, anaerobic digestion of pharmaceutical effluent, treatment of textile dye using bioaccumulation techniques, production of biosurfactants and their applications in bioremediation, biodegradable polymers, microbial fuel cell (MFC) technology, biodiesel from nonedible oil using a packed-bed membrane reactor, production of ecofriendly biodiesel from marine sources, pretreatment techniques for the enhancement of biogas production, and a review of source apportionment of air pollutants by receptor models. The book also covers information on newer biotechnological tools and techniques required for sustainable development. Various aspects covering efficient bioremediation methods have been included to project their importance in the 21st century.

This book can serve not only as an excellent reference material, but also as a practical guide for professors, research scholars, industrialists, biotechnologists, and workers in the applied field of environmental engineering.

Velmurugan Sivasubramanian

Editor

Velmurugan Sivasubramanian is an associate professor and former head of the Department of Chemical Engineering at National Institute of Technology Calicut (NITC), in Kozhikode, India. He worked at Tamil Nadu Industrial Explosives Ltd. as a senior plant engineer for five years after his graduation. Since 2001, he has been teaching fluidization engineering, environmental engineering, wastewater engineering, energy management, safety and hazards control, principles of chemical engineering, operations management, total quality management for master of technology degree in chemical engineering, petroleum refining and petrochemicals, pharmaceutical technology and environmental science and engineering; chemical reaction engineering, mechanical operations, mass transfer, chemical technology, environmental studies, safety in chemical process industries, energy management, process economics and industrial management, petrochemicals, downstream processing/bioseparations and operations research for a bachelor of technology in chemical engineering and biotechnology; and bioreactor engineering, fluidization engineering, biological wastewater treatment, environmental biotechnology, safety management in process industries, fire engineering and explosion control, and fire modeling and dynamics for the doctoral program.

He received the Dr. Radhakrishnan Gold Medal Award, Jawaharlal Nehru Gold Medal Award, and National Citizenship Gold Medal Award from Global Economic Progress and Research Association (Thiruvannamalai, Tamil Nadu) and the Universal Achievers Gold Medal Award from the Universal Achievers Foundation (Salem, Tamil Nadu) for his outstanding individual achievement in education and contribution to education and national development.

He is involved in the development and commercial exploitation of wastewater treatment using novel bioreactors such the inverse fluidized bed reactors (IFBRs), photobioreactors (PBRs) for the remediation of dyes, pharmaceuticals, and heavy metals. His team has installed a novel magnetic biocomposite for the treatment of dye effluent. Recently, his team successfully developed a 4 m^3 floating drum biodigester for the production of biogas, which is used as a fuel in the NITC hostel mess.

Sivasubramanian has published several research papers in international and national journals and conferences in the area of environmental and chemical engineering for the treatment of effluents from various industries. His team has earned five best paper awards at national and international conferences. He is editor for five peer-reviewed journals and he also serves as reviewer for more than 50 journals.

Contributors

Anant Achary
Department of Biotechnology
Kamaraj College of Engineering
  and Technology
Viruthunagar, India

N. Anu
Department of Civil Engineering
UKF College of Engineering
  and Technology
Kollam, Kerala, India

S. Bhuvaneshwari
Chemical Engineering Department
National Institute of Technology
  Calicut
Kozhikode, Kerala, India

P. Sankar Ganesh
Department of Biological Sciences
Birla Institute of Technology
  and Science Pilani
Hyderabad, India

George Philomin Doss Geoprincy
Centre for Nanoscience
  and Technology
Anna University
Chennai, India

Bhargavi Gunturu
Department of Biotechnology
Anna University
Chennai, India

Baskar Gurunathan
Department of Chemical
  Engineering
Anna University
Chennai, India

M. Jerold
Department of Chemical
  Engineering
National Institute of Technology
  Calicut
Kerala, India

Patchaimuthu Kalainila
Department of Biotechnology
Anna University
Chennai, India

J. Kanimozhi
Department of Chemical
  Engineering
National Institute of Technology
  Calicut
Kozhikode, Kerala, India

Ramachandran Kasirajan
Department of Applied Science
    and Technology
Anna University
Chennai, India

Uma Krishnakumar
Chemical Engineering
    Department
National Institute of Technology
    Calicut
Kozhikode, Kerala, India

B.G. Prakash Kumar
Chemical Engineering
    Department
Birla Institute of Technology
    and Science Pilani
Dubai, United Arab Emirates

**Dharmendira Kumar
Mahendradas**
Department of Applied Science
    and Technology
Anna University
Chennai, India

Vijay Mani
Department of Chemical
    Engineering
Annamalai University
Chidambaram, India

Krishnan Manigandan
Department of Mechanical
    Engineering
Podhigai College of Engineering
    and Technology
Tirupattur, India

Shyamasundari Murugan
Department of Biotechnology
Anna University
Chennai, India

Saravanan Panneerselvam
Department of Chemical
    Engineering
Anna University
Chennai, India

Kuravappullam Vedaiyan Radha
Department of Chemical
    Engineering
Anna University
Chennai, India

S. Ramachandran
Biotechnology Department
Birla Institute of Technology
    and Science Pilani
Dubai, United Arab Emirates

Sahadevan Renganathan
Department of Biotechnology
Anna University
Chennai, India

Arockiasamy Santhiagu
School of Biotechnology
National Institute of Technology
    Calicut
Kozhikode, Kerala, India

Arunachalam Bose Sathya
Department of Chemical
    Engineering
National Institute of Technology
    Calicut
Kozhikode, Kerala, India

N. Selvaraju
Department of Chemical
    Engineering
National Institute of Technology
    Calicut
Kozhikode, Kerala, India

Raja Sivashankar
Department of Chemical
    Engineering
National Institute of Technology
    Calicut
Kozhikode, Kerala, India

Velmurugan Sivasubramanian
Department of Chemical
    Engineering
National Institute of Technology
    Calicut
Kozhikode, Kerala, India

A. Surendhar
Department of Chemical
    Engineering
National Institute of Technology
    Calicut
Kozhikode, Kerala, India

Pandiyan Thyriyalakshmi
Department of Chemical
    Engineering
Anna University
Chennai, India

K. Vasantharaj
Department of Chemical
    Engineering
National Institute of Technology
    Calicut
Kozhikode, Kerala, India

Theresa Veeranan
Department of Biotechnology
Anna University
Chennai, India

Veerapadaran Velumani
Department of Mechanical
    Engineering
Podhigai College of Engineering
    and Technology
Tirupattur, India

D. Vidhyeswari
Chemical Engineering
    Department
National Institute of Technology
    Calicut
Kozhikode, Kerala, India

C. Vigneshwaran
Department of Chemical
    Engineering
National Institute of Technology
    Calicut
Kozhikode, Kerala, India

Balakrishnan Vinothraj
Department of Mechanical
    Engineering
Podhigai College of Engineering
    and Technology
Tirupattur, India

List of Abbreviations and Symbols

CHAPTER 1

AOP	advanced oxidation processes
BOD	biochemical oxygen demand
COD	chemical oxygen demand
GAC	granular activated carbon
IPS	intermediate pumping station
MF	microfiltration
MFC	microbial fuel cell
MLD	million liters per day
MPS	main pumping station
NF	nanofiltration
ORR	oxygen reduction reaction
PVC	polyvinyl chloride
RBC	rotating biological contactor
RO	reverse osmosis
STP	sewage treatment plant
TOC	total organic carbon
TOD	total oxygen demand
TTC-DHA	triphenyl tetrazoliumchloride dehydrogenase activity
UF	ultrafiltration
US	ultrasound
UV	ultraviolet
VOC	volatile organic carbon
VUV	vacuum ultraviolet
WAO	wet air oxidation

CHAPTER 2

Abbreviations

A	swept area
AC	alternating current
Au	gold
Bar	barometric pressure
BPPOdp	brominated poly(2,6-diphenyl-1,4-phenylene oxide)
BTX	benzene, toluene, and xylene
CdSe	cadmium selinide
CFM	cubic feet per minute
CNT	carbon nanotube
CO_2	carbon dioxide
Cp	power coefficient of wind turbine
CVD	chemical vapor deposition
DC	direct current
DD	dichlorodiphenyltrichloroethane
DP	deposition-precipitation
E_k	rate of kinetic energy change
HRSG	heat recovery and steam generation
HRTEM	high resolution transmission electron microscope
HTS	high temperature super conducting
ICMS	integrated cell-material sciences
IGCC	integrated gasification combined cycle
IPCC	Intergovernmental Panel on Climate Change
MTR	membrane technology research
MWCNTs	multiwalled carbon nanotubes
nm min^{-1}	nanometer per minute
NMP	N-methyl-2-pyrrolidone
NOM	natural organic matter
OTM	oxygen transport membrane
P	wind power
p-n junction	positive channel, negative channel junction
P3HT	poly(3-hexyl)thiophenezinc oxide
PAI	polyamide–imide
PCBs	polychlorinated biphenyls
PDMS	polydimethylsiloxane
PECVD	plasma-enhanced CVD
PEG-DME	polyethylene glycol-dimethylether

PEI–PAI   polyethylene imine–poly (amide–imide)
PIM-1   polymer with intrinsic microporosity
PM   permenant generators
PMSGs   permanent-magnet synchronous generators
POG   point of generation
Ppm   parts per million
Ppmv   parts per million by volume
PSiNWs   porous silicon nanowires
PU   poly(urethane)
PV   photovoltaic
PVAc   poly(vinyl acetate)
QDs   quantum dots
RO   reverse osmosis
SAED   selected area electron diffraction pattern
SAXS   small angle X-ray scattering
SBM   Santa Barbara Amorphous
SiNWs   silica nanowires
SM   steam generators
STP   standard temperature and pressure
SWNTs   single-wall carbon nanotubes
TCE   trichloroethylene
TEM   transmission electron microscopy
V   wind speed
V_1 and V_2   upwind and downwind
VALA   vacuum ultraviolet radiation-assisted laser ablation
VLS   vapor–liquid–solid
VOCs   volatile organic compounds
VUV   vacuum ultraviolet radiation
WEC   World Energy Council
WGSMR   water gas shift membrane reactor
WTGs   wind turbine generating systems
XANES   x-ray absorption near edge structure
XPS   x-ray photon spectroscopy
XRD   x-ray diffraction

Symbols

(Ω)   electrical resistance ohm
ρ   air density
Ωcm   ohm centimeter

%	percentage
°C	degree centigrade
°F	degree Fahrenheit
Al_2O_3	aluminum trioxide
Au	gold
BiMo	bismuth molybdenum
C_2H_3CN	vinyl cyanide
C_2H_4	ethene
C_2H_4O	acetaldehyde
C_3H_4O	cycloproponone
C_3H_6	propene
Ce	cerium
CeO_2	cerium oxide
cm s^{-1}	centimeter per second
Co	cobalt
Co_3O_4	cobalt oxide
Cu_2O	copper oxide
eV	electron volt
Fe_2O_3	ferric oxide
GW	gigawatt or one billion watts
H	hour
H_2SO_4	sulfuric acid
HCl	hydrochloric acid
in.	inch
kW	kilowatt
LiCl	lithium chloride
m	mass
m	meter
m	momentum
mg/L	milligram per liter
MgO	magnesium oxide
MJ/kg	mega joule per kilogram
mL	milliliter
MW	milliwatt
N	niobium
N_2	nitrogen
NH_3	ammonia

Nm	nanometer
O$_2$	oxygen
Si	silica
SiCl$_4$	silicon tetrachloride
SiH$_4$	silane
Te	tellurium
TiO$_2$	titanium dioxide
V	vanadium
wt%	weight percentage
ZnO	zinc oxide
ZnS	zinc sulphide
μm	micrometer

CHAPTER 3

°C	degree centigrade
AgNO$_3$	silver nitrate
AgNPs	silver nanoparticles
CdS	cadmium sulfide
cm	centimeter
CTAB	cetyltrimethyl ammonium bromide
CuNPs	copper nanoparticles
EGFR	epidermal growth factor receptor
FCC	face cubic center
FT-IR	Fourier transform infrared spectroscopy
KHz	kilohertz
mm	millimeter
MRI	magnetic resonance imaging
N$_2$H$_4$	hydrazine
Nd:YAG	neodymium yttrium aluminum garnet
nm	nanometer
O/W	oil in water
PWD	pulsed wire discharge
Rf	retention factor
SEM	scanning electron microscope
SPR	surface plasmon res
TLC	thin layer chromatography
UV	ultraviolet spectroscopy

W/O	water in oil
XRD	x-ray diffraction
ZnO	zinc oxide

CHAPTER 4

^{137}**Cs**	isotope of cesium
^{234,238}**U**	isotope of uranium
^{239}**Pu**	isotope of plutonium
^{90}**Sr**	isotope of strontium
AATC	American Association of Textile Chemists and Colourists
Ag	silver
Au	gold
BBR	Brilliant Blue R
BOD	biochemical oxygen demand
Cd	cadmium
COD	chemical oxygen demand
CV	crystal violet
CWs	constructed wetlands
Hg	mercury
MG	malachite green
Mn	manganese
Mo	molybdenum
NADH	nicotinamide adenine dinucleotide–hydrogen (reduced)
Ni	nickel
O$_2$	oxygen
Pb	lead
RR	Ramezol Red
TOC	total organic carbon
TOD	total oxygen demand
UK	United Kingdom
UV	ultraviolet
WHO	World Health Organization
Zn	zinc

CHAPTER 5

Cr(VI)	hexavalent chromium
mg/g	milli gram per gram
NMR	nuclear magnetic resonance
$\mathbf{q_{max}}$	maximum uptake

CHAPTER 7

AMD	acid mine drainage
APS	adenosine-5′-phosphosulfate
ATP	adenosine triphosphate
BSR	biological sulfate reduction
COD	chemical oxygen demand
DO	dissolved oxygen
FBR	fluidized bed reactor
GEMS	Global Environmental Monitoring System
HRT	hydraulic retention time
$\mathbf{k_d}$	decay coefficient
$\mathbf{K_s}$	saturation constant
NRB	nitrate reducing bacteria
ppm	parts per million
SOB	sulfide oxidizing bacteria
SRB	sulfate reducing bacteria
UASB	upflow sludge blanket reactor
UNEP	United Nations Environment Programme
US DHEW	United States Department of Health, Education and Welfare
US EPA	United States Environmental Protection Agency
WAS	waste activated sludge
$\mathbf{\mu_m}$	maximum specific growth rate

CHAPTER 8

% w/v	percent of weight of solution in the total volume of solution
AD	anaerobic digestion
ADBA	Anaerobic Digestion and Bioresource Association
BOD	biological oxygen demand
C	Celsius
COD	chemical oxygen demand
mg/L	milligram per liter
OLR	organic loading rate
pH	hydrogen ion concentration
SRT	sludge retention time
TDS	total dissolved solids
UASB	upflow anaerobic sludge blanket reactor

CHAPTER 9

AFM	atomic force microscopy
CaCl$_2$	calcium chloride
CR	Congo red
CV	crystal violet
FTIR	Fourier transform infrared spectroscopy
g	gram
g/L	gram per liter
H$_2$O$_2$	hydrogen peroxide
H$_2$SO$_4$	sulfuric acid
HCl	hydrochloric acid
HNO$_3$	nitric acid
MG	malachite green
mg	milligram
mg/g	milligram per gram
mg/L	milligram per liter
mL	milliliter
mm	millimeter
MnO	manganese oxide
Na$_2$CO$_3$	sodium carbonate
NaCl	sodium chloride
NaNO$_2$	sodium nitrite
NaOH	sodium hydroxide
O$_3$	ozone
rRNA	ribosomal ribonucleic acid
RSM	response surface methodology
SEM	scanning electron microscopy
TEM	transmission electron microscopy
UV	ultraviolet

CHAPTER 10

Symbols

%	percent
°C	degree Celsius
±	plus-minus

Notations

CMC	critical micelle concentration
CSTR	continuous stirred tank reactor
g/L, gL⁻¹	grams per liters
HPLC	high performance liquid chromatography
K	potassium
Mg	magnesium
mg/L	milligram per liter
mg/mL	milligram per milliliter
mN/m	millinewton per meter
Na	sodium
rpm	revolution per minute
SBR	sequencing batch reactor
sp.	species
vvm	volume of air per volume of liquid per minute

CHAPTER 11

DMT	dimethylterephthalate
EPS	extracellular polysaccharides
MMC	mixed microbial cultures
PBSA	polybutylene succinate-co-butylene adipate
PBSu	polybutylene succinate
PCL	polycaprolactone
PESu	polyethylene succinate
PHA	polyhydroxyalkanoate
PHB	polyhydroxybutyrate
PHBV	polyhydroxybutyrate-valerate
PLA	polylactic acid
ROP	ring opening polymerization
TBOT	tetrabutylorthotitanate
Tg	glass transition temperature in °C
Tm	melting temperature in °C

CHAPTER 12

Abbreviations

AEM	anode exchange membrane
BOD	biological oxygen demand

BPM	bipolar membrane
CEM	cation exchange membrane
COD	chemical oxygen demand
IEM	ion exchange membrane
MFC	microbial fuel cell
MFM	microfiltration membrane
PEM	proton exchange membrane
SEAs	separator electrode assemblies
SEM	scanning electron microscope
UFM	ultrafiltration membrane

Symbols

ε_E	energy efficiency
CE	coulombic (or current) efficiency
E	electrode potential
E 0	standard potential
E_{cell}	cell potential
E_{emf}	cell electromotive force
E_{emf}°	standard cell electromotive force
F	Faraday's constant
I	current
M	molecular weight of oxygen
n	number of electrons exchanged per mole
P	power
Ps	power density
R	resistance
R	universal gas constant
R_{ext}	fixed external resistor
T	absolute temperature
V	actual cell voltage
V_{Anode}	liquid volume of the anode chamber
ΔCOD	change in chemical oxygen demand (COD) over time
ΔG_r	Gibbs free energy for the specific conditions
ΔG_r°	standard Gibbs energy change
ΔHM	heat of combustion of organic substrate

CHAPTER 13

ASTM	American Society for Testing and Materials
FAME	fatty acid methyl ester

GC analysis   gas chromatography analysis
KOH   potassium hydroxide
MF   microfiltration
MG   monoglycerides
NaOH   sodium hydroxide
NF   nanofiltration
RO   reverse osmosis
SVO   straight vegetable oil
TG   triglycerides
TiO$_2$/Al$_2$O$_3$ titanium dioxide/aluminum oxide
UF   ultrafiltration

CHAPTER 14

ASTM   American Society for Testing and Materials
B100   biodiesel 100 percentage
C. sativa   *Camelina sativa*
CFPP   cold filter plugging point
CP   cloud point
DG   diglyceride
EN   European Standard
FAME   fatty acid methyl ester
FAP   fatty acid profile
FFA   free fatty acid
J. curcas   *Jatropha curcas*
ME   methyl ester
MG   monoglyceride
P. pinnata   *Pongamia pinnata*
sp. (or) spp.   species
TAN   total acid number
TG   triglyceride
US   United States

CHAPTER 16

Symbols
$V_{e_{jj}}$ effective variance least square
ε_{ij} error term consisting of all the variability in X_{ij}
χ^2 reduced chi-square
p pth source profile

σ_{ij}	uncertainty of species j measured on sample i
B	emission rate from source
D	number of days
D_k	dispersion factor
E	error of the multivariate model
e_{ij}	error of the multivariate model for the species j measured on sample i
F	source profile (source signature)
f_{kj}	concentration of species j in factor profile k
G	contribution of source (factor)
g_{ik}	relative contribution of factor k to sample i
Q	emission rate of pollutants, units/s
R	receptor concentration matrix
R^2	regression coefficient
X	concentration of species
x_{ij}	concentration of species j measured on sample i

Abbreviations

CMB	chemical mass balance
CMB-fmincon	chemical mass balance model using "fmincon" solver in MATLAB
CMB-GA	chemical mass balance model using "Genetic algorithm" in MATLAB
CMB8.2	chemical mass balance software version 8.2
EC	elemental carbon
FA	factor analysis
ISC	industrial source complexes
MATLAB	Matrix Laboratory
MDL	minimum detectable limit
MLRA	multilinear regression analysis
NCMB	nested chemical mass balance
OC	organic carbon
PAH	polycyclic aromatic hydrocarbon
PCA	principal component analysis
PM	particulate matter
$\mathbf{PM_{0.1}}$	particulate matters with aerodynamic size less than 0.1 μm
$\mathbf{PM_{10}}$	particulate matters with aerodynamic size less than 10 μm
$\mathbf{PM_{2.5}}$	particulate matters with aerodynamic size less than 2.5 μm

PMF	positive matrix factorization
SA	source apportionment
SVD	singular value decomposition
TSP	total suspended particulates
USEPA	United States Environmental Protection Agency

CHAPTER 1

Wastewater Treatment Technologies and Recent Developments

Raja Sivashankar, Arunachalam Bose Sathya, and Velmurugan Sivasubramanian

CONTENTS

Abstract

This chapter illustrates different water treatment technologies. Water treatment technologies can be organized into three general areas: physical, chemical and biological methods. The purpose of the chapter is to supply an outline of wastewater treatment, purification roles, and technology progression to help readers understand the consanguinity of the various technologies to the overall strategies employed in water treatment applications. Since the beginning of the Industrial Revolution, the production of wastes has increased rapidly and unceasingly as a consequence of human actions. Untreated wastewater generally contains high levels of organic material, numerous pathogenic microorganisms, and foods and toxic compounds, and thus entails environmental and health risk issues. And, accordingly, wastewater must be conveyed away from its generation sources and treated appropriately before final disposition. So, a contemporary tool for assessing wastewater and an improved technology is indispensable to prevent the ecosystem from being contaminated with waste and to treat effluent wastewater before discharge. This chapter provides an interdisciplinary assortment of topics pertaining to wastewater treatment and this collective information will be supportive to all budding researchers of different disciplines connected to wastewater.

1.1 SCOPE OF THIS CHAPTER

This chapter describes some water treatment technologies and recent approaches, and aims to provide information on developing wastewater treatment technologies. The chapter discusses the overall outlook on wastewater and treatment methods for wastewater recycling. Further, the chapter reviews technological details and applications of new technologies over the established treatment methods.

1.2 WATER RESOURCES AND CRISIS

The earth contains an enormous amount of water, roughly 1.4 billion km^3, but 97.5% of these resources are briny and unfit for consumption, including the waters of the oceans that comprise 96.5% of the total quantity and the reserves of highly mineralized underground brines that constitute about 1% (13 million km^3). Thus freshwater constitutes only 2.5% of the entire reservoir, and two-thirds of this is in frozen glaciers, mainly in the Artic and Antarctic regions. There is only 10.5 million km^3 of liquid freshwater on the globe. Most of this water is underground and currently difficult to access. Only 135,000 km^3 of freshwater, or 0.01% of the world's aquatic resources, is readily accessible. Figure 1.1 shows the distribution of water present on and in the earth. This includes water in rivers, lakes, marshes, and the soil, as well as water in the atmosphere,

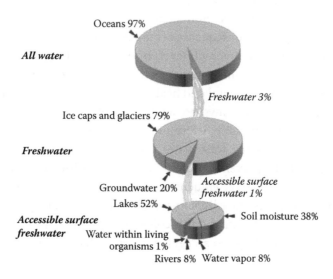

FIGURE 1.1   Distribution of water present on and in the earth.

which is in a state of continuous circulation (Shiklomanov, 1993; Gleick, 1996).

The total annual utilizable water resources in India is 1086 km³, which is only 4% of world's water resources. The total annual utilizable resources of surface water and groundwater are 690 and 396 km³, respectively (National Commission for Integrated Water Resource Development, 1999). Consequent to rapid growth in population and increasing water demand, stress on water resources in India is increasing, and per capita water availability reduces every day. In India, the per capita surface water availability in the years 1991 and 2001 were 2300 m³ (6.3 m³/day) and 1980 m³ (5.7 m³/day), respectively, and these are projected to reduce to 1401 and 1191 m³ by the years 2025 and 2050, respectively. Total water requirement of the country in 2050 is estimated to be 1450 km³, which is higher than the current availability of 1086 km³ (Kumar et al., 2005). Human impacts on freshwater systems are substantial in most populated parts of the world. Overextraction of freshwater, mainly for agriculture, has led to significant degradation of rivers, lakes, and aquifers. Liberation of water for the environment through substitution with wastewater has been promoted as a means of reducing anthropogenic impacts (Hamilton et al., 2005a,b).

1.3 SOURCES OF WASTEWATER

Any material, solid, liquid, or gas that is unwanted or unvalued, and discarded or discharged by its owner is called a waste. The wastes that are dissolved in water during or after processes are considered wastewater. Wastewater typically includes various pollutants, depending on what it is used for. Wastewater can be classified into following major categories based on the source:

- Domestic or sanitary wastewater—This comes from residential sources, including toilets, sinks, bathing, and laundry. It can have body wastes containing intestinal disease organisms.

- Industrial wastewater—This is discharged by manufacturing processes and commercial enterprises. Process wastewater can contain rinse waters, including such things as residual acids, plating metals, and toxic chemicals.

- Agricultural runoff—This carries fertilizer (e.g., phosphate) and pesticides, which constitute a major cause of eutrophication of lakes.

- Storm runoff—When in highly urbanized areas, it may cause significant pollution.

Usually wastewater, treated or untreated, that is discharged to a natural water body is referred to as receiving water. Receiving water is treated to remove pollutants. Figure 1.2 illustrates the imperative contaminants of wastewater.

1.4 STATUS OF WASTEWATER GENERATION AND TREATMENT IN INDIA

As per the latest estimate, out of 22,900 MLD (million liters per day) of wastewater generated in India, only about 5900 MLD (26%) is treated before letting out; the rest (i.e., 17,000 MLD) is disposed of untreated. Twenty-seven cities have only primary treatment facilities and forty-nine have primary and secondary treatment facilities. The level of treatment available in cities with existing treatment plants varies from 2.5% to 89% of the sewage generated. The total wastewater generated by the 299 class I cities is 16,662 MLD, approximately 81% of the water supplied. The state of Maharashtra alone contributes about 23%, and the Ganga river basin contributes about 31% of the waste generated. Only 74% of the total

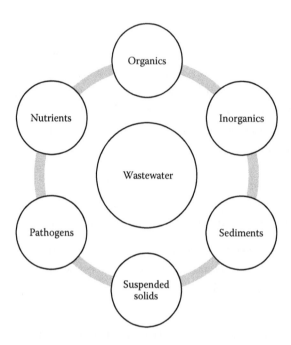

FIGURE 1.2   Essential contaminants of wastewater.

wastewater generated is collected. Out of 299 class I cities, 160 cities have sewerage coverage for more than 75% of the population and 92 cities have between 50% and 75% of population coverage. On the whole, 70% of the population of the class I cities are serviced by a sewerage facility. The type of sewerage system is either open or closed or piped (Dhote et al., 2012).

1.4.1 Status of Sewage Generated in Metropolitan Urban Centers in India

Discharge of untreated sewage in water sources, both surface waters and groundwaters, is the main water polluting source in India. Out of about 38,000 MLD of sewage generated, treatment capacity exists for only around 12,000 MLD. Thus, there is a big gap between generation and treatment of sewer water in India. Even the existing treatment capacity is not effectively utilized due to performance and maintenance problems. Cognitive process and maintenance of existing plants and sewage pumping stations is not satisfactory, as nearly 39% plants are not conforming to the general standards prescribed under the environment (protection) rules for discharge into streams as per the Central Pollution Control Board survey report. In a number of cities, the existing treatment capacity remains underutilized while a lot of sewage is discharged without treatment in the same metropolis. The auxiliary power backup facility is required at all intermediate and main pumping stations of all sewage treatment plants.

1.5 NEED FOR WASTEWATER TREATMENT

Wastewater treatment involves the breakdown, either physicochemically or by using microorganisms (biological treatment), of complex organic compounds in the wastewater into simpler compounds that are stable and nuisance free. The adverse environmental impact of allowing untreated wastewater to be discharged in groundwater or surface water bodies or lands include the following:

- The decomposition of the organic materials contained in wastewater can lead to the production of large quantities of malodorous gases.

- Untreated wastewater contains a large amount of organic matter that will consume dissolved oxygen for satisfying the biochemical oxygen demand (BOD) of wastewater when discharged into a river stream and thus deplete the dissolved oxygen of the stream, thereby causing fish kills and other undesirable effects.

- Wastewater may also carry nutrients, which can induce the development of aquatic plants and algal blooms, hence contributing to eutrophication of lakes and flows.

- Untreated wastewater usually contains numerous pathogenic, or disease-causing microorganisms and toxic compounds, that live in the human intestinal tract or may be present in certain industrial waste. These may contaminate the ground or the water body, where such sewage is disposed.

For these reasons the treatment and disposal of wastewater is necessary (Topare et al., 2011; Dhote et al., 2012). Wastewater treatment is a process to recover and purify the water, confiscate some or all of the contaminants, making it fit for reuse or discharge back to the environment. Appropriately treating wastewater guarantees that a tolerable overall water quality is sustained. In many parts of the world, health problems and diseases have often been caused by discharging untreated or inadequately treated wastewater. Such discharges are called water pollution, and result in the spreading of disease, fish kills, and destruction of other forms of aquatic life (Cushman and Carlson, 2004). The pollution of water has a serious impact on all living creatures, and can negatively affect the use of water for drinking, household needs, recreation, fishing, transportation, and commerce. Basic wastewater treatment facilities reduce organic and suspended solids to limit pollution to the environment (Corcoran et al., 2010). Advancement in needs and technology has necessitated the evolving of treatment processes that remove dissolved matter and toxic substances. At present, the progression of scientific knowledge and moral awareness has led to a reduction of discharges through pollution prevention and recycling, with the noble goal of zero discharge of pollutants. Treatment technology includes physical, biological, and chemical methods. Residual substances removed or created by treatment processes must be dealt with and reused or disposed of in a safe way. The purified water is discharged to surface water or groundwater. Residuals, called sludges or biosolids, may be reused by carefully controlled composting or land application. Sometimes they are incinerated (Viessman, 2003).

1.6 WASTEWATER TREATMENT TECHNOLOGIES

In general, wastewater management is composed of the following phases: wastewater collection, wastewater removal, wastewater treatment, and wastewater discharge (Babu et al., 2013). This chapter emphasizes wastewater treatments and methods. An understanding of the nature of the wastewater is fundamental for the design of appropriate wastewater treatment plants and the selection of effective treatment technologies (Topare et al., 2011). Wastewater originates predominantly from water usage by residences, commercial, and industrial establishments together with groundwater, surface water, and stormwater. Wastewater quality may be defined by its physical, chemical, and biological characteristics. Physical parameters include color, odor, temperature, and turbidity. Insoluble contents such as solids, oil, and grease also fall into this category. Solids may be further subdivided into suspended and dissolved solids as well as organic (volatile) and inorganic (fixed) fractions. Chemical parameters associated with the organic content of wastewater include biochemical oxygen demand (BOD), chemical oxygen demand (COD), total organic carbon (TOC), and total oxygen demand (TOD). Inorganic chemical parameters include salinity, hardness, pH, acidity, and alkalinity, as well as concentrations of ionized metals (such as iron and manganese) and anionic entities (such as chlorides, sulfates, sulfides, nitrates, and phosphates). Biological parameters include coliforms, fecal coliforms, specific pathogens, and viruses. Both constituents and concentrations vary with time and local conditions. Wastewater is classified as strong, medium, or weak, depending on its contaminant concentration. The effects of the discharge of untreated wastewater into the environment are diversified and depend on the types and concentrations of pollutants (Tchobanoglous and Burton, 1991).

Physical, chemical, and biological methods are used to remove contaminants from wastewater. In order to achieve different levels of contaminant removal, individual wastewater treatment procedures are combined into a variety of systems, classified as primary, secondary, and tertiary wastewater treatment. Different methods of water treatment and its unit operations are specified in Table 1.1. More rigorous treatment of wastewater includes the removal of specific contaminants as well as the removal and control of nutrients. Natural systems are also used for the treatment of wastewater in land-based applications. Sludge resulting from wastewater treatment operations is treated by various methods in order to reduce its water and organic content, and to make it suitable for final disposal and reuse. Wastewater treatment methods are broadly classifiable into physical, chemical, and biological processes (Tchobanoglous and Burton 1991; Water Environment Federation and the American Society of Civil Engineers/Environmental and Water Resources Institute, 1992; D.H.F Liu and Liptak, 1999; Economic and Social Commission for Western Asia, 2003).

1.6.1 Physical Treatment

Physical methods include procedures where no gross chemical or biological changes are carried out and strictly physical phenomena are used to improve or treat the sewer water. Among the treatment methods used were physical unit operations stands in which physical forces are applied to eliminate contaminants.

TABLE 1.1   Water Treatment Methods and Their Unit Operations

Physical Unit Operations	Chemical Unit Operations	Biological Unit Operations
Screening	Chemical precipitation	Activated sludge process
Comminution	Adsorption	Aerated lagoon
Flow equalization	Disinfection	Trickling filters
Sedimentation	Dechlorination	Rotating biological contactors
Flotation	Ion exchange	Pond stabilization digestion
Granular-medium filtration		Anaerobic digestion

1.6.1.1 Screening

Screening, one of the oldest treatment techniques, eliminates gross pollutants from the waste stream to protect downstream equipment from damage, avoid interference with plant operations, and prevent objectionable floating material from entering the primary settling tanks. The material retained from the manual or mechanical cleaning of bars, racks, and screens is referred to as "screenings," and is either disposed of by burial or incineration, or returned into the waste flow after grinding. This water treatment process traps and removes floating matter (such as paper and wood) by using automated mechanically raked bar screens.

Comminutors are used to macerate large floating material in the wastewater. They are installed where the handling of screenings would be impractical, generally between the grit chamber and the primary settling tanks. Comminutors are mainly used for reducing odors, flies, and unpleasantness. A comminutor may have either rotating or oscillating cutters. Rotating-cutter comminutors either engage a separate stationary screen alongside the cutters, or a combined screen and cutter rotating together.

1.6.1.2 Flow Equalization

Flow equalization is a technique used to develop the effectiveness of secondary and advanced wastewater treatment processes by leveling out operation parameters such as flow, pollutant levels, and temperature over a period of time. Variations are damp until a near-constant flow rate is achieved, minimizing the downstream effects of these parameters. Flow equalization may be applied at a number of locations within a wastewater treatment plant, for example, near the head end of the treatment works, prior to discharge into a water body and prior to advanced waste treatment operations.

1.6.1.3 Sedimentation

Sedimentation, an elementary and extensively used unit operation in wastewater treatment, involves the gravitational settling of heavy particles suspended in a mixture. This process is used for the removal of grit, particulate matter in the primary settling basin, biological floc in the activated sludge settling basin, and chemical flow when the chemical coagulation process is used. Sedimentation takes place in a settling tank, also referred to as a clarifier. Four types of settling occur, depending on particle concentration: discrete, flocculent, hindered, and compression. It is common for more than one type of settling to occur during a sedimentation operation.

1.6.1.4 Flotation

Flotation is a unit operation used to remove solid or liquid particles from a liquid phase by introducing a fine gas, usually air bubbles. The gas bubbles either adhere to the liquid or are trapped in the particle structure of the suspended solids, raising the buoyant force of the combined particle and gas bubbles. Particles that have a higher density than the liquid can thus be made to rise. In wastewater treatment, flotation is used mainly to remove suspended matter and to concentrate biological sludge. The chief advantage of flotation over sedimentation is that very small or light particles can be removed more completely and in a shorter time. Once the particles have floated to the surface, they can be skimmed out. Flotation, as currently practiced in municipal wastewater treatment, uses air exclusively as the floating agent. Furthermore, several chemical additives can be presented to enhance the removal procedure.

1.6.1.5 Granular Medium Filtration

The filtration of effluents from wastewater treatment processes is a relatively recent practice but has come to be widely used for the supplemental removal of suspended solids from wastewater effluents of biological and chemical treatment processes, in accession to the removal of chemically precipitated phosphorus. The complete filtration operation comprises two phases: filtration, and cleaning or backwashing. The wastewater to be filtered is passed through a filter bed consisting of granular material (sand, anthracite, and/or garnet), with or without added chemicals. Within the filter bed, suspended solids contained in the wastewater are removed by means of a complex process involving one or more removal mechanisms such as straining, interception, impaction, sedimentation, flocculation, and adsorption. The phenomena that happen during the filtration phase are essentially the same for all types of filters used for wastewater filtration. The cleaning/backwashing phase differs, depending on whether the filter operation is continuous or semicontinuous.

1.6.2 Chemical Treatment

Chemical processes used in effluent treatment are thought to handle some sort of change by means of chemical reactions. They are invariably utilized in combination with physical and biological operations. Broadly speaking, chemical processes possess an intrinsic drawback compared to physical operations in that they are additive processes.

1.6.2.1 Chemical Precipitation

Chemical coagulation of raw wastewater prior to sedimentation endorse the flocculation of finely divided solids into more readily settleable flocs, thereby enhancing the efficiency of suspended solid, BOD, and phosphorus removal as compared to plain sedimentation without coagulation. The precipitation of ions (heavy metals) and colloids (organic and inorganic) are mostly held in solution by electrical charges. By the increase of ions with opposite charges, these colloids can be destabilized and coagulation can be achieved by chemical or electrical methods. Chemical coagulants that are commonly used in wastewater treatment include alum $(Al_2(SO_4)_3 \cdot 18\ H_2O)$, ferric chloride $(FeCl_3 \cdot 6H_2O)$, ferric sulfate $(Fe_2(SO_4)_3)$, ferrous sulfate $(FeSO_4 \cdot 7H_2O)$, and lime $(Ca(OH)_2)$. Organic polyelectrolytes are sometimes used as flocculation aids. The advantages of coagulation include greater removal efficiency, the feasibility of using high overflow rates, and more uniform execution. On the other hand, coagulation results in a larger mass of primary sludge that is often more difficult to thicken and dewater. It also requires higher operational costs and demands greater care on the role of the manipulator.

1.6.2.2 Adsorption Using Carbon

Adsorption is a widely used method for the handling of industrial wastewater containing color, heavy metals, and other inorganic and organic impurities. The advantages of the adsorption process are simplicity in operation, inexpensive (compared to other separation processes), and no sludge formation. Activated carbon has been widely used as adsorbent in various industrial applications such as refinement of water in sewage facilities and filtration of air in toxicity treating factories (Bansal et al., 1998). Numerous efforts have been made to alter the chemical and physical properties of activated carbon for improving its adsorbing ability (Domingo-Garcia et al., 2000). Applications of sorption technologies in water treatment have sprung up from the original and still widely employed, use granular activated carbon (GAC) as a broad spectrum sorbent for removal of assortments of organic chemicals and heavy metals over wide concentration ranges (Isabel and Wayne, 2005).

1.6.2.3 Disinfection

Disinfection refers to the selective destruction of disease-causing microorganisms. This is a significant process in wastewater treatment due to the nature of wastewater, which harbors a bit of human enteric organisms that are affiliated with various waterborne diseases. The usual method of disinfection comprises the following: (a) physical agents such as warmth and light; (b) mechanical means, such as screening, sedimentation, and filtration; (c) radiation, mainly gamma rays; and (d) chemical agents, including chlorine and its compounds, bromine, iodine, ozone, phenol and phenolic compounds, alcohols, heavy metals, dyes, soaps and synthetic detergents, quaternary ammonium compounds, hydrogen peroxide, and various alkalis and acids. The most frequent chemical disinfectants are the oxidizing chemicals and of these chlorine is the most extensively applied. Disinfectants act through one or more number of mechanisms like damaging the cell wall, altering cell permeability, altering the colloidal nature of the protoplasm, and inhibiting enzyme activity. In applying disinfecting agents, various elements need to be considered: contact time, concentration and type of chemical agent, intensity and nature of physical agent, temperature, number of organisms, and nature of suspending liquid (Qasim, 1999).

1.6.2.4 Dechlorination
Dechlorination is the removal of free and total combined chlorine residue from chlorinated wastewater effluent before its reuse or discharge to receiving waters. Chlorine compounds react with many organic compounds in the effluent to produce undesired toxic compounds that cause long-term adverse impacts on the water environment and potentially toxic effects on aquatic microorganisms. Dechlorination may be brought about by the economic consumption of activated carbon or by the addition of a reducing agent, such as sulfur dioxide (SO_2), sodium sulfite (Na_2SO_3), or sodium metabisulfite ($Na_2S_2O_5$).

1.6.2.5 Ion Exchange
Ion exchange has been applied extensively to remove hardness, iron, and manganese salts in drinking water provisions. It has likewise been applied selectively to remove specific impurities and to recover valuable trace metals like chromium, nickel, copper, lead, and cadmium from industrial waste emissions. The procedure takes advantage of the ability of certain natural and synthetic materials to switch one of their ions. When water is too hard, it is difficult to clean and often leaves a gray residue. Calcium and magnesium are common ions that lead to water hardness. To soften the water, positively charged sodium ions are presented in the form of molten sodium chloride salt, or seawater. Hard calcium and magnesium ions exchange places with sodium ions, and free sodium ions are simply released into the water. Nevertheless, after softening a large quantity of

water, the softening solution may fill with extra calcium and magnesium ions, requiring the solution to be recharged with sodium ions.

1.6.3 Biological Treatment

The purpose of biological treatment is to remove dissolved organic contaminants (e.g., BOD) from a soluble form into suspended matter in the course of cell biomass, which can then be later removed by particle-separation processes (e.g., sedimentation). Biological unit processes are applied to win over the finely divided and dissolved organic matter in wastewater into flocculent settleable organic and inorganic solids. In these procedures, microorganisms, especially bacteria, convert the colloidal and dissolved carbonaceous organic matter into various gases and into cell tissue, which is then removed in sedimentation tanks. Biological processes are commonly applied in conjunction with physical and chemical processes, with the primary aim of cutting down the organic content (measured as BOD, TOC, or COD) and nutrient content (notably nitrogen and phosphorus) of wastewater. Biological processes used for wastewater treatment may be sorted out under five major headings: aerobic processes, anoxic processes, anaerobic processes, combined processes, and pond processes.

1.6.3.1 Activated-Sludge Process

The activated-sludge process is an aerobic, continuous-flow system containing a mass of activated microorganisms that are capable of stabilizing organic matter. The procedure consists of delivering clarified wastewater, after primary settling, into an aeration basin where it is combined with an active mass of microorganisms, primarily bacteria and protozoa, that aerobically degrade organic matter into carbon dioxide, urine, new cellphones, and other end products. After a specific retention time, the mixed liquor passes into the secondary clarifier, where the sludge is allowed to go down and a clarified effluent is produced for discharge. The process recycles a portion of the settled sludge back to the aeration basin to keep the required activated sludge concentration.

1.6.3.2 Aerated Lagoons

The microbiology involved in aerated lagoons is similar to that of the activated-sludge process. Nevertheless, disputes arise because the large surface area of a lagoon may cause more temperature effects than are normally encountered in conventional activated-sludge processes. Wastewater is oxygenated by surface, turbine, or diffused aeration. The turbulence created by aeration is applied to hold the contents of the basin in suspension. Most of these solids must be withdrawn in a settling basin before final effluent discharge.

1.6.3.3 Trickling Filters

The trickling filter is the most commonly encountered aerobic attached-growth biological treatment process employed for the removal of organic matter from wastewater. It consists of a bed of highly permeable medium to which organisms are attached, forming a biological slime layer, and through which wastewater is percolating. The organic material present in the wastewater is degraded by adsorption onto the biological slime layer. In the outer component of that stratum, organic material is taken down by aerobic microbeings. As the microorganisms grow, the thickness of the slime layer increases and the oxygen is used up before it has gone through the entire depth of the slime layer. An anaerobic environment is thus established near the surface of the filter medium. As the slime layer increases in thickness, the organic matter is degraded before it progresses to the microorganisms near the surface of the medium. After going through the filter, the treated liquid is accumulated in an under drain system, in concert with any biological solids that have become detached from the medium. The collected liquid then goes into a settling tank where the solids are separated from the treated effluent.

1.6.3.4 Rotating Biological Contactors

A rotating biological contractor is an attached growth biological process that consists of one or more basins in which large closely spaced circular disks mounted on horizontal shafts rotate slowly through wastewater. The disks, which are constructed of high-density polystyrene or polyvinyl chloride (PVC), rotate the bacteria exposing them alternately to wastewater, from which they adsorb organic matter and absorb oxygen. A final clarifier is needed to remove sloughed solids.

1.6.3.5 Stabilization Ponds

A stabilization pond is a relatively shallow body of wastewater contained in an earthen basin, using a completely mixed biological process without solids return. Mixing may be either natural (wind, heat, or fermentation) or induced (mechanical or diffused aeration). Stabilization ponds are usually classified, based on the nature of the biological activity that takes place in them, as aerobic, anaerobic, or aerobic anaerobic. The bacterial population oxidizes organic matter, making ammonia, carbon dioxide, sulfates, water, and other end products, which are afterward utilized by algae during daylight to get oxygen. Bacteria then use this supplemental oxygen and the oxygen provided by wind action to break down the remaining organic matter. Wastewater retention time ranges between 30 and 120 days. This is a treatment procedure that is very commonly found in rural areas because of its low construction and operating costs (Mara, 1992).

1.6.3.6 Anaerobic Digestion

Anaerobic digestion involves the biological conversion of organic and inorganic matter in the absence of molecular oxygen to end products including methane and carbon dioxide. A consortium of anaerobic organisms is used to degrade the organic sludges and wastes in three steps: hydrolysis of high molecular mass compounds, acidogenesis, and methanogenesis. The procedure takes place in an airtight reactor. Sludge is introduced continuously or intermittently, and retained in the reactor for varying periods of time. Following withdrawal from the reactor, the stabilized sludge is reduced in organic and pathogen content, and is nonputrescible. Anaerobic digesters are commonly used for the treatment of sludge and wastewaters with high organic content (Van Haandel and Lettinga, 1994). Table 1.2 provides the details of different unit operations used for excretion of several contaminants in wastewater.

1.7 DEGREES OF WASTEWATER TREATMENT PROCESSES

Stages of treatment are sometimes indicated by the use and in order of increasing treatment level, such as elementary, secondary, and tertiary treatment (Tchobanoglous and Burton, 1991; Tchobanoglous et al., 2003; Aziz Mojiri, 2014). Typical phases in the handling of wastewater are shown in Figure 1.3.

1.7.1 Preliminary Treatment

Preliminary treatment prepares wastewater effluent for further handling by cutting or getting rid of nonfavorable wastewater characteristics that might otherwise impede operation or excessively increase maintenance of downstream processes and equipment. These characteristics include large solids and rags, abrasive grit, odors, and, in certain cases, unacceptably high peak hydraulic or organic loadings. Preliminary treatment processes consist of physical unit operations, namely, screening and comminution,

TABLE 1.2   Unit Operations for Excretions of Various Contaminants of Wastewater

Function	Unit Operations
Removal of biodegradable material in wastewaters	• Activated sludge treatment • Trickling filters • Rotating biological contactors (RBCs) • Aerated lagoons • Anaerobic lagoons • Facultative lagoons • Anaerobic treatment
Removal of suspended solid in wastewaters	• Screening and comminution • Grit removal • Sedimentation • Flotation • Filtration and centrifugation • Coagulation/sedimentation
Removal of VOCs in wastewaters	• Adsorption • Absorption • Air stripping • Condensation • Freezing • Incineration • Combustion
Removal of nitrogen in wastewaters	• Biological nitrogen utilization in activated sludge process • Biological nitrification and denitrification • Air stripping of ammonia • Chlorination • Adsorption
Removal of phosphorus in wastewaters	• Biological phosphorus utilization in activated sludge process • Chemical additions (metal salts or polymers) • Lime addition • Biological/chemical treatment
Removal of organic priority pollutants in wastewaters	• Aerobic biological treatment • Anaerobic biological treatment • Biological treatment/activated • Carbon adsorption • Chemical oxidation • Adsorption
Removal of heavy metals and dissolved inorganic solids in wastewaters	• Chemical oxidation/reduction • Precipitation • Ion exchange • Reverse osmosis • Ultrafiltration

for the removal of debris and rags, grit removal for the elimination of coarse suspended matter, and flotation for the removal of oil and grease. Other preliminary treatment operations include flow equalization, septage handling, and odor control methods.

1.7.2 Primary Treatment

Primary treatment involves the partial removal of suspended solids and organic matter from the wastewater by means of physical operations such as screening and sedimentation. Preaeration or mechanical flocculation with chemical additions can be employed to enhance primary treatment. Primary treatment acts as a precursor for secondary treatment. It is targeted primarily at developing a liquid effluent suitable for downstream biological treatment and separating out solids as a ooze that can be convenient and economically treated before ultimate disposal. The effluent from primary treatment contains a good deal of organic matter and is characterized by a relatively high BOD.

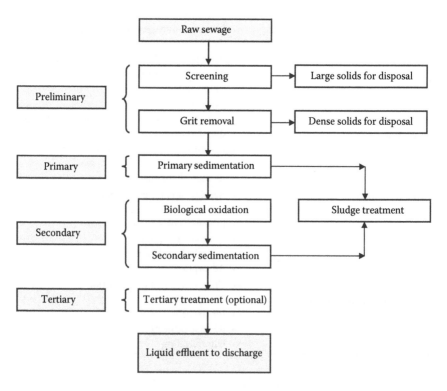

FIGURE 1.3   Typical stages in the conventional treatment of sewage/wastewater.

1.7.3 Secondary Treatment

The role of secondary treatment is the removal of soluble and colloidal organics and suspended solids that have escaped the principal treatment. This is typically done through biological processes, namely, treatment by activated sludge, fixed-film reactors, or lagoon systems and sedimentation.

1.7.4 Tertiary/Advanced Wastewater Treatment

Tertiary treatment goes beyond the level of conventional secondary treatment to remove significant amounts of nitrogen, phosphorus, heavy metals, biodegradable organics, bacteria, and viruses. In addition to biological nutrient removal processes, unit operations frequently used for this purpose include chemical coagulation, flocculation, and sedimentation, followed by filtration and activated carbon. Less frequently used processes include ion exchange and reverse osmosis for specific ion removal or for dissolved solids reduction.

1.8 RECENT DEVELOPMENTS IN WASTEWATER TREATMENT

1.8.1 Membrane Technology for Wastewater Treatment

The recent technological developments in the progression of wastewater treatment include membrane technology, which has emerged as a substantial innovation for treatment and rehabilitation for water reuse, considerably as a primary method in the betterment and expansion of wastewater treatment plants. In the past few years, there has been a rapid increment in the volume of wastewater that is treated with membranes to intensely high quality criteria, frequently for reuse purpose. At present, most municipal wastewater treatment amenities are using membrane technologies, since they are reliable. Likewise, the membranes may be an option because they help with the elimination of contaminants that other technologies cannot. In application of effluent treatment, membranes are presently functioning as a tertiary advanced treatment for the excretion of dissolved species, organic compounds, phosphorus, nitrogen, colloidal solids, suspended solids, and human pathogens, including bacteria, protozoan cysts, and viruses. Table 1.3 presents the membrane technologies for wastewater treatment.

The main technological issue with the utilization of membranes for wastewater treatment is the high possibility for fouling. Membrane fouling can be caused by colloids, soluble organic compounds, and microorganisms that are usually not properly removed by conventional pretreatment methods. Membrane fouling also increases feed pressure and entails frequent membrane cleaning. This decreases the efficiency and membrane life. Other technical obstructions may comprise the intricacy and expenditure of the concentrate (residuals) clearance from high-pressure membranes. However, membrane technologies can be advantageous for a landlocked situation such as urban, agricultural, and industrial reuse; groundwater recharge and salinity barriers; and augmentation of potable water supplies meeting very low effluent nutrient limits. Membrane processes by means of different types of membranes are gaining popularity for the production of drinking water from seawater, brackish water, wastewater, surface water, and groundwater. Table 1.4 shows the details of the pore size of different membranes used for production of drinking water.

1.8.2 Ultrafiltration

Ultrafiltration (UF) is a pressure-driven process that eliminates emulsified oils, metal hydroxides, colloids, emulsions, dispersed material, suspended solids, and other large molecular weight materials from water. UF membranes are identified by their molecular weight cutoff. The major opportunities for UF involve clarification of solutions containing suspended solids, removal of viruses and bacteria, or high concentrations of macromolecules. These include oil–water separation, fruit juice clarification, milk and whey production and processing, automotive electrocoat paint filtration,

TABLE 1.3 Membrane Technologies for Wastewater Treatment

Treatment Methods Using Membranes	Membrane Technologies
Membrane bioreactors	Microfiltration (MF) or ultrafiltration (UF) membranes immersed in aeration tanks (vacuum system) are implemented in external pressuredriven membrane units, as a replacement for secondary clarifiers and tertiary polishing filters.
Low pressure membranes	MF or UF membranes, either as a pressure system or an immersed system, offering a higher stage of suspended solids removal following secondary clarification. UF membranes are effective for virus removal.
High pressure membranes	Nanofiltration or reverse osmosis pressure systems for treatment and production of high quality product water suitable for indirect potable reuse and high purity industrial process water. Besides, recent research has demonstrated that microconstituents, such as pharmaceuticals and personal care products, can be removed by high pressure membranes.

TABLE 1.4   Types of Membrane Used for Treatment of Water

Membrane Type	Pore Size (nm)	Pressure (Bar)	Product
Reverse osmosis	<0.6	30–70	Pure water
Nanofiltration	0.6–5	10–40	Pure water and low molecular solutes
Ultrafiltration	5–50	0.5–10	Pure water, low molecular solutes, and macromolecules
Microfiltration	50–500	0.5–2	Colloids and all the above

purification of pharmaceuticals, polyvinyl alcohol and indigo recovery, potable water production, and secondary or tertiary wastewater reuse.

1.8.3 Reverse Osmosis

The membrane type with the smallest pores is reverse osmosis (RO), which entails a reversal of the osmotic process of a solution in order to drive water away from dissolved molecules. RO depends on ionic diffusion to achieve separation. A frequent application of RO is seawater and brackish water desalination.

1.8.4 Nanofiltration

Nanofiltration (NF) functions similarly to RO, however, it is commonly targeted to eliminate only divalent and larger ions. Monovalent ions such as sodium and chloride will pass through an NF membrane resulting in desalting of the process flow, thus affecting numerous uses of NF. NF membranes are used for hardness removal (in place of water softeners), pesticide elimination, and color reduction.

1.8.5 Microfiltration

Microfiltration (MF) includes considerable applications in simple dead-end filtration for water filtration, sterile bottling of fruit juices and wine, and aseptic uses in the pharmaceutical industry. The general applications of MF are clarification of whole cell broths and purification processes in which macromolecules must get separated from other large molecules such as proteins and/or cell debris (Bernardes et al., 2014; Pabby et al., 2015).

1.8.6 Nanotechnology for Wastewater Treatment

Nanomaterials are usually distinct materials smaller than 100 nm in at least one dimension. At this scale, materials often possess novel size-dependent properties different from their large equivalent, and therefore are highly explored for applications in water and wastewater treatment. Some of these applications utilize the smoothly scalable size-dependent properties of nanomaterials, which relate to the high specific surface area, such as fast dissolution, high responsiveness, and strong sorption. Others take advantage of their discontinuous properties, such as superparamagnetism, localized surface plasmon resonance, and quantum confinement effect. Nanotechnology for water remediation will play a crucial function in water security and therefore the food security of the macrocosm. The applications of nanotechnology in the cleanup of contaminated water could be (Smith, 2006)

- Nanoscale filtration techniques
- The adsorption of pollutants on nanoparticles
- The breakdown of contaminants by nanoparticle catalysts

The application of nanotechnology in wastewater treatment includes adsorption, membrane processes, photocatalysis, disinfection and microbial control, and sensing and monitoring (Qu et al., 2013). Table 1.5 shows the potential application of nanotechnology in wastewater treatment.

1.8.7 Fuel Cell Technology for Wastewater Treatment

Microbial fuel cells (MFCs) are a promising technology for sustainable wastewater treatment. An MFC is a device designed for the purpose of electricity generation in the process of wastewater handling. Therefore, it is a supreme resolution for a sustainable, nonrenewable source of energy. MFCs can be defined as electrochemical devices that change the chemical energy contained in organic matter into electricity by means of catalytic (metabolic) activity of the living microorganisms (Kim et al., 2002; H. Liu et al., 2004; Logan and Regan, 2006). MFCs have achieved a healthier importance in the last few decades due to their potential to generate energy, either as electricity or through hydrogen production, from renewable sources such as sewage waste and other similar waste sources.

In an MFC, biochemical reactions are carried out by electrogenic bacteria in an anaerobic anode chamber, generating electrons and protons through the degradation of the organic substrates entrenched in wastewater; concomitantly, electrochemical reactions take place in the aerobic cathode chamber, through which electrons and protons are accepted

TABLE 1.5   Application of Nanotechnology in Wastewater Treatment

Facilitating Operations	Supporting Technology	Assisting Nanomaterials
Adsorption	Contaminant preconcentration/detection, adsorption of recalcitrant contaminants	Carbon nanotubes
	Adsorptive media filters, slurry reactors	Nanoscale metal oxides
	Reactive nanoadsorbents	Nanofibers with core shell structure
Membranes and membrane processes	High permeability thin film nanocomposite membranes	Nanozeolites
	Anti-biofouling membranes	Nano-Ag, carbon nanotubes
	Aquaporin membranes	Aquaporin
	Reactive membranes, high performance thin film nanocomposite membranes	Nano-TiO_2
	Forward osmosis	Nanomagnetite
Photocatalysis	Photocatalytic reactors, solar disinfection systems	Nano-TiO_2, fullerene derivatives
Disinfection and microbial control	Water disinfection, anti-biofouling surface	Nano-Ag, carbon nanotubes, nano-TiO_2
	Decontamination	Nano-TiO_2
Sensing and monitoring	Optical detection	Quantum dots
	Optical and electrochemical detection	Noble metal, carbon nanotubes
	Sample preconcentration	Carbon nanotubes, magnetic nanoparticles
	Purification	Magnetic nanoparticles

by an oxygen reduction reaction (ORR). Anaerobic treatment of wastewaters is significantly less energy intensive than aerobic treatment; however, it takes longer to accomplish due to the exclusively slow growth process of anaerobic microorganisms. Thus, there has been little concern in applying anaerobic processes to dilute wastewaters (e.g., domestic wastewater).

There are two forms of microbial fuel cells: mediator and mediatorless. In mediator microbial fuel cells, the bacteria are electrochemically inactive and the microbes digest the organic thing and create electrons. The created electron transfer from microbial cells to the electrode is assisted by mediators such as thionine, humic acid, methyl viologen, methyl blue, and neutral red (Park and Zeikus, 2000; Rabaey et al., 2004). The majority of the mediators are expensive and lethal to bacteria.

Mediator-less microbial fuel cells do not involve a mediator except to exploit electrochemically active bacteria to transfer electrons to the electrode, i.e., the electrons are transmitted right away from the bacterial respiratory enzyme to the electrode (Bond and Lovley, 2003). Several bacteria that have pili on their external membrane are able to transmit their electrons produced through these pili. Microbes in mediator-less type MFCs usually include electrochemically active redox enzymes such as cytochromes on their outer membrane that can transfer electrons to external materials (Muralidharan et al., 2011). An estimated 2% of our worldwide power capacity is utilized to handle effluent, at an outlay of approximately $40 billion per year. Still, a new wastewater treatment process that uses microbial fuel-cell technology can create sufficient energy to operate the complete treatment process, trimming costs by 30% to 40% and cutting nearly 80% of leftover sludge.

The electrode surface area can be achieved with a maximum of 1.5 watts per meter squared in MFCs using batch mode operation. In case of continuous flow MFC, approximately 15.5 watts can be achieved per cubic meter of household wastewater flowing through it. The existing possible application of MFCs comprises brewery and domestic effluent treatment, desalination, hydrogen production, remote sensing, pollution remediation, and as a remote power source. As a result, MFC is a promising technology to treat wastewater equally as a traditional aeration process with immense potential as an energy-positive process, as it saves 100% of aeration energy with additional electricity production. Further, it also appreciably trims sludge production that might cut down the size of secondary clarifier and save the cost of sludge disposal.

1.8.8 Sonication for Wastewater Treatment

Ultrasound (US) treatment is one of various technologies that support hydrolysis, the rate-determining step in wastewater treatment. The underlying principle of US is based along the end of bacterial cells but difficult-to-degrade organics. The wastewater contains various substances and factors, which are compiled in the form of aggregates and flakes, including bacteria, viruses, cellulose, and starch. These aggregates are mechanically broken down and the constituent structure of the wastewater is altered by the energy produced during US treatment, thereby allowing the water to be separated more easily. The separation is achieved by hydrolytic reactions that are catalyzed by isoenzymes that are produced when the bacterial cell walls are intruded by the US. This results in an acceleration in the partitioning of organic stuff into smaller readily biodegradable fractions. The subsequent increase in biodegradable material improves bacterial kinetics resulting in lower wastewater quantities and increased biogas production in the case of anaerobic digestion (Mahvi, 2009).

US increases the degradation of different chemical pollutants such as herbicides and pesticides of agrochemical basis or hydrocarbon from industrial effluent. The US method is far sufficient to disintegrate the sewage sludge and for degradation or transformation of wastewater pollutants. Gogate and Pandit (2004) looked into the effect of application of US techniques for wastewater treatment. The advanced oxidation processes work on the basis of generation of free radicals and the ensuing attack of the same on contaminants and the H_2O_2 attack to oxidants in the presence of the US. The application of US irradiation on the enhancement of organic pollutant degradation in biological activated carbon membrane reactors was investigated by H. Liu et al. (2005). It was noted that ultrasonic treatment at 10 W for 24 hours, perhaps, will enhance organic load of the bioreactor and the removal efficiency of organic substances, and improve the level of 2,3,5 triphenyl tetrazoliumchloride dehydrogenase activity (TTC-DHA) in the biological activated carbon reactor. Transducers are generally utilized for the generation of US and they are liquid-driven or magneto-strictive or piezoelectric transducers. Generally, the transmissions of the sound waves in US occur during the rarefaction stage of the wave, and the liquid molecules are ruptured away from each other causing cavitation (voids formation) to take place. These cavities forcefully collapse in the next compression cycle. Consequently, transmission of ultrasonic waves in the liquid encourages the configuration of cavitation bubbles that are able to grow and implode in the periodic variation of the pressure fields. New developments in application of sonication in wastewater treatments include the following.

1.8.8.1 Catalytic Sonicative Degradation of Wastewater

Catalytic modifications are usually involved along with sonication, which improves the rates, since sonolysis is a slow process. Filtration processes are utilized, which supports separation of solid catalysts. Catalysts like Ni, V_2O_5, Pt, AgO, and Fenton reagent are assisted in the oxidation reactions.

1.8.8.2 Chemi-Sonication of Wastewater

Radical attacks degrade and oxidize a number of organic contaminants. This consistent extermination of compounds like CCl_4, CH_2Cl_2, and

$C_2H_2Cl_4$ from 100 to 1000 ppm is feasible in combination with strong oxidizing chemicals, such as H_2O_2, and US. The destruction process occurs around 1 to 4 hours for the aforementioned concentrations, depending on the quantity of the consigned gas and concentration of H_2O_2 used along with cavitation.

1.8.8.3 Assistance of Sonication with Conventional Methods

Pairing of US with ultraviolet (UV) radiation or electrostatic forces or with wet air oxidation (WAO) has led to treatment efficiency developments as discussed next in brief:

- Photocatalytic action with the US has resulted in higher degradation rates of the contaminants. This is due to the mechanical effects of cavitation involving photocatalyst surface cleaning and increased mass transfer of the polluting species to the powdered catalyst surface.

- Electrochemical effects. Effective oxidation rate of aromatics is significantly increased when employed with US. When a salt solution is present with an aromatic compound, it has been observed that the reaction becomes faster via formation of intermediates such as chlorinated hydrocarbons.

- Sonication followed by WAO gives an economic hybrid model and an efficient process. Sonication in the presence of the catalytic WAO including catalysts like $CuSO_4$ and $NiSO_4$ is more effective than without a catalyst. Sonication cans break down large molecular weight compounds, which are then easily oxidized by WAO.

The main cause of US efficiency may be the cavitation phenomena that accompanies the generation of local high temperature, pressure, and reactive radical species ($OH°$, $HO_2°$) via thermal dissociation of water and oxygen. The amount of hydroxyl radicals in the sonolysis system is directly related to the degradation efficiency because oxidation by hydroxyl radical is the main degradation pathway. Most experiments of the US technique were carried out on a laboratory scale due to costs. This technology in some aspects, such as algae removal, was used in full scale and had the highest efficiency. The cost of US techniques must be decreased, suggesting utilization of solar energy. There are various parameters that can affect to application and efficiency of

US in water treatment such as power density, frequency, and irradiation time (Doosti et al., 2012; Upadhyay and Khandate, 2012; Yaqub et al., 2012).

1.8.9 Advanced Oxidative Processes

Advanced oxidation processes (AOPs) exploit the hydroxyl radical (°OH) for oxidation, which has received growing attention in the research and development of wastewater treatment technologies in the last decades. AOPs have been effectively applied for the removal or degradation of toxic pollutants, or used as pretreatment to convert recalcitrants into biodegradable compounds, which are then treated by conventional biological methods. The effectiveness of AOPs depends on the production of reactive free radicals, the most important of which is the hydroxyl radical (°OH). These OH radicals attack the most organic molecules rapidly and nonselectively. The versatility of AOPs is also enhanced by the fact that they offer various alternative methods of hydroxyl radical production, thus allowing a better compliance with specific treatment requirements. The ecofriendly end product is the special feature of these AOPs, which are more efficient as they are capable of mineralizing a wide range of organic pollutants. AOPs such as ozonation (O_3); ozone combined with hydrogen peroxide (O_3/H_2O_2) and UV irradiation (O_3/UV) or both ($O_3/H_2O_2/UV$); ozone combined with catalysts ($O_3/catalysts$); UV/H_2O_2, Fenton, and photo-Fenton processes (Fe_2+/H_2O_2 and $Fe_2+/H_2O_2/UV$); and the ultrasonic process and photocatalysis have been successfully used for wastewater treatment. AOPs are generally a series of methods that produce highly oxidative radicals to degrade persistent molecules. Several techniques are involved and they are broadly classified into non-photochemical methods (methods that do not use radiation) and photochemical methods (methods that use radiation) (Munter, 2001; Wang and Xu, 2012; Muruganandham et al., 2014). Figure 1.4 shows the classification of AOPs.

1.8.9.1 Non-Photochemical Methods of Advanced Oxidative Processes

There are four well-known methods for generating hydroxyl radicals without using light energy. Three of the methods involve the reaction of ozone, while one uses Fe^{2+} ions as the catalyst. These methods are ozonation at elevated values of pH (>8.5), combining ozone with hydrogen peroxide, ozone + catalyst, and the Fenton system.

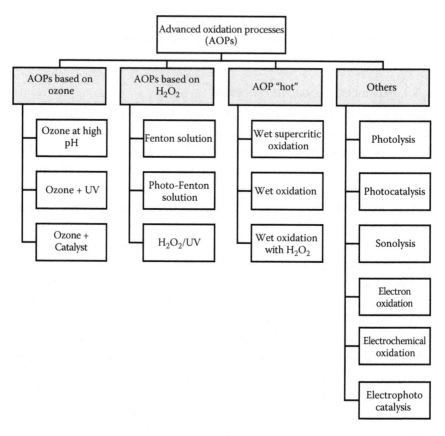

FIGURE 1.4   Classification of advanced oxidation processes.

1.8.9.1.1 Ozonation at Elevated pH   As the pH rises, the decomposition rate of ozone in water increases. Oxidation of organic species may occur due to a combination of reactions with molecular ozone and reactions with °OH radicals. The reaction between hydroxide ions and ozone leads to the formation of superoxide anion radical $O_2^{\circ-}$ and hydroperoxyl radical HO_2°. By the reaction between ozone and the superoxide anion radical, the ozonide anion radical $O_3^{\circ-}$ is formed, which decomposes immediately giving the °OH radical. Summarizing, three ozone molecules produce two °OH radicals:

$$3O_3 + OH^- + H^+ \rightarrow 2°OH + 4O_2 \qquad (1.1)$$

The rate of the attack by °OH radicals is typically 10^6 to 10^9 times faster than the corresponding reaction rate for molecular ozone. The major operating cost for the ozone oxidation process is the cost of electricity for ozone generation (Gottschalk et al., 2000).

1.8.9.1.2 O_3/H_2O_2 (Peroxone)   The addition of hydrogen peroxide to ozone can initiate the decomposition cycle of ozone, resulting in the formation of °OH radicals.

$$H_2O_2 \rightarrow HO_2^- + H^+ \tag{1.2}$$

$$HO_2^- + O_3 \rightarrow HO_2^\circ + O_3^{\circ -} \tag{1.3}$$

The reaction continues along the indirect pathway described earlier and °OH radicals are produced. The combination of different reaction steps shows that two ozone molecules produce two °OH radicals:

$$2O_3 + H_2O_2 \rightarrow 2°OH + 3O_2 \tag{1.4}$$

Hydrogen peroxide is a relatively inexpensive, readily available chemical oxidant. It is produced by electrolysis of ammonium bisulfate or by oxidation of alkyl hydroanthraquinones (Hoigne, 1982; Gottschalk et al., 2000).

1.8.9.1.3 Ozone + Catalyst   Another opportunity to accelerate ozonation reactions is to use heterogeneous or homogeneous catalysts. Several metal oxides and metal ions (Fe_2O_3, Al_2O_3–Me, MnO_2, Ru/CeO_2, TiO_2–Me, Fe^{2+}, Fe^{3+}, Mn^{2+}, etc.) have been studied and sometimes a significant acceleration in the decomposition of the target compound has been achieved, although the reaction mechanism in most cases remained unclear (Munter, 2001).

1.8.9.1.4 Fenton System   The Fenton process was reported by British chemist Henry John Horstman Fenton more than a hundred years ago for maleic acid oxidation:

$$Fe_2^+ + H_2O_2 \rightarrow Fe_3^+OH^- + °OH \tag{1.5}$$

The rate constant for the reaction of ferrous ion with hydrogen peroxide is high, and Fe(II) oxidizes to Fe(III) in a few seconds to minutes in the presence of excess amounts of hydrogen peroxide. It is believed that most waste destruction catalyzed by Fenton's reagent is simply a Fe(III)–H_2O_2 system catalyzed destruction process, and Fenton's reagent with an excess of hydrogen peroxide is essentially a Fe(III)–H_2O_2 process (known as a Fenton-like reagent). Thus, the ferrous ion in Fenton's reagent can be replaced with the ferric ion (Fenton, 1884).

1.8.9.2 Photochemical Methods

Photochemical oxidation is the reaction of a chemical change, initiated by light, in a substance that causes it to lose electrons. A common example is photochemical smog, which is caused by hydrocarbons and NOx reacting under the influence of UV light.

1.8.9.2.1 Vacuum UV (VUV) Photolysis   The vacuum ultraviolet (VUV) range is absorbed by almost all substances (including water and air). Thus it can only be transmitted in a vacuum. The absorption of a VUV photon causes one or more bond breaks. VUV photolysis can be achieved by the usage of either a monochromatic or polychromatic radiation source. These light sources have some limitations including high price and wavelength variations. For these reasons, application of VUV photolysis is limited (Bolton and Cotton, 2011).

1.8.9.2.2 Hydrogen Peroxide/Ultraviolet Process   The hydrogen peroxide/ultraviolet process (H_2O_2/UV) process is based on the direct photolysis of a hydrogen peroxide molecule by a radiation with a wavelength between the 200 and 300 nm region. Low-, medium-, and high-pressure mercury vapor lamps can be used for this process because they have significant emittance within 220 and 260 nm, which is the primary absorption band for hydrogen peroxide. Most UV light can also be absorbed by water. Use of low-pressure mercury vapor lamps can lead to usage of high concentrations of H_2O_2 for the generation of sufficient hydroxyl radical. However, high concentrations of H_2O_2 may scavenge the hydroxyl radical, making the H_2O_2/UV process less effective. Other variables such as temperature, pH, concentration of H_2O_2, and presence of scavengers affect the production of hydroxyl radicals (EPA, 1998; Bolton, 2001; Mandal et al., 2004; Azbar et al., 2005).

1.8.9.2.3 Ozone/Ultraviolet Process   Photolysis of ozone in water with UV radiation in the range of 200 to 280 nm can lead to a yield of hydrogen peroxide. Low-pressure mercury vapor UV lamps are the most common sources of UV irradiation used for this process. Many variables such as pH, temperature, scavengers in the influent, turbidity, UV intensity, lamp spectral characteristics, and pollutant type(s) affect the efficiency of the system. A number of articles provide details about laboratory, pilot, and full-scale applications of ozone/UV and hydrogen peroxide/UV processes (EPA, 1998; Azbar et al., 2005).

1.8.9.2.4 Ozone/Hydrogen Peroxide/Ultraviolet Process   The $O_3/H_2O_2/UV$ method is considered to be the most effective and powerful method that provides a fast and complete mineralization of pollutants. Similar to other ozone including AOPs, increasing of pH affects the hydroxyl radical formation. Additional usage of UV radiation also affects the hydroxyl radical formation. Efficiency of $O_3/H_2O_2/UV$ process is being much higher with the addition of hydrogen peroxide (Mokrini et al., 1997; Horsch, 2000; Contreras et al., 2001; Azbar et al., 2005). The main short mechanism of the $O_3/H_2O_2/UV$ process is

$$2O_3 + H_2O_2 \xrightarrow{\text{UV}} 2HO° + 3O_2 \qquad (1.6)$$

1.8.9.2.5 Photo-Fenton Process   The combination of the Fenton process with UV light, the so-called photo-Fenton reaction, had been shown to enhance the efficiency of the Fenton process. Some researchers also attributed this to the decomposition of the photo active $Fe(OH)^{+2}$, which lead to the addition of the HO° radicals. The short mechanism of the photo-Fenton reaction is

$$Fe(OH)^{+2} + hv \rightarrow Fe^{+3} + HO° \qquad (1.7)$$

with $Fe(OH)^{2+}$ being the dominant Fe(III) species in solution at pH 2–3. High valence Fe intermediates formed through the absorption of visible light by the complex between Fe(II) and H_2O_2 are believed to enhance the reaction rate of oxidation production (Sun and Pignatello, 1993; He and Lei, 2004).

1.9 CONCLUSION

Water treatment and recycling has been recognized as a key approach to mitigate water shortage, which has now become a worldwide issue.

Effective wastewater collection and treatment are of immense significance from the perspective of both environmental and public health. The wide range of research activity has led to considerable development and multifariousness of the process and techniques used for wastewater treatment. This chapter provides a brief summary of various technologies commonly used for wastewater treatment, which are generally classified as physical, chemical, and biological treatments. Today, the biological method is the most widely used method because of its simplicity and relatively low cost, but it is less successful when the effluent contains highly toxic organic pollutants. It occupies a large space and this could be a severe limitation. Conventional water and wastewater treatments are no longer capable of removing the ever-increasing complexity of water pollutants. Thus, for treatment of a highly contaminated and toxic effluent there is no option other than advanced wastewater technology like membrane technology, fuel cell technology, sonication, and advanced oxidation processes, and these techniques have been influential in driving water recycling forward. The advanced treatment technologies have moved in such a direction that any stage of treatment can be done with a suitable combination of a large number of processes. The physical and chemical treatments in combination with biological treatment are considered to be the only adequate technique for wastewater treatment. Though technologies are available, pollution will persist, for it is easier to pollute than to reclaim. Therefore, it is a high aspiration to shelter the water environment either by forbidding the discharge of pollutants into water bodies or by promoting the effluent quality standard.

Thus, concluding with the thought that "water is the driving force of all nature and pure water is the world's first and foremost medicine" magnifies that every individual must recognize the importance of water and treatment of wastewater, which protects water from contamination and promotes reclaiming.

REFERENCES

Azbar, N., K. Kestioğlu, and T. Yonar. 2005. Application of Advanced Oxidation Processes (AOPs) to Wastewater Treatment. Case Studies: Decolourization of Textile Effluents, Detoxification of Olive Mill Effluent, Treatment of Domestic Wastewater, Ed. A.R. BURK, Water Pollution: New Research, Nova Science Publishers, New York, 99–118.

Aziz, H.A., and A. Mojiri. 2014. *Wastewater Engineering: Advanced Wastewater Treatment System*. IJSR, Penang, Malaysia.

Babu, A., J.A. Jerald, S.S. Kumar, and P. Saravanan. 2013. Wastewater management—Contemporary position and strategies for the expectations. *International Journal of Life Sciences Biotechnology and Pharma Research* 2, no. 1:229–237.

Bansal, R.C., J.B. Donnay, and F. Stoeckli. 1998. *Activated Carbon*. Marcel Dekker, New York.

Bernardes, A.M., M.A.S. Rodrigues, and J.Z. Ferreira. 2014. *Electrodialysis and Water Reuse: Novel Approaches*. Springer, Berlin.

Bolton, J.R. 2001. *Ultraviolet Applications Handbook*. Bolton Photosciences Inc., Edmonton, AB, Canada.

Bolton, J.R., and C.A. Cotton. 2011. *The Ultraviolet Disinfection Handbook*. American Water Works Association, Denver.

Bond D.R., and D.R. Lovley. 2003. Electricity production by *Geobacter sulfurreducens* attached to electrodes. *Applied Environmental Microbiology* 69, no. 3:1548–1555.

Contreras, S., M. Rodriguez, E. Chamarro, S. Esplugas, and J. Casado. 2001. Oxidation of nitrobenzene by UV/O$_3$: The influence of H$_2$O$_2$ and Fe(III) experiences in a pilot plant. *Wat. Sci. Tech.* 44:39–46.

Corcoran, E., C. Nellemann, E. Baker, R. Bos, D. Osborn, and H. Savelli. 2010. Sick water? The central role of wastewater management in sustainable development. A Rapid Response Assessment. United Nations Environment Programme, UN-HABITAT, GRID-Arendal. Birkeland Trykkeri AS, Norway.

Cushman, R., and G. Carlson. 2004. Wastewater treatment (Pollution A to Z). In *Encyclopedia.com*.

Dhote, J., S. Ingoleb, and A. Chavhan. 2012. Review on wastewater treatment technologies. *International Journal of Engineering Research & Technology* 1, no. 5:1–10.

Domingo-Garcia, M., F.J. Lopez-Garzon, and M. Perez-Mendoza. 2000. Effect of some oxidation treatment on the textural characteristics and surface chemical nature of an activated carbon. *Journal of Colloid and Interface Science* 222:233–240.

Doosti, M.R., R. Kargar, and M.H. Sayadi. 2012. Water treatment using ultrasonic assistance: A review. *Proceedings of the International Academy of Ecology and Environmental Sciences* 2, no. 2:96–110.

Economic and Social Commission for Western Asia. 2003. Wastewater treatment technologies: A general review. United Nations, New York.

EPA, 1998. Handbook on Advanced Photochemical Oxidation Process. US. EPA, Washington, DC.

Fenton, H.J. 1884. Oxidative properties of the H2O2/Fe2+ system and its application. *J. Chem. Soc.* 65:889–899.

Gleick, P.H. 1996. Water resources. In *Encyclopedia of Climate and Weather*, edited by S.H. Schneider. Oxford University Press, New York.

Gogate, P.R., and B.A. Pandit. 2004. A review of imperative technologies for wastewater treatment I: Oxidation technologies at ambient conditions. *Advances in Environmental Research* 8:501–551.

Gottschalk, C., J.A. Libra, and A. Saupe. 2000. *Ozonation of Water and Waste Water.* Wiley-VCH, Verlag GmbH.

Hamilton, A.J., A.-M. Boland, D. Stevens, J. Kelly, J. Radcliffe, A. Ziehrl, P. Dillon, and B. Paulin. 2005a. Position of the Australian horticultural industry with respect to the use of reclaimed water. *Agricultural Water Management* 71:181–209.

Hamilton, A.J., M. Mebalds, R. Aldaoud, and M. Heath. 2005b. A survey of physical, agrochemical and microbial characteristics of wastewater from the carrot washing process: Implications for re-use and environmental discharge. *Journal of Vegetable Science* 11:57–72.

He, F., and L. Lei. 2004. Degradation kinetics and mechanisms of phenol in photo-fenton process. *Journal of Zhejiang University Science* 5, no. 2:198–205.

Hoigne, J. 1982. Mechanisms, rates and selectivities of oxidations of organic compounds initiated by ozonation of water. In *Handbook of Ozone Technology and Applications.* Ann Arbor Science Publ., Ann Arbor, MI.

Horsch, F. 2000. Oxidation Eines Industriellen Mischabwassers mit Ozon und UV/H_2O_2. *Vom Wasser* 95:119–130.

Isabel, L., and E. Wayne. 2005. Utilization of turkey manure as granular activated carbon: Physical, chemical and adsorptive properties. *Waste Management* 25:726–732.

Kim, B.H., T. Ikeda, H.S. Park, H.J. Kim, M.S. Hyun, K. Kano, K. Takagi, and H. Tatsumi. 2002. Electrochemical activity of an Fe (III)-reducing bacterium, *Shewanella putrefaciens* IR-1, in the presence of alternative electron acceptors. *Biotechnology Techniques* 13:475–478.

Kumar, R., R.D. Singh, and K.D. Sharma. 2005. Water resources of India. *Current Science* 89:794–811.

Liu, D.H.F., and B.G. Liptak. 1999. *Wastewater Treatment.* Boca Raton, Lewis.

Liu, H., Y.H. He, X.C. Quan, Y.X. Yan, X.H. Kong, and A.J. Lia. 2005. Enhancement of organic pollutant biodegradation by ultrasound irradiation in a biological activated carbon membrane reactor. *Process Biochemistry* 40:3002–3007.

Liu, H., R. Ramnarayanan, and B.E. Logan. 2004. Production of electricity during wastewater treatment using a single chamber microbial fuel cell. *Environmental Science and Technology* 38:2281–2285.

Logan, B.E., and J.M. Regan. 2006. Microbial fuel cells challenges and applications. *Environmental Science and Technology* 40:5172–5180.

Mandal, A., K. Ojha, A.K. De, and S. Bhattacharjee. 2004. Removal of catechol from aqueous solution by advanced photo-oxidation process. *Chemical Engineering Journal* 102:203–208.

Mahvi, A.H. 2009. Application of ultrasound technology for water and wastewater treatment. *Iranian Journal of Public Health* 38:1–17.

Mara, D.D. 1992. *Waste Stabilization Ponds: A Design Manual for Eastern Africa.* Lagoon Technology International, Leeds.

Mokrini, M., D. Oussi, and S. Esplugas. 1997. Oxidation of aromatic compounds with UV radiation/ozone/hydrogen peroxide. *Water Science Technology* 35:95–102.

Munter, R. 2001. Advanced oxidation processes—Current status and prospects. *Proceedings of the Estonian Academy of Sciences, Chemistry* 50, no. 2:59–80.

Muralidharan, A., O.K. Ajay Babu, K. Nirmalraman, and M. Ramya. 2011. Impact of salt concentration on electricity production in microbial hydrogen based salt bridge fuel cells. *Indian Journal of Fundamental and Applied Life Sciences* 1, no. 2:78–184.

Muruganandham, M., R.P.S. Suri, Sh. Jafari, M. Sillanpaa, Gang-Juan Lee, J.J. Wu, and M. Swaminathan. 2014. Recent developments in homogeneous advanced oxidation processes for water and wastewater treatment. *International Journal of Photoenergy* 1:1–21.

National Commission for Integrated Water Resource Development (NCIWRD). 1999. Integrated water resource development: A plan for action, report of the NCIWRD. Volume I. Ministry of Water Resources, Government of India.

Pabby, A.K., S.S.H. Rizvi, and A.M. Sastre. 2015. *Handbook of Membrane Separations: Chemical, Pharmaceutical, Food and Biotechnological Applications.* CRC Press/Taylor & Francis Group, Boca Raton.

Park, D.H., and J.G. Zeikus. 2000. Electricity generation in microbial fuel cells using neutral red as an electronophore. *Applied Environmental Microbiology* 66:1292–1297.

Qasim, S.R. 1999. *Wastewater Treatment Plants: Planning, Design, and Operation.* 2nd ed. Technomic, Lancaster, PA.

Qu, X., P.J.J. Alvarez, and Q. Li. 2013. Applications of nanotechnology in water and wastewater treatment. *Water Research* 47:3931–3946.

Rabaey, K., N. Boon, S.D. Siciliano, M. Verhaege, and W. Verstraete. 2004. Biofuel cells select for microbial consortia that self-mediate electron transfer. *Applied and Environmental Microbiology* 70:5373–5382.

Shiklomanov, I.A. 1993. World fresh water resources. In *Water in Crisis*, edited by P.H. Gleick, chap. 2. Oxford University Press, New York.

Smith, A. 2006. Nanotech—The way forward for clean water? *Filtration and Separation* 43, no. 8:32–33.

Sun, Y., and J.J. Pignatello. 1993. Photochemical reactions involved in the total mineralisation of 2, 4-D by $Fe^{+3}/H_2O_2/UV$. *Environ. Sci. Technol.* 27, no. 2:304–310.

Tchobanoglous, G., and F.L. Burton. 1991. *Wastewater Engineering: Treatment, Disposal, and Reuse.* 3rd ed. McGraw Hill, New York.

Tchobanoglous, G., F.L. Burton, and H.D. Stensel. 2003. *Wastewater* Topare, Niraj S., S.J. Attar, and Mosleh M. Manfe. 2011. Sewage/wastewater treatment technologies: A review. *Scientific Reviews and Chemical Communications* 1, no. 1:18–24.

Upadhyay, K., and G. Khandate. 2012. Ultrasound assisted oxidation process for the removal of aromatic contamination from effluents: A review. *Universal Journal of Environmental Research and Technology* 2, no. 6:458–464.

Van Haandel, A.C., and G. Lettinga. 1994. *Anaerobic Sewage Treatment: A Practical Guide for Regions with a Hot Climate.* John Wiley, Chichester.

Viessman, W. 2003. Wastewater treatment and management (Water: Science and Issues). In *Encyclopedia.com.*

Wang, J.L., and L.J. Xu. 2012. Advanced oxidation processes for wastewater treatment: Formation of hydroxyl radical and application. *Critical Reviews in Environmental Science and Technology* 42, no. 3:251–325.

Water Environment Federation and the American Society of Civil Engineers/ Environmental and Water Resources Institute. 2010. *Design of Municipal Wastewater Treatment Plants: WEF Manual of Practice No. 8* (ASCE Manuals and Reports on Engineering Practice No. 76). McGraw-Hill Education, New York.

Yaqub, A., H. Ajab, M.H. Isa, H. Jusoh, M. Junaid, and R. Farooq. 2012. Effect of ultrasound and electrode material on electrochemical treatment of industrial wastewater. *Journal of New Materials for Electrochemical Systems* 15:289–292.

CHAPTER 2

Nanotechnology in Environmental Applications

Kuravappullam Vedaiyan Radha
and Pandiyan Thyriyalakshmi

CONTENTS

Abstract

Environmental nanotechnology is the creation of ecofriendly products for cleaning existing pollution and improving manufacturing methods to reduce pollution, making alternative energies more cost effective. The chapter discusses issues relating to low pollution levels during manufacturing processes using silver nanoparticles and production of energy using solar cells of silicon nanowires, improvement in windmills through silicon-epoxy nanomaterials. The chapter discusses novel nanomaterials developed for the treatment of surface water, groundwater, wastewater, and other environmental materials contaminated by organic, inorganic, and toxic metal ions using carbon nanotube (CNT) membranes. Capturing carbon dioxide in power plant exhausts using nanostructured membranes is one of the cutting-edge tasks of environmental nanotechnology.

A catalyst, composed of porous manganese oxide embedded with gold nanoparticles, has been demonstrated to break down volatile organic compounds (VOCs) from air. Storing hydrogen for fuel cells using grapheme layers and biosynthesis of nanoparticles are considered to be clean, nontoxic, and environmentally accepted green chemistry procedures. Most of the studies reported relate to green technology and a greener environment. Hence, environmental nanotechnology products have wide applications in the field of biomedical, industrial, and environmental remediation.

2.1 INTRODUCTION

This chapter discusses the latest developments in technologies adopted in production of energy using solar cells using silicon nanowires; improvement in windmills through silicon-epoxy nanomaterials; and novel nanomaterials developed for the treatment of surface water, groundwater, wastewater and other environmental materials contaminated by organic, inorganic, and toxic metal ions using carbon nanotube (CNT) membranes. Capturing carbon dioxide in power plant exhaust using nanostructured membranes is one of the cutting-edge tasks of environmental nanotechnology. A catalyst, composed of porous manganese oxide embedded with gold nanoparticles, has been demonstrated to break down volatile organic compounds (VOCs) from air. The nanomaterials used, their processes, merits, demerits, and future scope of the work in the related research area are discussed.

2.2 PRODUCTION OF ENERGY USING NANOMATERIALS

Solar cells, called photovoltaic (PV) cells by scientists, convert sunlight directly into electricity. The process of converting light (photons) to electricity (voltage) is called the PV effect. The PV effect was discovered in 1954, when scientists at Bell Telephone discovered that silicon (an element found in sand) created an electric charge when exposed to sunlight.

Semiconductor nanostructures are promising building blocks for next-generation solar cells with higher energy conversion efficiencies and lower cost. Silicon-based nanostructured solar cells have several advantages, including the natural abundance of silicon, lack of toxicity, and compatibility with mature integrated-circuit fabrication techniques.

Recently, porous silicon nanowires (PSiNWs) with mesopores were prepared by the metal-assisted chemical etching method. PSiNWs have the combined physical properties of SiNWs and porous silicon. They have promising energy-harvesting applications, and are basic components for future nanoelectronic, catalysis, and conversion.

Silicon solar cells based on nanowires have much shorter p-n junctions that thin film solar cells. In the nanowire structure, photo-excited electrons and holes (carriers) travel very short distances before being collected by the electrodes. This results in higher carrier-collection efficiency in the core-shell nanowire structure. This advantage leads to a higher tolerance for material defects and allows the use of lower-quality silicon. The core-shell nanowire structure addresses the carrier-collection issue, one of the key factors that determine the overall efficiency of a solar cell. However, the efficiency of photon capture in the nanowire structures, another very important factor, has not yet been determined.

2.3 METHODS ADOPTED FOR THE FABRICATION OF SILICON NANOWIRE

2.3.1 Vapor–Liquid–Solid (VLS) Mechanism

The vapor–liquid–solid (VLS) mechanism, first proposed by Wagner and Ellis in the mid-1960s, is the key mechanism for silicon-wire growth. The proposed VLS mechanism is based on two observations: that the addition of certain metal impurities is an essential prerequisite for growth of silicon nanowires in experiments and that small globules of the impurity are located at the tip of the wire during growth. From this, Wagner and Ellis deduced that the globule at the wire tip must be involved in the growth of the silicon wires by acting as a preferred sink for the arriving Si atoms or, perhaps more likely, as a catalyst for the chemical process involved (Figure 2.1). When Au, for example, is deposited on silicon substrate and this substrate is then heated to temperatures above about 363°C, small liquid Au–Si alloy droplets will form on the substrate surface.

Exposing such a substrate to a gaseous silicon precursor, such as silicon tetrachloride ($SiCl_4$) or silane (SiH_4) precursor molecules, will crack on the surface of the Au–Si alloy droplets, whereupon Si is incorporated into the droplet. The silicon supply from the gas phase causes the droplet to become supersaturated with Si until silicon freezes out at the silicon–droplet interface. The continuation of this process then leads to the growth of a wire with the alloy droplet riding atop the growing wire (Putnam et al. 2009).

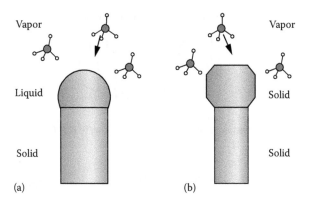

Vapor

Liquid

Solid

(a)

Vapor

Solid

Solid

(b)

FIGURE 2.1 Construction of nanowire by VLS mechanism. (Courtesy of Lensch-Falk, J.L. et al., *Journal of Material Chemistry*, 19:849–857, 2009.)

2.3.2 Chemical Vapor Deposition (CVD)

In chemical vapor deposition (CVD), a volatile gaseous silicon precursor, such as silane (SiH_4) or silicon tetrachloride ($SiCl_4$), serves as the silicon source. It is transported to the deposition surface at which the precursor reacts and is cracked into its constituents.

Originally, CVD was devised for the deposition of high purity films. Contaminations such as gold particles, however, were found to cause anisotropic growth of silicon, that is, the growth of silicon wires (Zhou and Biswas 2008). CVD allows epitaxial growth of SiNWs, with the growth velocity varying from about 10^{-2} to 10^{-3} nm min^{-1}, depending on the temperature and type of Si precursor used.

Furthermore, CVD offers broad possibility of modifying the properties of the silicon wires in a controlled fashion. A variety of derivatives of the CVD method exist. These can be classified by parameters such as the base and operation pressure or the treatment of the precursor. Since silicon is known to oxidize easily if exposed to oxygen at elevated temperatures, it is crucial to reduce the oxygen background pressure in order to be able to epitaxially grow uniform silicon nanowires. In particular, when oxygen-sensitive catalyst materials are used, it turns out to be useful to combine catalyst deposition and nanowire growth in one system, so that growth experiments can be performed without breaking the vacuum in between (Figure 2.2).

In any case, it is useful to lower the base pressure of the CVD reactor down to high or even ultrahigh vacuum, which reduces unwanted contamination

Metal-assisted nanowire growth mechanisms

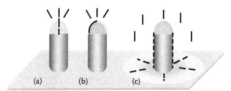

Different diffusion paths potentially contributing to NW formation

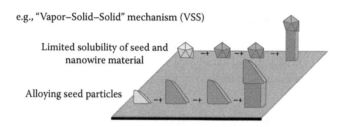

FIGURE 2.2 Mechanism of chemical vapor deposition. (a) The Si liquid droplet. (b) The flowing gases dissolve into the Si droplet. (c) Synthesized Si whisker. (Courtesy of Christian, B.S., *One-Dimensional Oxide Nanostructures via Chemical Vapor Deposition and Synthesis of Heterometallic Transition Metal Alkoxides [Co, Mn and Fe]*, Thesis, Saarland University, Germany, 2008.)

and enables growth at lowered temperatures. The pressure during growth mainly depends upon the gaseous silicon precursor and its cracking probability at the catalyst surface. The low growth pressures allow the combination of CVD with transmission electron microscopy (TEM), enabling in situ observation of the nanowire growth. In contrast to that, silane partial pressures required for wire growth are about five orders of magnitude higher.

By modifying the precursor before reacting with the sample surface, the temperature budget of the substrate can be lowered. In cases where the thermal load is critical or where a high supersaturation of the droplet is necessary, nanowire growth can be enhanced using plasma-enhanced CVD (PECVD).

Another advantage of CVD as a bottom-up synthesis method is its variability concerning the intended wire size. Wire diameters range from below 10 nm up to several hundred micrometers. Since surface diffusion only plays a minor role in CVD, the length of the wires can also be tuned accordingly by simply extending or decreasing the growth time. Thus, to summarize, a large range of length and diameter configurations can be fabricated. With CVD, not only the wire size but also its properties can be modified.

2.3.3 Hybrid Solar Cells

To overcome the problems of conventional Si-based solar cells, hybrid solar cells were developed. These cells are a combination of a polymer and a nano organic or inorganic semiconductor. Hybrid solar cells combine the excellent electronic properties of inorganic molecules with the reduced cost and flexibility of amorphous substrates. In order to achieve high hybrid solar cell performance, the electron and hole mobility should be balanced and optimized. Cadmium selenide (CdSe) is a promising photovoltaic material because of its high absorption coefficient and nearly optimum band gap energy for the efficient absorption of light and conversion into electrical power.

Hybrid solar cells are similar to conventional solar cells, but in this case, nanoscale CdSe acts as the electron receptor and P3HT acts as the hole receptor. P3HT was chosen due to the fact that it has the highest hole mobility in conducting plastics available today, while the inorganic semiconductor CdSe has a high electron mobility. High mobilities ensure fast charge transport and separation of the electron hole pair, reducing current losses due to recombination and increasing the efficiency of the cell. Electrons and holes are exchanged at the interaction surface between CdSe and P3HT.

Since an electron's position and direction can only be described by a probability, or wave function, the particular path that an electron will take is indeterminate. In an effort to direct the electron's path, the shape of the CdSe is long and narrow, like a wire, providing a directed, low-resistance path for the electron to follow. The nanorods may also be tuned to absorb different frequencies of light. The diameter of the rod determines the band gap, where smaller diameters absorb higher frequencies of light and larger diameters absorb lower frequencies. Alternatively, the length of the rods is proportional to the efficiency of the cell.

It was found that electron transport exhibited in the shorter nanoparticles is dominated by inefficient hopping, while band conduction is more prevalent in the longer particles. The size of CdSe nanoparticles plays a very important role in fluorescence. The band gap and also the light emitted varies with the change in size of CdSe nanoparticles. Hence, the particles to be chosen are quantum dots or rods of different lengths; this can be decided depending on the application and its requirement.

Although CdSe/P3HT solar cells have shown promising efficiencies, Cd is a toxic heavy metal and Se is also toxic in large quantities. To solve this problem, various hybrid solar cells have been reported using zinc oxide (ZnO) and TiO_2 nanocrystals. The hybrid film consists of TiO_2 nanocrystals and P3HT with a different concentration of TiO_2 nanocrystals and different solvents. ZnO and TiO_2 are nontoxic materials and have wide band gaps that only absorb deep violet; therefore, the efficiencies of ZnO/polymer or TiO_2/polymer solar cells are not as good as CdSe/polymer solar cells (Law et al. 2005).

Colloidal semiconductor quantum dots (QDs), made of CdSe capped with a thin layer of higher band-gap material, zinc sulfide (ZnS), have unique optical properties. Emission characteristics of CdSe/ZnS core-shell QDs strongly depend upon the size of the CdSe core. A change in size and size distribution of the QDs also changes the luminescence intensity and optical properties of the synthesized nanocrystals. CdSe quantum dots can also be used for supplement of dye or other quantum dots such as CdS to increase the efficiency of photovoltaic cells. CdSe have recently been reported for possible use as a photosensitizer absorbing photons on TiO_2 but the efficiency was low compared to desensitized solar cells.

2.3.4 TiO_2 Hybrid Solar Cells

Recently, nanocrystals have paved a new route for photovoltaic application due to its high surface area to volume ratio and tunable light absorption. One such application of nanocrystals is titanium dioxide (TiO_2) nanoparticle-based solar cell. This cell utilizes high surface-area-to-volume ratio TiO_2 nano crystals to increase the exciton dissociation area. These are n-type semiconductors that have a strong absorption peak in the ultraviolet but are transparent to visible light. They have a wide band gap of 3.05 to 3.5 eV. Single crystal titanium dioxide TiO_2 has a resistivity of about 10^{13} Ωcm at room temperature and about 10^7 Ωcm at 250°C.

Additionally, TiO_2 nanostructures may be cheaply synthesized in a modest chemistry lab via sol gel processing. TiO_2 nanoparticles are also effective photocatalysts for treatment of air and water pollutants. Inorganic semiconducting nanoparticles are preferred in photovoltaics because of improved light gathering properties as compared to conventional thin film PV cells. The total surface area of a collection of nanoparticles embedded in a polymer matrix is much larger than the contact area of the p-n junction in conventional devices. Additionally, the spherical geometry of particles allows for absorption of light with any angle of incidence, whereas conventional thin film systems have a maximum efficiency at incident angles close to the normal of the junction.

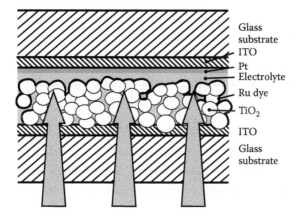

Glass substrate
ITO
Pt
Electrolyte
Ru dye
TiO₂
ITO
Glass substrate

FIGURE 2.3 TiO₂ hybrid solar cells. (Courtesy of Deshpande, V., and Seyezhai, R., *International Journal of Innovative Research in Computer and Communication Engineering*, 2:3387–3392, 2014.)

Several researchers have shown interest in the optical properties and low cost of TiO_2 nanoparticles for use in photovoltaic systems (Figure 2.3). A group at the Ecole Polytechnique University in Lausanne, Switzerland, has achieved 10% efficiency with a new form of photovoltaic cell where a porous network of TiO_2 nanoparticles is sandwiched between two electrodes. Since TiO_2 nanoparticles are transparent, they do not absorb visible light efficiently. Therefore a photosensitive liquid dye, known as a sensitizer, is added to the network to improve absorption in the visible range. Additionally, an electrolyte is required to return the electrons from the cathode to the dye molecules.

2.3.5 Advantages of Silicon Nanowires

- Increased surface area emits more power for each cell, increasing efficiency of energy yields per panel.

- Can be employed with nanoparticles as well as nanowires, broadening the prospective applications across several different profitable markets.

- Strong, structurally sound construction, decreasing future replacement costs.

- Easily and affordably maintained by the addition of a simple polymer, increasing product lifespan at little to no additional cost.

- The absorption of more radiation allows for an enhanced capitalization of free energy sources.

- Addition of polymers can act synergistically with the cells, also contributing to a potential increase in power emissions.

2.3.6 Disadvantages of Silicon Nanowires

A major drawback to silicon nanowires is that silicon forms an unavoidable passivation layer of silicon dioxide. The problem with this is that it affects the semiconducting properties of silicon nanowires.

2.3.7 Future Scope of Silicon Nanowires

Silicon nanowires are the most promising building block for future electronics devices. Tremendous progress has been made in the synthesis and fabrication methods of silicon nanowires. It also has a wide range of opportunities in interdisciplinary areas such as chemical and biological sciences.

2.4 IMPROVEMENT OF WINDMILLS THROUGH NANOMATERIALS

Windmills are used to convert kinetic energy into mechanical energy. The use of wind energy to generate electricity first appeared in the late 19th century but did not gain ground owing to the then control of steam turbines in electricity generation. Initially, wind energy started to gain popularity in electricity generation to charge batteries in remote power systems, residential scale power systems, isolated or island power systems, and utility networks. These wind turbines themselves are generally small (rated less than 100 kW) but could be made up to a large wind farm (rated 5 MW or so).

It was not until the early 1990s that the wind projects really took off the ground, primarily driven by the governmental and industrial initiatives. It was also in 1990s when there seemed a shift of focus from onshore to offshore development in major wind development countries, especially in Europe. Offshore wind turbines were first proposed in Germany in 1930s and first installed in Sweden in 1991 and in Denmark in 1992. By July 2010, there were 2.4 GW of offshore wind turbines installed in Europe. Compared to onshore wind energy, offshore wind energy has some appealing attributes such as higher wind speeds, availability of larger sites for development, lower wind shear, and lower intrinsic turbulence intensity.

Windmill power generators are improved by introducing nanomaterials due to their large surface volume ratio, higher reactivity, light weight, and so on. Incorporation of advanced composites materials made of a polymer matrix in the windmill blades improves rotation efficiency. Thermosetting epoxies reinforced with high-performance fibers have been widely used in windmill blades. These epoxy composites materials possess better properties such as stiffness, high specific strength, superior corrosion resistance, and excellent manufacturability (Lee and Neville 1967). However, these epoxy composites have the disadvantage of brittleness due to the unique hierarchical fiber architecture embedded in a brittle polymer matrix resulting in unavoidable crack nucleation and growth in polymer composites when subjected to external loads. From the physical strength, all the damage and failure processes identified in conventional polymer composites are thermodynamically irreversible, that is, the material properties always gradually degrade with time.

2.4.1 Development of High-Performance Materials

Since the early time of developing wind turbines, the three-phase synchronous machines were utilized. AC synchronous wind turbine generators (WTGs) scan work under the constant or DC excitations from either permanent magnets (PMs) or electromagnets. Hence, they are known as permanent magnet synchronous generators (PMSGs) and electrically excited synchronous generators (EESGs), respectively. The rotor operated by using wind turbine generates a three-phase power. The power is generated in the stator windings, which are connected to the grid through transformers and power converters. In the case of fixed-speed synchronous generators, the rotor speed must be kept at exactly the synchronous speed. If the power exceeds the limit or lowers the loss of synchronism, then it leads to the failure of the process.

2.4.1.1 Working Principle

Synchronous generators are known as machine technology. Synchronous generators have better performance for power generation and are widely accepted. Synchronous generators are easily controlled by adjusting the reactive power characteristics in the field circuit for electrical excitation. The tower shading effects and natural resonance of the components are the results of adopting fixed-speed synchronous generators. The random wind speed fluctuations and periodic disturbances lead to the formation of tower shading effects and natural resonances of components.

Speed effects can be passed onto the power grid and utilized. The low damping effect was also developed by using synchronous WTGs. This may not allow drive train transients to be absorbed electrically. So an additional damping element such as flexible coupling in the drive train or the gearbox assembly mounted on springs and dampers have been incorporated to the synchronous WTGs to get better results. This may cause delicate operations and be more complex to operate, have a high cost, and also be more prone to failure than induction generators.

Synchronous machines operated using electromagnetic fields were controlled by voltage supply. The working principle of PM generators is similar to that of synchronous generators. But PM generators can be operated asynchronously. PMSGs have more advantages such as elimination of the commutator, and slip rings and brushes so that the machines are rugged, reliable, and simple. PMs remove the field winding and lead to associated power losses. Field control is also impossible, and the high cost and tedious maintenance are the disadvantages of PM generators. In the case of PMSGs, power grid incorporation is required due to the variable wind speed.

The power grid is connected through AC/DC-AC conversion by power converters. The conversion of AC power with variable frequency and magnitude is converted to DC current after rectification. The DC current is again converted into AC power with fixed frequency and magnitude. This is one of the major uses of permanent magnet machines for direct drive application. PMSGs are physically larger and have more pole numbers and have rated geared machines. The high-temperature superconducting (HTS) generator is known as the potential variant in synchronous generators. The PMSG machine is composed of stator back iron, stator copper winding, HTS field coils, rotor core, rotor support structure, rotor cooling system, cryostat and external refrigerator, electromagnetic shield and damper, bearing, shaft, and housing. In the machine design, the arrangements of the stator, rotor, cooling, and gearbox may pose particular challenges in order to keep HTS coils in cooling temperature and better operational conditions.

2.4.1.2 Preparation of Epoxy Resins
To reinforce the epoxy resin the following were used:

- Silicon carbide (80 μm)

- Zirconium diboride (100 μm)

- Mullite-like oxide crystals $3Al_2O_3 \cdot 2SiO_2$ (diameter 2–8 μm, length 80–200 μm)

- Basalt powder (100 μm)

- Diluvium powder (60 μm); chemical composition: SiO_2, Al_2O_3, Fe_2O_3, CaO, MgO

The prepared mixture was briquetted in a graphite mold. Synthesis was performed at 1850 ± 25°C for 2.5 h vacuum electric furnace. Cooling was performed under vacuum over 10 h (Chikhradzea et al. 2015).

The permanent magnet excited machines were controlled by voltage supplies by changing their converter circuit. PM generators have high power density and low mass and are successfully used in wind turbine applications. The synchronous machines (SM) and PM generator machines are known as the permanent magnet synchronous generators. These are used in small wind turbine generators. Uninterrupted power supply has been achieved by using rugged PMs and are installed on the rotor to produce a constant magnetic field. The electricity generated has been passed from the armature (stator) via the use of the commutator, slip rings, or brushes. The advanced developments in PMs integrated into a cylindrical cast aluminum rotor are used to get power supply and reduce costs.

Instead of using copper wires, superconducting coils were used. Superconducting coils were used in the power supply circuits, which carry the current efficiently 10 times better than conventional copper wires. Another advantage of superconducting coil is the negligible resistance and conductor losses. Superconducting coils also eliminate all field circuit power loss. Superconducting coils are able to withstand high current density and allow high magnetic fields. This leads to a significant reduction in mass and size for wind turbine generators. Therefore, superconducting generators are useful in high capacity and weight reductions, and are suited for wind turbines rated 10 MW or more. The world's first superconducting wind turbine generator was established by Siemens in 2005 (Sieros et al. 2012). The superconducting 4 MW synchronous generator was launched for its efficient power generation and smaller size. However, the disadvantages faced by this system are the long-life, low-maintenance of wind turbine systems. For instance, cryogenic systems need to be maintained to keep the conditions cool and restoring operations following a stoppage is an additional issue.

The design life of modern wind turbine blades is generally 20 years and they normally rotate 10^9 times during their lifetime. The design of wind turbine blades depends on the materials used. The materials must be strong, stiff, and light to achieve cost-effective production.

In 1946 Betz applied simple momentum theory to the windmill established by Froude (1889) for a ship propeller (Maurya et al. 2012). In that work, the retardation of wind passing through a windmill occurs in two stages: before and after its passage through the windmill rotor. Provided that a mass (m) is air passing through the rotor per unit time, the rate of momentum change is $m(V_1 - V_2)$, which is equal to the resulting thrust. Here, V_1 and V_2 represent upwind and downwind speeds at a considerable distance from the rotor; the power absorbed (P) can be expressed as

$$P = m(V_1 - V_2)\bar{v} \tag{2.1}$$

On the other hand, the rate of kinetic energy change in wind can be expressed as

$$E_k = 1/2\,m\left(V_1^2 - V_2^2\right) \tag{2.2}$$

The expressions in Equations 2.1 and 2.2 should be equal. So the retardation of the wind, $V_1 - \bar{v}$, before the rotor is equal to the retardation, $\bar{v} - V_2$, behind it, assuming that the direction of wind velocity through the rotor is axial and that the velocity is uniform over the area A. Finally, the power extracted by the rotor is

$$P = \rho A\bar{v}(V_1 - V_2)\bar{v} \tag{2.3}$$

Furthermore,

$$P = \rho A\bar{v}^2(V_1 - V_2) = \rho A(V_1 + V_2/2)^2(V_1 - V_2) \tag{2.4}$$

and

$$P = \rho A\bar{v}_1^3/4[(1+\alpha)(1-\alpha^2)] \tag{2.5}$$

where $\alpha = V_2/V_1$.

Differentiation shows that the power P is a maximum when $\alpha = 1/3$, that is, when the final wind velocity V_2 is equal to one-third of the upwind velocity V_1. Hence, the maximum power that can be extracted is $\rho AV_1^3 8/27$, as compared with $\rho AV_1^3/2$ in the wind originally; that is, an ideal windmill could extract 16/27 or 0.593 of the power in the wind.

The selection of the correct material is a tough task and is based on different factors such as the properties of materials, performance, reliability, safety, effects on the environment, availability, recyclability, and, most important, economics. Conventional materials for the preparation of windmill blades are listed next.

- Wood—Wood is a composite material of cellulose and lignin, and potentially used to manufacturing windmill materials due to its low density and environmental friendliness.

- Steel and nickel alloy—Steel is an alloy of iron and carbon. Nickel alloy steel has also been used for blade manufacturing due to good thermal and chemical properties such as low corrosion.

- Aluminum—Aluminum has a lower density and lower cost than steel. It is a silvery white metal with good thermal conductivity and ductility.

- Fiber—Fiber is a class of materials that are continuous filaments or are in discrete elongated pieces, similar to lengths of threads. These materials are stiff and strong. Apart from their stiffness and strength, they cannot be solely used in the manufacturing of wind turbine blades. However, they are combined in composite materials. The fibers are divided into different subclasses, for example, glass fiber, carbon fiber, aramid, polyethylene, and cellulose.

    Fiberglass materials are oxides of silicon and aluminum and some other oxides as well. The glass fiber is an amorphous solid and has properties like stiffness and thermal expansion. The glass fibers are made with different chemical compositions. The glass fibers are prepared by pulling the fibers from molten glass by the spinnerets and kept into huge bundles. The diameters of glass fiber are in the range of 10 to 20 μm. Their surfaces are coated with a polymer sizing that is used to guard them against cracking and enhance the bonding force with polymer matrix.

- Carbon fibers—Carbon fibers are composed of carbon, which forms a crystallographic lattice with a hexagonal shape called graphite. The atoms are bounded by a strong covalent bond. Carbon fibers provide higher strength than glass fibers and they are more useful in handling the fatigue.

2.4.2 Merits

- Silicon epoxy resins have reduced peak exotherm during curing for wind blade production (Chikhradzea et al. 2015).

- Provide extended pot life, fast curing, and shortened cycle times.

- Fabricators are able to control resin-curing properties.

- Optimize cycle times without compromising the strength and integrity of the blade structure.

2.4.3 Demerits

Wood

- Availability; in terms of large quantities it is difficult to conduct large-scale economical manufacturing with wood.

- Low stiffness limits to producing large rotor blades because large blades suffer from high elastic deflections.

Steel and nickel alloy

- High weight

- High cost

- Low fatigue level

Aluminum

- Low fatigue level

- Low stiffness

Fiber

- Low fatigue resistance

- Low elastic modulus and higher density than carbon fibers

2.4.4 Future Work

The growth in wind turbine installations around the world is expected to continue and actually accelerate. A study conducted by the World Energy Council (WEC) projected a worldwide wind capacity of 13 gigawatts (GW) by 2000 (actual installed capacity was 13.6 GW by the end of 1999), increasing to 72 GW by 2010 and 180 GW by 2020. WEC also considered an "environmentally driven scenario" that has much faster growth if national policies were adjusted. That scenario projected 470 GW of wind power by 2020.

Another technology shift is occurring in the drive train. In some cases the gearbox is being eliminated by employing variable speed generators and solid-state electronic converters that produce utility quality alternating current (AC) power. This trend began in small machines and is now being incorporated in turbine sizes from 100 kW to 3 MW.

2.5 NANOMATERIALS FOR WATER TREATMENT

Nanomaterials are approximately within the range of 100 nanoscale size with unique optical, magnetic, electrical, electronic, and other properties. Nanomaterials have greater surface area and quantum effects, which make them an effective material in different fields of application such as optical, electrical, medical, and environmental.

Traditional water treatment technologies remain ineffective for providing adequate safe water due to increasing demand of water coupled with stringent health guidelines and emerging contaminants. Nanotechnology-based multifunctional and highly efficient processes are providing affordable solutions to water/wastewater treatments that do not rely on large infrastructures or centralized systems.

Various nanostructured materials have been fabricated with features such as high aspect ratio, reactivity, and tunable pore volume, electrostatic, hydrophilic and hydrophobic interactions, which are useful in adsorption, catalysis, sensing, and optoelectronics. Silver nanoparticles are effective in disinfecting biological pollutants such as bacteria, viruses, and fungi. Titanium nanoparticles have been used in micropollutants transforming redox reactions. Photocatalyst nano-TiO_2 can degrade phenolic recalcitrant compounds, microbial, and odorous chemicals into harmless species. Gold and iron nanoparticles are especially suitable for removing inorganic heavy metals from surface waters and wastewaters. Nanomaterials are not only effective in disintegrating various pollutants, but they are also active in bimetallic coupling with other metals and metal oxides, which synergistically improve pollution catalysis (Saravanan et al. 2008).

Various nanomaterials can be used to make composite membranes. This enhances salt retention ability, and curtails costs, land area, and energy for desalination. For example, zeolite nanoparticles are mixed with polymer matrix to form a thin film RO membrane. It increases water transport and salt retention ability. Silica nanoparticles were doped with RO polymer matrices for water desalination. It improved polymeric networks, pore diameters, and transport properties. CNTs and graphene were used for adsorption-based desalination because of their extraordinary adsorption capacities.

2.5.1 Fabrication of Poly(Amide-Imide) (PAI) Hollow Fiber Membrane

Dry-jet wet spinning technique was used to fabricate the poly(amide-imide) (PAI) hollow fiber substrate. N-Methyl-2-pyrrolidone (NMP) and LiCl were used as solvents, in which NMP acts as an additive and LiCl was used as a pore former. In a jacket flask, 56 g of PAI and 14 g of LiCl were added to 330 g of NMP solvent. The mixture was stirred under 70°C for 3 days to dissolve the polymer and filtered then degassed. The resultant dope solution was incorporated with multiwall CNTs (MWCNTs) by vacuum filtration. The ends of each fiber were glued by epoxy into a 3 cm polypropylene tube (1/2 in. to 3/8 in.). The resultant fibers were immersed in MWCNT solution (0.31 mg/L) and kept under vacuum at 3.5 min. Finally MWCNT-incorporated PAI fibers were washed with deionized water and centrifuged at 1200 rpm for 20 s to remove the unbound CNTs (Figure 2.4).

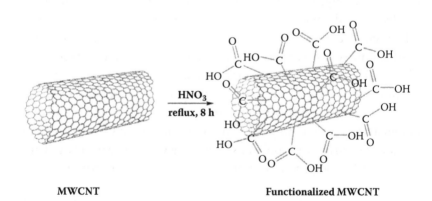

MWCNT **Functionalized MWCNT**

FIGURE 2.4 Fabrication of polyamide-imide (PAI) hollow fiber membrane with MWCNT. (Courtesy of Goh, K. et al., *Journal of Membrane Science* 446: 244–254, 2013.)

2.5.2 Fabrication of Polyethyleneimine–Poly(Amide-Imide) (PEI–PAI) Hollow Fiber Membrane

Posttreatment to incorporate polyethyleneimine (PEI) to PAI incorporated MWCNTs was prepared by immersing the fibers into a 500 mL PEI solution of 1 wt% concentration at a temperature of 70°C for 90 min. After treatment, the fibers were rinsed three times using deionized water to remove excess of PEI (Goh et al. 2013).

2.5.3 Treatment of Salinated Water through Carbon Nanotube Membranes

CNTs are used as nanoadsorbents that have attributed the antimicrobial effects of CNTs to their unique physical, cytotoxic, and surface functionalizing properties; their fibrous shape; the size and length of the tubes; and number of layers (single or multiwalled) (Figure 2.5). CNTs are very effective in removing bacterial pathogens and the aforementioned construction would work best for the treatment. CNTs possess antimicrobial characteristics against a wide range of microorganisms including bacteria such as *E. coli* and *Salmonella* and viruses. The adsorption of cyanobacterial toxins on CNTs is also higher when compared with carbon-based adsorbents mainly due to large specific surface area, external diameter of CNTs, and large composition of mesoporous volume (Noy et al. 2007). Nanoporous activated carbon fibers and CNTs are used for the adsorption of organic contaminants such as benzene, toluene, xylene, ethyl benzene, 1,2-dichlorobenzene, trihalomethanes, chlorophenols and herbicides, and DDTs. MWCNTs functionalized with Fe nanoparticles have been used as an effective sorbent for aromatic compounds like benzene and toluene.

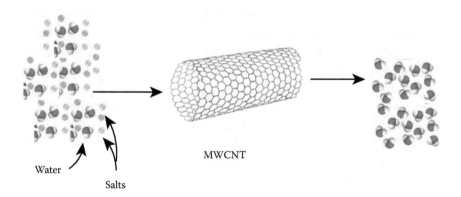

FIGURE 2.5 Treatment of salinated water through CNT membrane.

Filtration membranes containing radially aligned CNTs are very effective in removing both bacteria and viruses in a very short time due to size exclusion and depth filtration, and thus enable such filters to be used as cost-effective and point-of-use water disinfection devices (Das et al. 2014). CNTs can also reduce membrane biofouling and a nanocomposite membrane of single-walled CNTs and polyvinyl-N-carbazole showed high inactivation of bacteria upon direct contact. Another example of controlling the biofouling in thin film nanocomposite membranes is the single-walled CNTs covalently bonded to a thin film composite membrane surface that has exhibited moderate antibacterial properties.

Among the other nanomaterials for microbial disinfection, bifunctional ferrous oxide (Fe_3O_4) with Ag and Fe_3O_4 with TiO_2 nanoparticles were employed successfully against pathogenic bacteria including *E. coli*, *Staphylococcus epidermidis*, and *Bacillus subtilis*. Magnesium oxide nanoparticles were also used effectively as biocides against both Gram-positive and Gram-negative bacteria and bacterial spores. Nano tungsten oxide and palladium-incorporated ZnO nanoparticles have shown good antibacterial properties at removal of *E. coli* from water. Nitrogen-doped ZnO and zirconium oxide nanoparticles also have proved to be good antibacterial nanostructured materials.

Different types of nanomaterials used as a nanosorbents for the removal of heavy metals from wastewater, such as CNTs, zeolites, and dendrimers, are used to adsorb Cd^{2+}, Cr^{3+}, Pb^{2+}, and Zn^{2+} and metalloids such as arsenic compounds. Composites nanomaterials are also used for the metal adsorption, such as CNTs incorporated with Fe and cerium oxide. Cerium oxide nanoparticles supported on CNTs are used effectively to adsorb arsenic. Effective treatment of Cr(VI) to Cr(III) are reported for titania nanoparticles and palladium nanoparticles. The iron oxide nanoparticles are used for the arsenic removal, and polymer-grafted Fe_2O_3 nanocomposite is effectively used to remove divalent heavy metal ions from copper, nickel, and cobalt. Bisphosphonate-modified magnetite nanoparticles are used to remediate the radioactive metal toxins such as uranium dioxide (Hristovski et al. 2009). The zero-valent iron or iron nanoparticles are very effective for the transformation of heavy metal ions such as As(III), As(V), Pb(II), Cu(II), Ni(II), and Cr(VI). Chitosan nanoparticles are used for the removal of lead from the contaminated water.

The nanofilter membranes of self-assembled positive poly(allyl amine hydrochloride) and negative poly(styrene sulfonate) onto porous alumina exhibited a high retention of the salts of calcium and magnesium. The iron nanoparticles composite membranes prepared from incorporating the iron hydroxide nanoparticles into porous carbon materials are used for the removal of both inorganics and organics and utilized for point-of-use applications.

The metal oxide nanomaterials such as TiO_2 incorporated CeO_2 are used for the complete degradation of organic pollutants in ozonation processes (Patkar and Laznow 1992). Photocatalysts like TiO_2 nanoparticles are used to treat organic pollutants such as polychlorinated biphenyls (PCBs), benzenes, and chlorinated alkanes. Nanocomposites such as nitrogen and Fe (III) doped with TiO_2 nanoparticles are used for the degradation of azo dyes and phenol, and TiO_2 nanoparticles deposited onto porous alumina are used for the removal of TOC of wastewater. The silica incorporated TiO_2 nanocomposites are used to remove the aromatic pollutants. The TiO_2 nanotubes are used to degrade organic pollutants such as toluene.

Nanostructured ZnO nanomaterials are useful for the treatment of hospital wastewater containing organic contaminants such as 4-chlorocatechol (Vlazan et al. 2015). Ag and amidoxime nanocatalyst fibers are used for the degradation of organic dyes. The Pd-based nanoparticles are reported for the degradation of halogenated organic compounds and the bimetallic nanocomposite of Pd-Cu/-alumina was used for the reduction of nitrate. Pd nanoparticles and bimetallic Pd/Au (gold) nanoparticles and the immobilized nanoparticles of metallo-porphyrinogens are used effectively for the degradation of hydrodechlorination of trichloroethylene (TCE). MnO_2 nanoparticles are reported for the removal of organic pollutants.

Cellulose acetate membrane incorporated with zero-valent iron is useful for treating hospital wastewater for the dechlorination of TCE. The bimetallic zero-valent iron and platinum nanoparticles are used to reduce TCE. Bimetallic Fe/Ni and Fe/Pd nanoparticles incorporated in zero-valent iron as polymer–inorganic porous composite membranes have been successfully used for the reductive degradation of halogenated organic solvents. Polyvinylidene fluoride film containing Pd and Pd/Fe are used for the dechlorination of PCB. Alumina-zirconia-titania ceramic membrane coated with Fe_2O_3 nanoparticles are reported for the reduction of the dissolved organic carbon and enhancing the degradation of NOM (Maximous et al. 2010). The ceramic composite membranes incorporated with TiO_2 and CNTs enhance the membrane permeability and improve the photocatalytic activity. The ceramic membrane incorporated with Al_2O_3, Fe, and Mn nanoparticles are useful for the selective adsorption of dye wastewater. The integrated membranes incorporated with the silica compounds increase the membrane quality with improved antifouling performance and flux increase.

The ultrafiltration membranes are improved with the incorporation of nanomaterials and are used to treat the organic and inorganic pollutants in industrial discharge. The dendritic nonmaterials loaded polymers are used as water-soluble ligands for radionuclides and inorganic anions. The composite membranes of alumina and citrate stabilized gold nanoparticles are reported for the complete reduction of 4-nitrophenol. The ultrafiltration membranes loaded with metal oxide nanoparticles such as silica, TiO_2, alumina, and zeolites to polymeric ultrafiltration membranes are efficient for reducing fouling.

Nanomaterials, nanofibers, nanomembranes, and composite nano-structures membranes can help to degrade a wide range of organic and inorganic contaminants in the field of wastewater treatment. The natural attenuation by the nanoparticles and the synthesized nanoparticles are useful for the degradation of wastewater containing metal ions, dye, and other contaminants. The nanocomposites membranes are reported for their improved performance in the field of ultrafiltration, nanofiltration, and the degradation of radioelements and harmful metal ions, toxicants from herbicide, and hospital wastewater.

2.5.3.1 Merits of CNT Membranes

- Frictionless water flow
- Retention of a broad spectrum of water pollutants
- The inner hollow cavity of CNTs resulted an improved desalinating water
- The high aspect ratios, smooth hydrophobic walls, and inner pore diameter of CNTs allow ultraefficient transport of water molecules

The smooth and hydrophobic inner core of hollow CNTs allowed the uninterrupted and spontaneous passage of water molecules and retained the pollutants. The purified water only passed away and the impurities were adsorbed on the adsorbent (Das et al. 2014). The membrane allowed the high velocities of feed from 9.5 to 43.0 cm s^{-1}/bar, which is 4 to 5 times faster than the conventional membrane flow velocities.

2.5.3.2 Demerits of Nanomaterials for Water Treatment Process

Nanomaterials for water treatment also have the drawbacks, which include

- Thermal instability
- Requirement of high pressure
- Fouling
- Pollutant precipitation
- Pore blocking
- Low influx
- Slow reaction
- Formation of toxic intermediates
- Aggregation on storage

2.5.4 Future Scope

CNT-based membranes have remarkable accomplishments in terms of water permeability, desalination capacity, solute selectivity, robustness, antifouling, energy savings, and scalability. CNT membranes can be used at all levels from the point of generation (POG) to the point of use (POU) treatments. CNT membrane filters have the potential to give potable water instantly and serve as next-generation universal water filters. CNT membranes can also be used to scavenge heavy metal ion pollutants. CNT membranes also resist membrane fouling followed by coagulation and pore blocking; hence, they can be used in an antibacterial filter. CNT membranes reduce biofouling and are able to self-clean. The small size and soft condensed architecture of CNT membranes help them autoarrange themselves without any external input. High water flux with selective salt rejection has raised the demand for water desalination.

2.6 CAPTURING CARBON DIOXIDE USING NANOSTRUCTURE MEMBRANES

Increasing CO_2 emissions in the environment leads to global warming, which is an issue of great concern today. CO_2 is the major contributor for global warming among other greenhouse gases and contributes approximately 55% of the observed global warming. CO_2 alone is responsible for about 64% of the enhanced greenhouse effect.

Currently, coal-fired power plants emit about 2 billion tons of CO_2 per year. The Intergovernmental Panel on Climate Change (IPCC) predicts that by the year 2100 the CO_2 presence in the atmosphere may be up to 570 ppm, causing a rise of mean global temperature of around 1.9°C and an increase in mean sea level of 3.8 m. The fuel thermal power plant, including coal, oil, and gas, provides 85% of total world demanded energy. Nearly, 40% of total CO_2 emission is from fossil fuel power plants and coal-fired power plants are the main contributors among them. The capture of CO_2 from power plant flue gas accounts for three-quarters of the total cost of carbon capture and storage. Nowadays, more international attention is focused on global warming.

There are several technologies available for CO_2 capture, including absorption, adsorption, gas separation, membrane, and cryogenic separation. In membrane technology for CO_2 separation, the required energy per unit mass of CO_2 captured is least in the range of 0.5–6 MJ/kg of CO_2 as compared to other existing technologies.

Carbon captured (CC) membrane gas separation, especially CO_2 separation, has been known since the 1980s. Membrane separation is used both in the petrochemical industry and in the cleaning of natural gas. The membranes used in these cases were to recover hydrogen and carbon dioxide from methane. These processes proceeded at high pressure, and the use of highly selective materials, especially polymers, has allowed membrane preparation with selectivity ranges between 200 and 400.

2.6.1 Membrane Technology

A novel capture of CO_2 is the use of selective membranes to separate certain components from a gas stream from natural gas (natural gas processing). Membranes are semipermeable barriers able to separate substances by solution/diffusion, adsorption/diffusion, molecular sieve, and ionic transport mechanisms, and are available in different forms either organic (polymeric) or inorganic (carbon, zeolite, ceramic, or metallic), and can be either porous or nonporous (Ma et al. 2005). The membrane processes for CO_2 removal are classified into two types:

- Gas separation membrane

- Gas absorption membrane

2.6.1.1 Gas Separation Membrane
Gas separation membranes (Figure 2.6) operate on the principle of referential permeation of mixture constituents through the pores of the

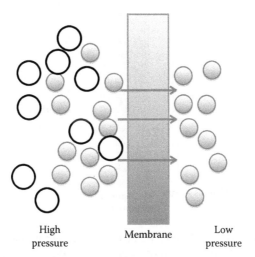

FIGURE 2.6 Gas absorption membrane. (Courtesy of International Energy Agency Working Party for Fossil Fuels, 2002, Solutions for the 21st century: Zero emissions technologies for fossil fuels.)

membrane causing one component to diffuse through the membrane faster than the others, as observed by Mondal et al. (2012). The main design and operational parameters of membrane are selectivity and permeability.

In this process, the gas stream containing CO_2 is introduced at elevated pressure into the membrane separator, and CO_2 selectively permeates through the membrane and is recovered at reduced pressure on the shell side of the separator. Various types of gas separation membranes including ceramic, polymeric, and a combination of both materials or mixed matrix membranes are available.

Figure 2.6 indicates the gas absorption membrane consists of microporous solid membrane. The CO_2 to be separated from flue gas is diffused through the membrane, and then is recovered by liquid absorbent by absorption. Gas adsorption membrane process gives higher removal rate than gas separation membrane due to high driving force at any instant compact than a conventional membrane separator.

According to Mondal et al. (2012), porous membrane processes (Figure 2.7) have the high capability of maintaining product purity even though the capacity is reduced to 10% of the initial design value. Moreover, it is a low-cost process for separating gases and extremely reliable with respect to the on-stream factor.

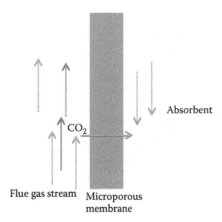

Absorbent

CO_2

Flue gas stream   Microporous
membrane

FIGURE 2.7 Porous membrane. (Courtesy of International Energy Agency Working Party for Fossil Fuels, 2002, Solutions for the 21st century: Zero emissions technologies for fossil fuels.)

2.6.1.2 Integrated Gasification Combined Cycle Systems with CO_2 Capture

Nowadays, the integrated gasification combined cycle (IGCC) acts as an efficient power generation technology with the lowest carbon dioxide emissions among coal power plants and is a good candidate for CO_2 capture (Descamps et al. 2008).

2.6.1.3 Systems Using Hydrogen Selective Membranes

To increase power plant efficiency Rezvani et al. (2009) developed the IGCC system configured with a water gas shift membrane reactor (WGSMR) and an oxygen transport membrane (OTM) system. The OTM utilizes the remaining combustibles in the outcoming gas from WGSMR. According to Kotowicz et al. (2010), the generated hydrogen in the retentive side permeates through the hydrogen selective membrane. A steam flow coming from the heat recovery at a temperature of 400°C and steam generation (HRSG) is used to sweep the hydrogen from the permeate side to the gas turbine. The remaining incombustible gas constituents are combusted in an oxygen selective membrane and feedback to the gas turbine. After the flue gas condensates, the CO_2 is compressed for the transport through the pipeline. The power requirement for the compression train is reduced by the high pressure of CO_2. The CO_2 capture efficiency is around 93%, higher than using physical absorption, which is nearly 91%. The hydrogen selective membrane used in this investigation is based on a palladium silver alloy (Pd/Ag) on a silicon wafer with a perm selectivity of around 4000.

According to Rezvani et al. (2009) in general, Pd/Ag membranes achieve a very high permselectivity. However, they are associated with poor chemical and thermal stability. Thus the IGCC systems fabricated with hydrogen selective membranes for CO_2 capture resulting in low energy penalties and minimized CO_2 avoidance costs.

Gas transport behavior of PIM-1/silica nanocomposite membranes is by the addition of filler enhanced gas permeability. This shows that the trend is related to the true silica volume and void volume fraction (Ahn et al. 2010). The incorporation of silica nanoparticles in PIM-1 significantly increased overall gas permeability.

CNTs are very effective in reinforcing polymeric materials, but it is unknown whether they degrade the membranes' gas separation performance. Cong et al. (2007) by the addition of the single-wall CNTs (SWNTs) and multiwall CNTs (MWNTs) formed polymeric nanocomposite membranes with brominated poly (2,6-diphenyl-1,4-phenylene oxide) BPPOdp. Here, the CO_2 permeability increased with increasing the CNT content and reached a maximum of 155 barrer at 9 wt% of SWNTs. The addition of CNTs to polymeric membranes improved mechanical strength without deteriorating the gas separation performance of the membranes. High CNT concentrations degraded the CO_2 permeability and selectivity. It was found that MWNTs were more effective at increasing gas permeability than SWNTs.

A poly(ethylene oxide)–poly(butylene terephthalate) multiblock copolymer used as membrane material was reported by Yave et al. (2010a). Here, additives such as polyethylene glycol dibutyl ether were incorporated as spacers or fillers increasing the CO_2 permeability (750 barrer) up to fivefold without the loss of selectivity for producing nanostructured materials. The membranes present outstanding performance for CO_2 separation, and the measured CO_2 flux is extremely high ($>2m^3m^{-2}$ h^{-1} bar^{-1}) making them attractive for CO_2 capture.

Carbon dioxide selective membrane materials from a commercially available poly(amide-ethylene oxide) (Pebax®, Arkema) blended with polyethylene glycol ethers were fabricated by Yave et al. (2010b). Poly(ethylene glycol) dimethyl ether (PEG-DME) is well known as a physical solvent for acidgas absorption. It is used under the trade names Genosorb® and Selexol®. The combination of absorbent (PEG-DME) with multiblock copolymer resulted in mechanically stable films with superior CO_2 separation properties was observed by Reijerkerk et al. (2010). Here the incorporation of 50 wt% PEG-DME to the copolymer resulted in an eightfold increase of the carbon dioxide permeability. The diffusivity as well as solubility of carbon dioxide is strongly increased by the addition of PEG ethers.

Nanocomposite membranes consisting of SWNTs embedded in a poly(imide siloxane) copolymer was evaluated by Kim et al. (2006) for their gas transport properties. The siloxane segment enhanced the interfacial contact and the polyimide component imparted mechanical integrity. A poly(imide siloxane) was synthesized using an aromatic dianhydride, an aromatic diamine, and amine-terminated PDMS for the siloxane block. The permeability of CO_2 increased 166.02 ± 1.76 to 190.67 ± 0.59 after the addition of 2 wt% CNTs.

Ahn et al. (2008) prepared glassy polysulfone mixed-matrix membranes by introducing nonporous nanosized silica particles. The addition of silica significantly enhances the gas permeability of polysulfone with increasing silica content that results from an increase in free volume because of the inefficient chain packing as well as the presence of extra void volume at the interface between polymer and silica clusters. Consequently, the incorporation of fumed silica did not result in an overall improvement in performance of permeability versus selectivity in relation to the upper bound. However, it is notable that nonporous nanosized silica disrupts polymer chain packing leading to an increase in free volume in low-free-volume glassy polymers. Furthermore, the relative enhancement in permeability is significantly greater than in high-free-volume glassy polymers.

According to the reports by Ghalei and Semsarzadeh (2007) poly(urethane)/poly(vinyl acetate) (PU/PVAc) blend membranes prepared by the solvent evaporation process had a higher permeability for CO_2 than the pure PU and a lower permeability to the other gases.

In membrane science and technology, the fabrication of ultrathin polymer films (defect-free) made from a CO_2 philic polymer material on square meter scale with uniform thickness (<100 nm) show high permeances for CO_2, i.e., >5 m^3 (STP) m^{-2} h^{-1} bar^{-1}. Hence it finds application in the capturing of CO_2 from flue gas. Yave et al. (2010b) demonstrated that ultrathin film membranes manufactured from the commercial Polyactive (poly1.5*k*) are attractive for CO_2– capture from flue gas of coal-fired power plants.

Reijerkerk et al. (2010) reported the blend membranes made from the commercially available PEBAX® MH1657 and a PEG-based additive, showed an increase in CO_2 permeability by a factor of 5 to approximately 530 barrer at 50 wt% PDMS-PEG loading. The results show the blend has strong potential to increase the gas permeability.

For the purpose of carbon dioxide separation from nitrogen, Xomeritakis et al. (2009) fabricated amine-derivatized and nickel-doped sol–gel silica membranes Membralox-type commercial ceramic supports (Figure 2.8). Pure silica exhibited flux decline in the presence of trace SO_2 gas in the feed. Doping the membranes with nickel (II) nitrate salt was effective for higher permeance and higher separation factor of the doped membrane compared to the pure (undoped) silica membrane after 168 h exposure to simulated flue gas conditions.

2.6.2 Advantages of Membranes Used for CO_2 Separation

According to Kim et al. (2006), the energy needs of the CCS installation result from the power demand for CO_2 separation, and compression reduces the efficiency of the power plant. For a pressure of 150 bar, CCS installation requires 31.6% of the total power of the plant, and the efficiency drops by 15.4 percentage points. Lowering CO_2 temperature before compression to 79.5°C means that the CCS installation will use 21% of the total plant power, with an overall drop in efficiency of 10.26 percentage points.

2.6.3 Disadvantages of Membranes Used for CO_2 Separation

There are some limitations in the use of membranes for the CO_2 capture. But low removal efficiency and low purity of CO_2 make this process ineffective, and its feasibility is decreased when concentration of CO_2 in feed stream is below 20%.

Usually large quantities of gases will need to be processed. Since the high temperatures of flue will damage the membrane and the temperature of the flue should be cooled below 100°C. Also, the membrane needs to be chemically resistant to corrosive chemicals present in the flue gases that were subjected to membrane separation. In addition, the required pressure difference leads to large power consumption in the power plant.

Since, the membranes are sensitive to sulfur compounds and other traces, multiple stages or recycling are necessary. Thus, much attention is needed to improving the transport properties of membrane.

2.6.4 Future Developments

Currently, Membrane Technology and Research, Inc. (MTR), have fabricated, MTR polaris membrane, with high flux, that shows a permeance 10 times higher than cellulose acetate.

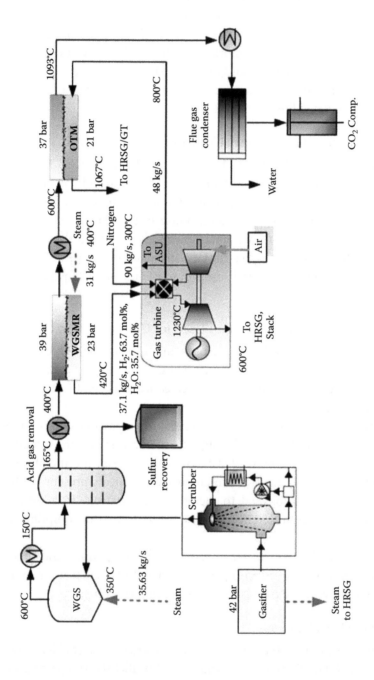

FIGURE 2.8 Schematic operation of CO$_2$ removal using membrane technologies. (From Rezvani, S. et al., *Fuel* 88, no. 12:2463–2472, 2009.)

Researchers in Japan from the Institute for Integrated Cell-Material Sciences (ICMS) and the University of Cambridge created an advanced membrane capable of rapidly separating gases with advanced features efficient in removing harmful greenhouse gases from the atmosphere. Polymers of intrinsic microporosity (PIM-1) have highly permeable membranes compared with commercially available ones. The main disadvantages of the membrane is that it is flimsy and its starting selectivity. These drawbacks are overcome by thermal oxidation of PIM-1 at temperatures ranging from 120°C to 450°C in the presence of oxygen. The resulting, improved PIM-1 was found to be twice as selective for carbon dioxide. It allows air 100 times faster compared with commercially available polymers. Also, it is affordable and long lasting, and reduces the cost of carbon dioxide capture by as much as 1000 times.

2.7 NANOPARTICLES FOR TREATING VOLATILE ORGANIC COMPOUNDS

Volatile organic compounds (VOCs) are emitted as gases from certain solids or liquids. Concentrations of many VOCs are higher indoors than outdoors. In recent years, VOCs have contributed drastically to air pollution. VOCs such as acetone, formaldehyde, benzene, toluene, ethylbenzene, xylene (BTEX), dichlorobenzene, and chloroform are currently receiving much attention due to their global greenhouse effect and their severe effect on the human health. The sources of VOC emission are as follows: (1) outdoor plants, factories, incinerators, office equipment such as copiers and printers, pesticides, and means of transport; and (2) indoor furnishings, paints, cigarettes, cosmetics, aerosol sprays, room fresheners, and adhesives. The common health effects caused by VOCs are eyes, nose, and throat irritation; headaches; loss of coordination; and nausea. Some organics can cause cancer in animals; some are suspected to cause cancer in humans.

There are various methods to remove VOCs, such as adsorption, absorption, biofiltration, thermal incineration (combustion), and catalyst incineration, among which catalytic combustion is believed to be one of the most efficient pathways (Ruddy and Carroll 1993). Catalysts can be used to enable a chemical reaction at lower temperatures or make the reaction more effective. Nanotechnology can improve the performance and cost of catalysts used to transform vapors escaping from cars and industrial plants into harmless gases or pollutants emitted from buildings. That is because catalysts made from nanoparticles have a greater surface area to interact with the reacting chemicals than catalysts made from larger particles, which makes the catalyst more effective.

2.7.1 Catalytic Oxidation System

Catalytic oxidation systems could directly combust VOCs at a lower temperature of about 700°F to 900°F. This is made possible by the use of catalysts that reduce the combustion energy requirements (Figure 2.9). The incoming gas stream is heated, most often in a recuperative heat exchanger followed by additional input from a burner if needed, and passed through a honeycomb or monolithic support structure coated with catalysts.

Catalyst systems can be designed to handle a capacity of 1000 to 100,000 cfm and VOC concentration ranges from 100 to 2000 ppm. They are often used for vent controls where flow rates and VOCs content are variable. Lower operating temperatures, combined with a recuperative heat exchanger, reduce the startup fuel requirement. Catalyst systems can produce secondary combustion wastes.

Halogens and sulfur compounds are converted to acidic species by the catalytic combustion process; these are treated by using acid–gas scrubbers. Also, the spent catalyst materials can require disposal as a hazardous waste, but they are not recyclable. Catalyst materials can be sensitive to poisoning by non-VOC materials such as sulfur, chlorides, and silicon.

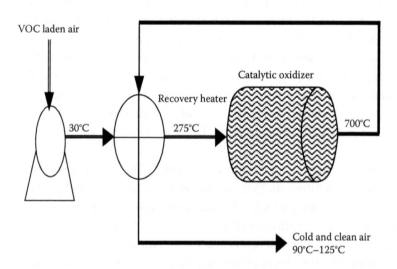

FIGURE 2.9 Schematic diagram of catalytic oxidizer. (Courtesy of Khan F.I., and Ghoshal, A.K., *Journal of Loss Prevention in the Process Industries*, 13:527–545, 2000.)

Many catalyst manufacturers have overcome sensitivity to some of these substances, but every catalyst has susceptibilities that must be considered at the process selection stage.

For example, some catalysts are sensitive to deactivation by high molecular weight hydrocarbons or polymerizing materials. Also, the catalyst support may become deformed at high temperatures and high concentrations. Researching these issues should be part of the process selection activity if catalytic oxidation is under consideration (William and Lead 1997).

Catalytic oxidation for control of trace VOCs can be carried out in two ways: (1) a direct contact with an open flame to "preheat" the gas stream of the catalyst or (2) the open flame both preheats the gas stream to an elevated temperature so catalytic oxidation can take place and actually accomplishes some measure of VOC oxidation. This preheat flame may accomplish a significant portion of the overall observed oxidation (Palazzolo et al. 1985). The chemical mechanism of open-flame oxidation involves free-radical-induced homogeneous reactions and is fundamentally different from heterogeneous catalytic oxidation, which involves activated complexes formed on the catalyst surface. The second method involves only a catalyst bed, over which the gas stream passes, usually after some indirect preheating. The difference between these two configurations is the presence of an open flame, but these differences can be important because the mechanism of oxidation on a catalyst in close proximity to a flame may be different from that on a catalyst by itself (Figure 2.10).

Industrial catalytic oxidation reactions and catalysts (Satterfield 1980):

$$C_2H_4 + 1/2\,O_2 \rightarrow C_2H_4O \qquad \text{Ag on } \alpha\text{-alumina (catalyst)}$$
$$C_3H_6 + O_2 \rightarrow C_3H_4O + H_2O \qquad \text{BiMo/SiO}_2, Cu_2O \text{ (catalyst)}$$
$$C_3H_6 + NH_3 + 3/2\,O_2 \rightarrow C_2H_3CN + 3H_2O \quad \text{Te, Ce oxides on silica (catalyst)}$$

An interesting approach for producing porous manganese oxide embedded with gold nanoparticles was provided by Sinha et al. (2007). γ-MnO$_2$ was synthesized through a surfactant-assisted wet-chemistry route and presented an exceptionally high surface area (300 m^{2}g^{-1}). Then, gold nanoparticles were deposited on this oxide by a vacuum ultraviolet radiation (VUV)-assisted laser ablation (VALA) method to include lattice defects and strong metal-support interactions. By this way, Au

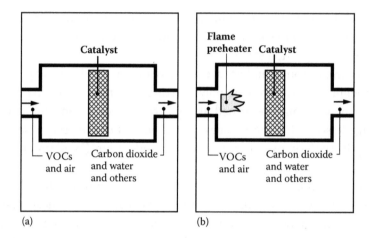

FIGURE 2.10 Catalytic oxidation systems (a) without preheater and (b) with direct flame preheater. (Courtesy of Bond, G.C. et al., Complete catalytic oxidation of volatile organics, in Catalysis—A review of recent literature, *Royal Society Chemistry*, 8, 204, ISBN: 0851865941, 9780851865942, 1989.)

nanoparticles of 3 to 6 nm in size were well dispersed and clearly seen to be embedded in the support lattice. Some Au nanoparticles were observed inside mesoporous channels.

XANES (x-ray absorption near edge structure) measurements confirmed the presence of completely metallic particles. However, XPS (x-ray photoelectron spectroscopy) showed a slight shift in Au^o binding energies indicating a partial positive charge on Au sites resulting from the strong electronic interaction between the metal and the support. This material was tested for VOCs removal and it was found very active. Three different classes of VOCs, specifically acetaldehyde toluene and *n*-hexane, were chosen to reveal the efficacy of the mesoporous γ-MnO_2/nanogold catalysts.

Manganese oxide, without gold doping, was able to completely remove acetaldehyde within 1 h at room temperature, with efficiency that is about 2 to 3 times better than those of conventional materials used to remove VOCs. However, nondoped material was not able to convert *n*-hexane in a mixture of toluene, acetaldehyde, and *n*-hexane, and toluene conversion was only at around 60%. On the contrary, γ-MnO_2/Au showed a toluene conversion of 99% and *n*-hexane conversion of 30%. Moreover, an increase in conversion was obtained by raising the temperature up to 85°C, with 99% toluene removal and 77% *n*-hexane removal.

A different method for preparing a 3D mesoporous chromium oxide material was demonstrated by Sinha et al. (2005). Chromium oxide material was synthesized by a neutral templating route and it confirmed an exceptionally high removal/oxidation ability for VOCs. XRD (x-ray power diffraction) analysis of template-free calcined sample confirmed the presence of trivalent Cr. However, XPS revealed the presence of Cr species with different oxidation states.

The template free sample, which obtained a surface area of 212 m^2g^{-1} and decreases to 78 m^2g^{-1} after calcinations at 500°C, was tested for VOC removal and it was found very efficient. The sample was able to remove the highest amount of toluene among all conventional materials, with 52% toluene removal and 11 ppm CO_2 formation at room temperature after 25 h. Further, an increase in toluene removal ability was obtained by increasing the temperature to 85°C, with 65% toluene removal and 66 ppm CO_2 formation.

Porous Co_3O_4 nanoflower clusters fabricated by a simple low-temperature hydrothermal method were reported by Yan et al. (2012). The XRD pattern of the sample, prepared by calcining the cobalt precursor at 300°C for 5 h, implied a good crystallinity of pure Co_3O_4 phase. The SAED (selected area electron diffraction) pattern indicated that the material is of polycrystals nature. Further, XPS confirmed the presence of Co_3O_4 by the O 1s peak at 531.7 eV.

The catalytic activities of prepared Co_3O_4 nanoflower clusters were compared with Co_3O_4 blocks, namely, Co-T_1, Co-T_2, and Co-T_3, and they were found much superior under the same conditions. Catalytic activity of Co_3O_4 blocks at 100% conversion were in the order Co-T_1 > Co-T_2 > Co-T_3 and revealed that the toluene removal rate of all the three samples are less than 40%. The Co_3O_4 nanoflower clusters sample was able to remove 90% toluene after 50 h at 250°C.

Synthesis of mesoporous silica with gold nanoparticles for catalytic oxidation of VOCs was demonstrated by Wu et al. (2013). Silica is used as a support and fabrication of Au/SiO_2 catalyst was explained by various methods, for example, the colloidal deposition method, doping oxides, premodification, one-pot synthesis, two-step in situ method, and gold cationic complex precursor. Colloidal deposition method was able to aggregate large gold nanoparticles on the surface of amorphous silica with irregular shapes.

Doping metal oxides, namely, Co_3O_4, CeO_2, TiO_2, FeO_x, and CuO_x, onto the silica support confirmed an increase in the binding effect. By this way, gold nanoparticles of 3 to 5 nm were well dispersed. The premodification method functionalized the negatively charges silica support with cationic organosilane, which facilitated uniform distribution of gold nanoparticles. One-pot synthesis effectively synthesized gold nanoparticles of 2 to 5 nm in mesoporous silica. The two-step in situ method encapsulated gold nanoparticles of in the size range of 2 to 10 nm in mesoporous silica support. Gold cationic complex precursor used $Au(en)_2Cl_3$ as the gold precursor and encapsulated gold nanoparticles in the silica matrix upon high temperature treatment about 800°C to 900°C.

Zinc-oxide-based gold nanoparticles were fabricated by Wu et al. (2011) for catalytic oxidation of benzene, toluene, and xylene (BTX). The zinc oxide nanoparticles were prepared by the sol-gel method. The gold catalyst was prepared by a colloidal deposition method. The Au/ZnO catalyst was fabricated by adding support ZnO nanoparticles to the gold solution. EDX (energy-dispersive x-ray spectroscopy) confirmed the presence of gold nanoparticles. High-resolution transmission electron microscopy (HRTEM) illustrated that the Au nanoparticles are hemispherical in shape and highly dispersed in the pores of ZnO nanoparticles.

Au/ZnO was tested for catalytic oxidation of BTX and it was found very effective compared to Au/Al_2O_3 and Au/MgO. However, the Au/ZnO with lower BET area and Au loading, the catalytic activity is still 2.5 and 5.2 times higher than that of Au/Al_2O_3 and Au/MgO at 150°C. Catalytic activity of the Au/ZnO catalyst was in the following order: benzene > toluene > p-xylene The Au/ZnO catalyst was able to remove more than 80% benzene at 150°C.

Total oxidation of volatile organic compounds on Au/FeO_x catalysts supported on mesoporous SBA-15 silica was reported by Bonelli et al. (2011). Support, mesoporous SBA-15 was fabricated by use of commercially available triblock poly(ethylene oxide)–poly(propylene oxide)–poly(ethylene oxide) copolymers in acid media. Au/FeO_x/SBA-15 material prepared by bimetallic carbonyl cluster deposition method was compared against that prepared by the deposition-precipitation (DP) method.

The small-angle X-ray scattering (SAXS) pattern showed a slight shift in peak from 0.92 to 0.95° 2θ indicating the presence of metallic carbonyl cluster-derived catalysts, and a decrease in d-spacing from 9.6 nm to 9.3 nm confirmed the presence of metal particles inside the pore channels of SBA-15. This material was tested for VOC removal and it showed the best performance. Au/FeO_x/SBA-15 material prepared by bimetallic carbonyl cluster deposition method was able to remove above 90% methanol at 150°C and a similar performance was shown by Au/FeO_x/SBA-15 material prepared by the DP method. However, the activity of bare SBA-15 was not very active in methanol removal.

Catalytic properties of silicate MCM-41 material modified with gold, vanadium, and niobium were described by Sobczak et al. (2008). MCM-41 materials containing Au and V, or Au, V, and Nb were fabricated by the hydrothermal method with the use of H_2SO_4 and HCl. XRD patterns of MCM-41 materials containing Au and V, or Au, V, and Nb with the use of H_2SO_4 revealed a well-ordered hexagonal peak at $2\theta = 2.2°$ and confirmed the presence of gold particles. However, XRD patterns of MCM-41 materials containing Au and V with the use of HCl revealed the disordering of the hexagonal structure in the long range.

These materials were tested for methanol oxidation. The basicity of AuMCM-41 was reduced by introducing V and/or Nb together with Au into MCM-41. AuVNbMCM-41(H_2SO_4) catalyst was more selective to formaldehyde in methanol oxidation. The mesoporous materials prepared by the use of HCl appeared to be a high total oxidation of methanol to CO_2. In addition, the total oxidation activity decreased with an increase in the acidic nature of the catalyst.

Total oxidation of VOCs on gold embedded on cobalt doped mesoporous UVM-silica was studied by Solsona et al. (2012). Cobalt doped mesoporous UVM-silica was prepared by introducing Co-atrane complex to the silatrane-containing solution. Then, gold was deposited using a postsynthesis deposition-precipitation method. XRD, TEM, and N_2 adsorption-desorption isotherms confirmed the mesoporous nature of material. Atomic absorption analysis established the percentage weight of gold deposited. This material was tested for VOC oxidation and it gave a good performance in the total oxidation of propane and toluene. Bimetallic Au/Co-UVM catalyst was more active than Au/UVM catalyst. Moreover, this catalyst has shown a high stable activity during oxidation of propane and toluene for at least 50 h at 350°C.

2.7.2 Advantages of Nanocatalysts in Catalytic Oxidation

- Increased selectivity and activity, low energy consumption, and long lifetime of catalysts by controlling pore size.

- A catalyst tailored at the nanoscale improves chemical reactivity and reduces process costs.

- Catalysts can remove unwanted molecules from gases very efficiently by controlling the pore size and characteristics.

- In catalytic oxidation system, maximum energy recovery is possible up to 70% by nanocatalysts.

2.7.3 Disadvantages of Nanocatalysts in Catalytic Oxidation

- Certain compounds can poison the catalyst.

- May require additional control equipment downstream.

2.7.4 Future Scope

Research in Utrecht makes use of structured nanoporous materials that were made by coating micelles. This results in structured pore systems of SiO_2 with uniform 3 nm pores that have a total surface area of 1000 nm² per gram of catalytic material. Using electron tomography (3D-TEM), single-pore analysis of occupancy with nanoparticles is carried out.

Molecular catalytic systems, a new strategy developed for the synthesis, development, and testing of catalysts, holds the promise of new improvements. This is now possible by improved molecular understanding of catalytic systems, derived from research in the past decade. The systems approach aims to understand and design catalytic systems in their full complexity. Components or parameters are combined rather than isolated, and the function and interaction of various components is studied within the system. The focus is on the interaction of components, giving rise to new properties that are only present in the system, not in its parts.

A modern approach of colloid chemistry is presently being developed to increase the reactivity of nanoparticles in a limited size up to 10 nm, using several types of stabilizers as well as several types of support for their heterogenization. It appears that a fundamental understanding of the surface properties of such metallic nanoparticles is needed to get more efficient and selective nanocatalysts in the future.

REFERENCES

Ahn, J., Chung, W., Pinnau, I., and Guiver, M. 2008. Polysulfone/silica nanoparticle mixed-matrix membranes for gas separation. *Journal of Membrane Science* 314, no. 1–2:123–133. doi:10.1016/j.memsci.2008.01.031.

Ahn, J., Chung, W., Pinnau, I., Song, J., Du, N., Robertson, G., and Guiver, M. 2010. Gas transport behavior of mixed-matrix membranes composed of silica nanoparticles in a polymer of intrinsic microporosity (PIM-1). *Journal of Membrane Science* 346, no. 2:280–287. doi:10.1016/j.memsci.2009.09.047.

Bond, G.C., Webb, G., Abrahams, R.J., Colbourn, E.A., Evans, J., Grant, G.H., Malinowski, S., Marczewski, M., and Spivey, J.J. 1989. Complete catalytic oxidation of volatile organics in: Catalysis—A review of recent literature. *Royal Society Chemistry* 8, 204. ISBN: 0851865941, 9780851865942.

Bonelli, R., Lucarelli, C., Pasini, T., Liotta, L., Zacchini, S., and Albonetti, S. 2011. Total oxidation of volatile organic compounds on Au/FeOx catalysts supported on mesoporous SBA-15 silica. *Applied Catalysis A: General* 400, no. 1–2:54–60. doi:10.1016/j.apcata.2011.04.014.

Chikhradzea, N.M., Marquise, F.D.S., and Abashidzeb, G.S. 2015. Hybrid fiber and nanopowder reinforced composites for wind turbine blades. *Journal of Materials Research and Technology* 4, no. 1:60–67.

Christian, B.S. 2008. One-dimensional oxide nanostructures via chemical vapor deposition and synthesis of heterometallic transition metal alkoxides (Co, Mn and Fe), Ph.D. Thesis, Saarland University, Germany.

Cong, H., Zhang, J., Radosz, M., and Shen, Y. 2007. Carbon nanotube composite membranes of brominated poly (2, 6-diphenyl-1, 4-phenylene oxide) for gas separation. *Journal of Membrane Science* 294, no. 1–2:178–185. doi:10.1016/j.memsci.2007.02.035.

Das, R., Ali, M.E., Hamid, S.B.A., Ramakrishna, S., and Chowdhury, Z.Z. 2014. Carbon nanotube membranes for water purification: A bright future in water desalination. *Desalination* 336:97–109.

Descamps, C., Bouallou, C., and Kanniche, M. 2008. Efficiency of an integrated gasification combined cycle (IGCC) power plant including CO_2 removal. *Energy* 33, no. 6:874–881. doi:10.1016/j.energy.2007.07.013.

Deshpande, V., and Seyezhai, R. 2014. *International Journal of Innovative Research in Computer and Communication Engineering* 2:3387–3392.

Ghalei, B., and Semsarzadeh, M. 2007. A novel nano structured blend membrane for gas separation. *Macromolecular Symposia* 249–250:330–335. doi:10.1002/masy.200750354.

Goh, K., Setiawan, L., Wei, L., Jiang, W., Wang, R., and Chen, Y. 2013. Fabrication of novel functionalized multi-walled carbon nanotube immobilized hollow fiber membranes for enhanced performance in forward osmosis process. *Journal of Membrane Science* 446:244–254.

Hristovski, K.D., Nguyen, H., and Westerhoff, P.K. 2009. Removal of arsenate and 17α-ethinyl estradiol (EE2) by iron (hydr)oxide modified activated carbon fibers. *Journal of Environmental Science and Health—Part A Toxic/Hazardous Substances and Environmental Engineering* 44, no. 4:354–361.

International Energy Agency Working Party for Fossil Fuels. 2002. Solutions for the 21st century: Zero emissions technologies for fossil fuels.

Khan, F.I., and Ghoshal, A.K. 2000. Removal of volatile organic compounds from polluted air. *Journal of Loss Prevention in the Process Industries* 13:527–545.

Kim, S., Pechar, T., and Marand, E. 2006. Poly(imide siloxane) and carbon nanotube mixed matrix membranes for gas separation. *Desalination* 192, no. 1–3:330–339. doi:10.1016/j.desal.2005.03.098.

Kotowicz, J., Chmielniak, T., and Janusz-Szymanska, K. 2010. The influence of membrane CO_2 separation on the efficiency of a coal-fired power plant. *Energy* 35, no. 2:841–850. doi:10.1016/j.energy.2009.08.008.

Law, M., Greene, L.E., Johnson, J.C., Saykally, R., and Yang, P. 2005. Nanowire dye–sensitized solar cells. *Nature Materials* 4:455–459.

Lee, H., and Neville, K. 1967. *Handbook of Epoxy Resins.* McGraw-Hill, New York.

Lensch-Falk, J.L., Hemesath E.R., Perea, D.E., and Lauhon, L.J. 2009. Alternative catalysts for VSS growth of silicon and germanium nanowires. *Journal of Material Chemistry* 19:849–857.

Ma, X., Dai, J., Zhang, H., and Reisner, D. 2005. Protonic conductivity nanostructured ceramic film with improved resistance to carbon dioxide at elevated temperatures. *Surface and Coatings Technology* 200, no. 5–6:1252–1258. doi:10.1016/j.surfcoat.2005.07.099.

Maurya, V.K., Jalan, R., Pal, D., Abdi, S.H., Tripathi, G., Siddrath, A.M., Singh, B., and Kumar, P. 2012. Role of wind energy in Indian electricity system. *International Journal of Geology, Earth and Environmental Sciences* 2, no. 2:121–128.

Maximous, N., Nakhla, G., Wong, K., and Wan, W. 2010. Optimization of Al_2O_3/PES membranes for wastewater filtration. *Separation and Purification Technology* 73, no. 2:294–301.

Mondal, M., Balsora, H., and Varshney, P. 2012. Progress and trends in CO_2 capture/separation technologies: A review. *Energy* 46, no. 1:431–441. doi:10.1016/j.energy.2012.08.006.

Noy, A., Park, H.G., Fornasiero, F., Holt, J.K., Grigoropoulos, C.P., and Bakajin, O. 2007. Nanofluidics in carbon nanotubes. *Nano Today* 2:22–29.

Palazzolo, M.A., Steinmetz, J.I., Lewis, D.L., and Beltz, J.F. 1985. Parametric evaluation of VOCIHAP destruction via catalytic incineration. EPA-600/2-85-041. National Technical Information Service (NTIS no. PB85 191187).

Patkar, A. N., and Laznow, J. 1992. Hazardous air pollutant control technologies. *Hazmat World* 2:78.

Putnam, M.C., Turner-Evans, D.B, Kelzenberg, M.D., Boettcher, S.W., Lewis, N.S., and Atwater, H.A. 2009. 10 μm minority–carrier diffusion lengths in Si wires synthesized by Cu–catalyzed vapor–liquid–solid growth. *Applied Physics Letters* 95:163116.

Reijerkerk, S., Knoef, M., Nijmeijer, K., and Wessling, M. 2010. Poly(ethylene glycol) and poly (dimethyl siloxane): Combining their advantages into efficient CO_2 gas separation membranes. *Journal of Membrane Science* 352, no. 1–2:126–135. doi:10.1016/j.memsci.2010.02.008.

Rezvani, S., Huang, Y., McIlveen-Wright, D., Hewitt, N., and Mondol, J. 2009. Comparative assessment of coal fired IGCC systems with CO_2 capture using physical absorption, membrane reactors and chemical looping. *Fuel* 88, no. 12:2463–2472. doi:10.1016/j.fuel.2009.04.021.

Ruddy, E.N., and Carroll, L.A. 1993. Select the best VOC control strategy. *Chemical Engineering Progress* 7:28.

Saravanan, P., Gopalan, R., and Chandrasekaran, V. 2008. Synthesis and characterization of nanomaterials. *Defence Science Journal* 58, no. 4:504–516.

Satterfield, C.N. 1970. Mass transfer in heterogeneous catalysis. MIT Press, Cambridge, MA.

Satterfield, C.N. 1980. *Heterogeneous Catalysis in Practice.* McGraw-Hill, New York.

Sieros, G., Chaviaropoulos, P., Sorensen, J., Bulder, B., and Jamieson, P. 2012. Upscaling wind turbines: Theoretical and practical aspects and their impact on the cost of energy. *Wind Energy* 15, no. 1:3–17.

Sinha, A., and Suzuki, K. 2005. Three-dimensional mesoporous chromium oxide: A highly efficient material for the elimination of volatile organic compounds. *Angewandte Chemie* 117, no. 2:275–277. doi:10.1002/ange.200461284.

Sinha, A., Suzuki, K., Takahara, M., Azuma, H., Nonaka, T., and Fukumoto, K. 2007. Mesostructured manganese oxide/gold nanoparticle composites for extensive air purification. *Angewandte Chemie (International Edition)* 46, no. 16:2891–2894. doi:10.1002/anie.200605048.

Sobczak, I., Kieronczyk, N., Trejda, M., and Ziolek, M. 2008. Gold, vanadium and niobium containing MCM-41 materials—Catalytic properties in methanol oxidation. *Catalysis Today* 139, no. 3:188–195. doi:10.1016/j.cattod.2008.05.029.

Solsona, B., Pérez-Cabero, M., Vázquez, I., Dejoz, A., García, T., Álvarez-Rodríguez, J., El-Haskouri, J., Beltrán, D., and Amorós, P. 2012. Total oxidation of VOCs on Au nanoparticles anchored on Co doped mesoporous UVM-7 silica. *Chemical Engineering Journal* 187:391–400. doi:10.1016/j.cej.2012.01.132.

Vlazan, P., Ursu, D.H., Moisescu, C.I., Miron, I., Sfirloaga, P., and Rusu, E. 2015. Structural and electrical properties of TiO2/ZnO core–shell nanoparticles synthesized by hydrothermal method. *Materials Characterization* 101:153–158.

Wagner, R.S., and Ellis, W.C. 1964. Vapor-liquid-solid mechanism of single crystal growth. *Applied Physics Letters* 4:89–90.

William, J.C., and Lead, P.E. 1997. VOC control strategies in plant design. In *Chemical Processing: Project Engineering Annual*, 44.

Wu, H., Pantaleo, G., Venezia, A., and Liotta, L. 2013. Mesoporous silica based gold catalysts: Novel synthesis and application in catalytic oxidation of CO and volatile organic compounds (VOCs). *Catalysts* 3, no. 4:774–793. doi:10.3390/catal3040774.

Wu, H., Wang, L., Zhang, J., Shen, Z., and Zhao, J. 2011. Catalytic oxidation of benzene, toluene and p-xylene over colloidal gold supported on zinc oxide catalyst. *Catalysis Communications* 12, no. 10:859–865. doi:10.1016/j.catcom.2011.02.012.

Xomeritakis, G., Tsai, C., Jiang, Y., and Brinker, C. 2009. Tubular ceramic-supported sol gel silica-based membranes for flue gas carbon dioxide capture and sequestration. *Journal of Membrane Science* 341, no. 1–2:30–36. doi:10.1016/j.memsci.2009.05.024.

Yan, Q., Li, X., Zhao, Q., and Chen, G. 2012. Shape-controlled fabrication of the porous Co3O4 nanoflower clusters for efficient catalytic oxidation of gaseous toluene. *Journal of Hazardous Materials* 209–210:385–391. doi:10.1016/j.jhazmat.2012.01.039.

Yave, W., Car, A., Funari, S.S., Pereira Nunes, S. and Peinemann, K. 2010a. CO2-philic polymer membrane with extremely high separation performance. *Macromolecules* 43, no. 1:326–333. doi:10.1021/ma901950u.

Yave, W., Car, A., Wind, J., and Peinemann, K.-V. 2010b. Nanometric thin film membranes manufactured on square meter scale: Ultra-thin films for CO2 capture. *Nanotechnology* 21, no. 39:395301. doi:10.1088/0957-4484/21/39/395301.

Zhou, D., and Biswas, R. 2008. Photonic crystal enhanced light–trapping in thin film solar cells. *Journal of Applied Physics* 103:93–97.

CHAPTER 3

Green Synthesis of Ecofriendly Nanoparticles and Their Medical Applications

Sahadevan Renganathan,

George Philomin Doss Geoprincy,

and Patchaimuthu Kalainila

CONTENTS

Abstract

Silver nanoparticles are nontoxic to humans but most effective against bacteria, fungi, and viruses, even at low concentrations. Moreover, several salts of silver and their derivatives are commercially manufactured as antimicrobial agents. Microbes may produce resistance against antibiotics by producing repeated mutations to protect themselves from further attack. This is why microbes that are initially sensitive to certain chemotherapeutic agents eventually become resistant upon subsequent dosages of antibiotics. But, with respect to the activity of silver nanoparticles, there exists a complexity for microorganisms to offer resistance. This is because the metal nanoparticles attack a broad range of targets, which makes it difficult for the organism to develop mutations. Thus silver nanoparticles can be used a potential antimicrobial agent. The silver nanoparticles, synthesized using the leaf extracts of *Lucas aspera*, exhibit significant antibacterial and antifungal activity. *Lucas aspera*, commonly known as thumbai, is found to be rich in phytochemical constituents, which offers pharmacological effects. The medicinally active compounds present in the leaf extract reduce the silver ions to form silver nanoparticles, which can be further used for therapeutic applications.

3.1 INTRODUCTION

Our environment is undergoing great damage, and a large amount of hazardous and unwanted chemicals, gases, and substances are released due to rapid industrialization and urbanization. So, we need to learn about the secrets that are present in nature and its products that will lead to the growth of advancements in the synthesis processes of nanoparticles. Nanoparticle application is highly suitable for biological materials because of their choice properties. Nanotechnology is one of the interesting areas of research in the modern field of material science and biological science (Kaviya et al. 2008).

Most chemical methods used for the synthesis of nanoparticles are too expensive and also involved with the use of toxic and hazardous chemicals that are responsible for various biological risks. Nevertheless, in the case of biological methods, nanoparticle synthesis using plant extracts is the most adopted method, because it is ecofriendly, the green production of nanoparticles, it can act as a source of several metabolites, it is much safer to handle, and easily available (Ankamwar et al. 2005).

Also, most of the antibiotics are less sensitive toward many organisms. Hence, many researchers started to seek a new potential antimicrobial agent (Kim et al. 2007) with this respect, and silver and silver-based compounds have a strong bactericidal and fungicidal activity (Spadaro et al. 1974; Zhao and Stevens 1998). Nanoparticles that have a larger surface-area-to-volume ratio tend to possess higher antimicrobial activity (Lok et al. 2007). Also, silver has a lower tendency to stimulate microbial resistance than many other antimicrobial agents (Silver et al. 2007). Based on these properties, green synthesis of silver nanoparticles has been investigated. Several experiments have been performed on the synthesis of silver nanoparticles using medicinal plants such as *Zea mays, Saccharum officinarum, Helianthus annus, Sorghum bicolour,* and *Oryza sativa.* Silver nanoparticles using a methanolic extract of *Eucalyptus hybrida* have also been reported (Jeong et al. 2005).

The development of green processes of nanoparticles is evolving into an important branch of nanotechnology. The research on synthesized nanomaterials and their characterization is an emerging field of nanotechnology from the past two decades due to their huge application in the fields of physics, chemistry, biology, and medicine.

3.2 NANOPARTICLES

Nanoparticles are a group of atoms or molecules in the size range of 1 to 100 nm. The use of nanoparticles is gaining ground in the present century, as they possess defined optical, chemical, and mechanical properties (Gong et al. 2007). The metallic nanoparticles are most promising; they have good antibacterial properties due to their large surface-area-to-volume ratio, which is of much interest to researchers due to the growing microbial resistance against metal ions, antibiotics, and the development of resistant strains. Different types of nanomaterials, including zinc, copper, silver, titanium (Retchkiman-Schabe et al. 2006), magnesium, gold (Gu et al. 2003), alginate (Ahmad et al. 2005), and silver, have been researched, but silver nanoparticles have proven to be most effective, as they have good antimicrobial efficacy against microorganisms such as viruses, bacteria, and other eukaryotic microorganisms.

Silver nanoparticles show effective antimicrobial property compared to other nanoparticles due to their extremely large surface area, which provides better contact with microorganisms. The nanoparticles get attached to the cell membrane and also penetrate inside the bacteria. The bacterial membrane contains sulfur-containing proteins, and the silver nanoparticles interact with these proteins in the cell as well as with the phosphorus-containing compounds like DNA. The nanoparticles preferably attack the respiratory chain and cell division finally leading to cell death. The nanoparticles release silver ions in the bacterial cells, which enhance their bactericidal activity (Feng et al. 2000).

Nanoparticles smaller than 10 nm interact with bacteria and produce electronic effects, which improve the reactivity of nanoparticles. Thus, it is proven that the bactericidal effect of silver nanoparticles is size dependent (Morones et al. 2005; Raimondi et al. 2005). Sastry et al. (2003) reported the biosynthesis of nanoparticles using plant leaf extracts and their potential application. Researchers have also studied bioreduction of chloraurate ions and silver ions by extracts of geranium (Shankar et al. 2003) and neem leaf (Shankar et al. 2004a).

Biological methods of nanoparticle synthesis using microorganisms, enzymes, fungus, and plants have been suggested as possible ecofriendly alternatives to chemical and physical methods. Sometimes the synthesis of nanoparticles from plants can prove advantageous over other biological

TABLE 3.1   Sources Utilized for Synthesis of Nanoparticles

Source	Type of Nanoparticle	Reference
Catharanthus roseus	Silver	Ponarulselvam et al. 2012
S. tricobatum, S. cumini, C. asiatica	Silver	Logeswari et al. 2015
Capparis zylanica leaf	Copper	Saranyaadevi et al. 2014
Polyalthia longifolia leaf	Silver	Kaviya et al. 2008
Emblics officinalis fruit	Silver, gold	Ankamwar et al. 2005
Geranium leaf	Silver	Shankar et al. 2003
Azadirachta indica leaf	Silver, gold	Shankar et al. 2004a
Dioscorea bulbifera	Silver	Sougata et al. 2012
Papaya fruit	Silver	Jain et al. 2009
Vitex negundo	Silver	Mohzen et al. 2011

methods by eliminating the elaborate process of maintaining microbial cultures. There are various sources utilized for synthesis of nanoparticles (see Table 3.1).

3.3 TYPES OF NANOPARTICLES

Many nanoparticles are synthesized including as silver, copper, iron, gold, zinc, platinum, and palladium. A few of these are discussed next. These nanoparticles have been utilized for biomedical applications.

3.3.1 Silver Nanoparticles

Nanoparticles have been known to be used for numerous physical, chemical, biological, and pharmaceutical applications. Silver nanoparticles (Ag NPs) are used as antimicrobial agents. It is a well-known fact that silver ions and silver-based compounds are highly toxic to microorganisms (Slawson et al. 1992; Zhao and Stevens 1998). This aspect of silver makes it an excellent choice for multiple roles in the medical field. Generally, silver is used in the nitrate form to induce antimicrobial effect, but with the use of Ag NPs, there is a large surface area available for the microbe to be exposed. Thus Ag NPs find use in many antibacterial applications.

3.3.2 Copper Nanoparticles

Copper nanoparticles (Cu NPs) are very attractive due to their heat transfer properties, such as high thermal conductivity. Cu NPs have a low production cost, high surface-area-to-volume ratio, antibacterial potency, optical properties, catalytic activity, and magnetic properties as compared to precision metals such as gold, silver, or palladium (Borkow 2010; Borkow et al. 2009). Copper nanoparticles have great interest due to their catalytic, optical, mechanical, and electrical properties. Copper is an excellent alternative material for noble metals such as Au and Ag, as it is highly conductive and much more economical (Athanassiou et al. 2006). Copper plays an important role in electronic circuits because of its excellent electrical conductivity. Copper nanoparticles are highly inexpensive and their properties can be controlled depending on the synthesis method (Kantam et al. 2007).

3.3.3 Gold Nanoparticles

Elemental gold has many unique properties that have attracted interest in the biomedical field. It has been used for many decorative, ceremonial, colorful, and religious artifacts. Properties of gold nanoparticles are different from gold's bulk form, because bulk gold is yellow solid and it is inert in nature, whereas gold nanoparticles are wine red in color in solution and are reported to be antioxidant. Gold nanoparticles also exhibit different shapes such as spherical, suboctahedral, octahedral, decahedral, icosahedral multiple twined, multiple twined, irregular shape, tetrahedral, nanotriangles (Shankar et al. 2004b), nanoprisms, hexagonal platelets, and nanorods. Gold nanoparticles are widely used in biomedical science including tissue or tumor imaging, drug delivery, photothermal therapy, and immune chromatographic identification of pathogens in clinical specimens due to the surface plasmon resonance (SPR) (Chithrani et al. 2010).

3.3.4 Iron Nanoparticles

The synthesis of superparamagnetic nanoparticles has been intensively developed not only for fundamental scientific interest but also for many technological applications such as magnetic storage media, biosensing applications, and medical applications, such as targeted drug delivery. Superparamagnetic iron oxide nanoparticles with appropriate surface chemistry can be used for numerous in vivo applications (Gupta and Gupta 2005), such as MRI contrast enhancement, tissue repair, immunoassay, detoxification of biological fluids, hyperthermia, drug delivery, and cell separation. All of these biomedical applications require that the nanoparticles have high magnetization values, a size smaller than 100 nm, and a narrow particle size distribution. These applications also need particular surface coating of the magnetic particles, which has to be nontoxic and biocompatible and must also allow for a targetable delivery. Such magnetic nanoparticles can bind to drugs, proteins, enzymes, antibodies, or nucleotides, and can be directed to an organ, tissue, or tumor using an external magnetic field (Petri-Fink et al. 2005).

The most common methods including coprecipitation, thermal decomposition, hydrothermal synthesis, microemulsion, sonochemical synthesis, and sonochemical synthetic route can all be directed to the synthesis of high-quality iron oxide NPs. In addition, these NPs can also be prepared by other methods such as electrochemical synthesis, laser pyrolysis techniques, and microorganism or bacterial synthesis (Roh et al. 2006).

3.3.5 Zinc Oxide Nanoparticles

Zinc oxide is an inorganic compound with the molecular formula ZnO. It appears as a white powder and is nearly insoluble in water. The powder ZnO is widely used as an additive in numerous materials and products including ceramics, glass, cement, rubber (e.g., car tires), lubricants, paints, ointments, adhesives, plastics, foods (source of Zn nutrient), batteries, ferrites, and fire retardants.

ZnO semiconductor has several unique properties such as good transparency, high electron mobility, wide band gap, and strong room temperature luminescence. These properties account for its applications in transparent electrodes, in liquid crystal display, and in energy-saving or heat-protecting windows and other electronic applications. The wide band gap and large excitonic binding energy have made zinc oxide important for both scientific and industrial applications (Wang et al. 2004).

3.3.6 Cadmium Sulfide Nanoparticles

The wet chemical method is used for the fabrication of materials, typically a metal oxide starting either from a chemical solution or colloidal particles to produce an integrated network particle. Typical precursors are metal alkaloxide and metal chloride, which undergo hydrolysis in polycondensate reactions to form a colloid, a system composed of solid particles (size ranging from 1 nm to 100 nm) dispersed in a solvent.

Cadmium sulfide is a chemical compound with the formula CdS. Cadmium sulfide is yellow in color and is a semiconductor (2.42 eV). It exists in nature as two different minerals: green kite and hawleyite. Cadmium sulfide is a direct band gap semiconductor (gap 2.42 eV) and has many applications, for example, in light detectors. It forms thermally stable pigments ranging from deep red to yellow.

CdS nanoparticles have been extensively studied due to their potential applications, such as field effect transistors, light-emitting diodes, photo catalysis, and biological sensors. Many synthetic methods have been employed to prepare CdS nanoparticles including soft chemical reaction,

TABLE 3.2 Nanoparticles Utilized for Biomedical Applications

Source	Type of Nanoparticle	Application	Reference
Paederia foetida leaf	Silver	Antibacterial activity	Lavanya et al. 2013
Mimosa pudica leaf	Silver	Antibacterial activity	Akash Raj et al. 2014
Annona squamosa leaf	Silver	In vitro cytotoxicity effect on MCF-7 cell	Viveka et al. 2012
Erythrina indica lam root	Silver	Antimicrobial activity and cytotoxic activity	Rathi Sre et al. 2015
Carica papaya	Copper oxide	Photocatalytic dye degradation	Sankar et al. 2014
Commelina nudiflora L.	Gold	Antibacterial and antioxidant activity	Kuppusamy et al. 2015
Acalypha indica Linn.	Silver and gold	Cytotoxic effects against MDA-MB-231	Krishnaraj et al. 2014
Acorus calamus	Gold	Antibacterial and UV blocking	Ganesan et al. 2014
Ocimum sanctum leaf	Iron oxide	Spectroscopic and microscopic studies	Balamurughan et al. 2014
Rosa damascena petals	Silver	Anticancer activity	Venkatesan et al. 2014

solid state reaction, sol gel process, microwave heating, photo etching, and reverse micelle. There are various nanoparticles utilized for biomedical applications (Table 3.2).

3.4 METHODS AVAILABLE TO SYNTHESIZE SILVER NANOPARTICLES

Nanoparticles can be produced using many different techniques, typically classified as bottom-up or chemical methods, and top-down or physical methods (Bali et al. 2006). In the bottom-up approach, the structure of nanoparticles is constructed by atoms, molecules, or clusters. In the top-down approach, a bulk piece of a required material is reduced to nanosized dimensions using cutting, grinding, and etching techniques, i.e., nanomaterials are prepared from larger entities without atomic-level control. Chemical reduction, microemulsion (colloidal) techniques, sonochemical reduction, electrochemical, microwave-assisted, and hydrothermal syntheses are the main techniques for the synthesis of nanoparticles through the chemical approach. Biological or biosynthesis techniques are also considered as bottom-up or chemical processes. Physical methods for nanoparticles synthesis are laser (pulse) ablation, vacuum vapor deposition, pulsed wire discharge (PWD), and mechanical milling. A wide range of nanoparticles can be produced using physical methods with little modification for different metals, but the main disadvantages of these methods are the quality of the product, which is less as compared to nanoparticles produced by chemical methods. Usually these methods require costly vacuum systems or equipment to prepare nanoparticles. During the chemical synthesis of nanoparticles, the morphology and growth can be controlled by optimizing reaction conditions, such as concentration, temperature, surfactant precursor, capping/stabilizing agent, and the type of solvent (Chen et al. 2006). Using these optimum reaction conditions, a narrow size distribution can be achieved during chemical synthesis. These methods for the production of nanoparticles are appropriate for laboratory-scale synthesis but are not economical for a large-scale or commercial setup.

3.4.1 Chemical Methods

There are many methods are available to synthesize nanoparticles by chemical routes such as chemical reduction, microemulsion, sonochemical, electrochemical, and solvothermal decomposition. A few techniques are discussed next.

3.4.1.1 Chemical Reduction Method

The chemical reduction method is the simplest, easiest, and the most commonly used synthetic method for nanoparticles. In fact, the production of nanoparticles with good control of sizes and morphologies using chemical reduction methods can be achieved (Huang et al. 1997).

In the chemical reduction techniques, nanoparticles are reduced by a reducing agent such as sodium borohydride, hydrazine (N_2H_4), ascorbate, polyol, isopropyl alcohol with cetyl trimethyl ammonium bromide (CTAB) as well as glucose (Panigrahi et al. 2006).

3.4.1.2 Microemulsion Method

Microemulsion is a technique for the synthesis of nanoparticles in which two immiscible fluids such as water in oil (W/O) or oil in water (O/W) or water in supercritical carbon dioxide (W/Sc. CO_2) become a thermodynamically stable dispersion with the aid of a surfactant. An emulsion is a single phase of three components: water, oil, and a surfactant (Kapoor et al. 2002). Normally oil and water are immiscible but with the addition of a surfactant, the water and oil become miscible because the surfactant is able to bridge the interfacial tension between the two fluids (Kitchens and Roberts 2004).

Microemulsion consists of surfactant aggregates that are in the ranges of 1 nm to 100 nm. The location of the water, oil, and surfactant phases affects the geometry of aggregate. If the microemulsion is to be oil in water, the water is the bulk fluid and the presence of oil is in less quantity with small amounts of surfactant. Similarly, if it is water in oil, the oil is the bulk fluid and water is present in less quantity. The creation of oil in water and surfactant is called micelles, which is an aggregate formed to reduce free energy. The W/O microemulsion carries oil or organic solvent as bulk then the system is thermodynamically stable and called reverse micelles (Dadgostar 2008).

3.4.1.3 Sonochemical Method

Acoustic cavitation is a physical phenomenon that is responsible for sono-chemical reaction. This method was initially proposed for the synthesis of iron nanoparticles. Nowadays, this method is used to synthesize different metals and metal oxides. The main advantages of the sonochemical method are its simplicity, operating conditions (ambient conditions), and easy control of the size of nanoparticles by using precursors with different concentrations in the solution (Suslick et al. 1996).

Sonochemical reactions of volatile organometallics have been exploited as a general approach to the synthesis of various nanophase materials by changing the reaction medium. There are many theories presented by different researchers that have been developed to explain the mechanism of breakup of the chemical bond under 20 KHz ultrasonic radiations. They have explained the sonochemistry process in these theories, i.e., how bubble creation, growth, and its collapse is formed in the liquid.

3.4.2 Physical Methods

3.4.2.1 Pulsed Laser Ablation/Deposition

The laser ablation method is a commonly used technique for the preparation of nanoparticles in colloidal form in a variety of solvents. Nanoparticles are prepared in colloidal form to avoid oxidation. The process of pulsed laser ablation takes place in a vacuum chamber and in the presence of some background/inert gas. In pulsed laser deposition, a laser beam is focused inside a vacuum chamber in which a high-power pulse strikes a target in the material creating plasma, which then is converted into a colloidal solution of nanoparticles. Mostly second harmonic generation (Nd:YAG) lasers are being used to prepare the nanoparticles. There are many factors that affect the final product, such as the number of pulses, pulsing time, type of laser, and type of solvent (Song et al. 2007).

3.4.2.2 Mechanical/Ball Milling Method

Milling is a solid-state processing technique for the synthesis of nanoparticles. This technique was first used by Benjamin for the production of superalloys. In the milling process, raw material of micron size is fed to undergo several changes. Different types of mechanical mills are available that are commonly used for the synthesis of nanoparticles. These mills are characterized according to their capacities and applications.

Mechanical mills that are commonly used for the synthesis of nanoparticles are vibratory, planetary, attritor, and uniball. It is very difficult to produce ultrafine particles, due to mechanical limitations and long length of time. However, simple operation, low cost of production of nanoparticles, and the possibility to produce large quantities are the main advantages of mechanical milling. The important factors affecting the quality of the final product are the type of mill, container, milling speed, time, atmosphere, temperature, size and size distribution of the grinding medium, process control agent, and weight ratio of ball to powder and extent of filling the vial (Suryanarayana 2001).

3.4.2.3 Pulsed Wire Discharge Method

Pulsed wire discharge (PWD) is a physical technique to prepare nanoparticles. Compared to the other previously mentioned methods, synthesis of metal nanoparticles by the PWD technique follows a completely different mechanism. In PWD, a metal wire is evaporated by a pulsed current to create a vapor, which is then cooled by using an ambient gas to form nanoparticles. Preparations of metal, nitride, and oxide nanoparticles by PWD have been reported. This method has a high-energy efficiency, high production rate, and a simple apparatus consisting of a vacuum chamber to be used for the nanoparticle preparation using a powder collection filter and a discharging circuit. This process is not used conventionally for common industrial purposes, because it is not only very expensive but also impossible to use explicitly for different metals. It is mainly useful for high electrical conductivity metals that are easily available in the thin wire form (Jiang and Yatsui 1998).

Muraia et al. (2007) found that copper nanoparticles covered with organic matter can also be successfully prepared by evaporation of copper wire in an oleic acid vapor/mist, with the thickness of the coating layer up to a few nano meters. They prepared nanoparticles of size 10 to 25 nm using the pulsed wire technique.

3.4.3 Biological Synthesis

There is a need for biosynthesis of nanoparticles design, as the physical and chemical processes are costly and hazardous. Therefore, in the search for cheaper pathways, scientists used microorganisms and plant extracts for synthesis for nanoparticles. Nature has planned various processes for the synthesis of nano and microlength-scaled inorganic materials, which have contributed to the development of relatively new and largely unexplored areas of research based on the biosynthesis of nanomaterials. Biosynthesis of nanoparticles is also considered to be a bottom-up technique, where the oxidation/reduction is the main reaction that occurs during the production. Metals are usually reduced into their respective nanoparticles because of the microbial enzymes or the plant phytochemicals with antioxidant or reducing properties.

In the biosynthesis of nanoparticles, three important parameters are (1) the choice of the solvent medium used, (2) the choice reducing agent, and (3) the choice of a nontoxic material for the stabilization of the nanoparticles. The use of bacteria, actinomycetes, organisms, plants, and fungi to synthesize nanoparticles is being practiced (Bali et al. 2006). The biosynthesis of nanoparticles involves easy preparation protocols and less toxicity, and includes a wide range of applications according to their morphology. The size of nanoparticles can be controlled using this technique but not in a complete manner. The field of nanobiotechnology needs more research focused on the mechanism of nanoparticles formation, which may lead to fine-tuning of the process; ultimately it is very important to the synthesis of nanoparticles with a strict control over the size and shape parameters.

Certain plants are known to accumulate higher concentrations of metals compared to others and such plants are termed as hyperaccumulators. *Brassica juncea* had better metal-accumulating ability and also integrated the metal as nanoparticles (Cioffi et al. 2005).

3.5 CHARACTERIZATION TECHNIQUES

3.5.1 Ultraviolet-Visible Spectrometry

Samples for ultraviolet-visible (UV-Vis) spectrophotometry are most often liquids and gases, and even solids can be measured. Samples are always placed in a cuvette. Cuvettes are typically rectangular in shape, generally with an internal width of 1 cm.

UV-Vis absorption spectra of the colloidal dispersions were recorded using the Ultro spec 2100 spectrophotometer. The distribution of the particle size was measured by the Zeta-Sizer system (Malvern Instruments). The biosynthesized silver nanoparticles from mangosteen leaf has a resolution of 1 nm between 300 and 700 nm and possess a scanning speed of 300 nm/min, which was determined by UV-Vis absorption. Synthesis of silver nanoparticles from plant extract shows that maximum absorbance occurs at 430 nm, which increases as a function of reaction time. There is no evidence of absorbance for the UV-Vis spectra range between 400 nm 800 nm for the pure *Solanum torvum* plant extract, but when the plant extract gets exposed to $AgNO_3$ solutions, maximum absorbance was found at 434 nm, due to the formation of nanoparticles. The width and frequency of the surface plasmon absorption depends on the size and shape of the metal nanoparticles as well as on the dielectric constant of the metal itself and the surrounding medium (Sastry et al. 1997).

3.5.2 Fourier Transform Infrared Spectroscopy

In Fourier transform infrared spectroscopy (FTIR), infrared radiation is passed through a sample. The infrared radiation is absorbed by the sample and some of it is passed through the sample or transmitted. Finally, the resulting spectrum represents the molecular absorption and transmission, creating a molecular fingerprint of the sample. Elemental analysis of the synthesized silver nanoparticles from *Lucas aspera* was studied by FTIR.

FTIR analysis is mainly for determining functional groups present in the compound. FTIR absorption spectra of *Dioscorea bulbifera* tuber extract show a strong peak at 3300 cm^{-1} representing the O-H bond. But after bioreduction, it is not seen in the extracts of *D. bulbifera*. The absorbance bands at 2931 cm^{-1}, 1625 cm^{-1}, 1404 cm^{-1}, and 1143 cm^{-11} are associated with respect to the stretch vibrations of alkyl C=C, conjugated C-C with a benzene ring, bending of C-O-H and C-O stretch in saturated tertiary or secondary highly symmetric alcohol in *D. bulbifera*. The presence of peaks at 3749 cm^{-1} and 1523 cm^{-1} indicate the NH_2 symmetric stretching and N-O bonds in nitro compounds (Sougata et al. 2012).

FTIR measurements for *Gliricidia sepium* shows the absorption peak at around 1020 cm^{-1} can be assigned as absorption peaks of -C-O-C- or C-O-. The absorption spectra at 1638 cm^{-1} result from stretching of vibration of -C=C-. The peak at around 1640 cm^{-1} indicates the amide I bonds of proteins. The bonds or functional groups such as -C-O-C-, C-O-, and -C=C- are derived from heterocyclic compounds. The amides I bond derived from the proteins are the capping ligands of the nanoparticles (Raut et al. 2009).

3.5.3 X-Ray Diffraction

For x-ray diffraction (XRD), the Collidge tube is arranged and a power of 1.2 kw is produced from 40 kv and 30 mA to produce the x-ray of wavelength 1.5406 Å. The prepared sample is mounted on the focusing circle. The x-rays are allowed to fall on the sample and the counter is initially set at an angle of 0°. The program is set to run such that the detector counter moves through the required angle at specific counts and scans the sample. The angle range is between 10° and 70°.

As the x-ray beam is diffracted by the sample and detected at different angles, the output peak is generated on the computer screen. The output is a graph with different peaks corresponding to different planes of the crystal and the graph drawn was between 2θ in the x-axis and intensity in the y-axis. The obtained peak is compared with the data in the Joint Committee on Powder Diffraction Standards (JCPDS) tool, which is a standard. From this even the particle size of the crystal was calculated using the formula

$$D = 0.9\lambda/\beta\cos\theta$$

Formation of silver nanoparticles from papaya fruit extract shows three intense peaks that range from 10° to 80°. The average size of the particles was measured as 15 nm (Devendra et al. 2009). The XRD patterns of Ag/ *Vitex negundo* indicate the face-centered cubic structure of silver nanoparticles. The silver nanoparticles showed XRD peaks at 38.17°, 44.31°, 64.44°, 77.34°, and 81.33°, corresponding to the face-centered cubic planes (111, 200, 220, 311, and 222, respectively) of the silver crystals (Mohzen et al. 2011).

3.5.4 Scanning Electron Microscope

A scanning electron microscope (SEM) is a type of electron microscope that images a sample by scanning it with a high-energy beam of electrons in a raster scan pattern. The electrons interact with the atoms that make up the sample producing signals that contain information about the sample surface topography, composition, and other properties such as electrical conductivity. SEM is extensively used to study the detailed images of the sample taken for investigation.

Green synthesis of silver nanoparticles from *Cleome viscosa* was analyzed by SEM. SEM observations were established using the ZEISS EVO 40 EP electron microscope. SEM analysis shows that the silver nanoparticles are uniformly distributed on the surface. But it does not indicate that all the nanoparticles are bound to the surface. This may mean that the particles dispersed in the solution may also be deposited onto the surface (Bharathi et al. 2012).

3.6 BIOMEDICAL APPLICATIONS OF NANOPARTICLES

3.6.1 Antimicrobial Activity of Silver Nanoparticles

It is known that silver ions and nanoparticles are highly toxic and hazardous to microorganisms. Silver nanoparticles have many inhibitory and bactericidal effects and so their application is extended as an antibacterial agent. The antibacterial activity of silver nanoparticles is estimated by the zone of inhibition. Silver nanoparticles not only interact with the surface of the membrane but can also penetrate inside the microorganism. Recently, nanoparticles have gained significance in the field of biomedicine. The most distinguishing and significant property of nanoparticles is that they exhibit a large surface-area-to-volume ratio. When the surface area of the nanoparticles gets increased, their surface energy also increase, and hence their biological effectiveness increases (Srivastava et al. 2011). Smaller nanoparticles with a larger surface-area-to-volume ratio provide a more effective antibacterial activity even at a very low concentration. Silver nanoparticles of many different shapes (spherical, rod, truncated, triangular nanoplates) were developed by various synthetic routes. The triangular shape of silver nanoplates were found to show the strongest antibacterial activity. This property could be due to their larger surface-area-to-volume ratios and their crystallographic surface structures.

Silver nanoparticle is an effective and a fast-acting fungicide against a broad spectrum of common fungi including genera such as *Aspergillus*, *Candida*, and *Saccharomyces*. The standard well diffusion method was used to assay the antibacterial activity against human pathogenic bacteria such as *Pseudomonas aeruginosa*, *Escherichia coli*, *Bacillus subtilis*, and *Klebsiella pneumonia* (Aditi et al. 2011). In vitro antibacterial activity of the prepared nanoparticles was studied using the Kirby-Bauer technique, which confirmed the recommended standards of the National Committee for Clinical Laboratory Standards (NCCLS), now known as the Clinical and Laboratory Standards Institute (CLSI). The agar well diffusion method was used to assess the antibacterial activity of synthesized Ag nanoparticles. The zone of inhibition produced by various antibiotics was compared with the inhibitory zone produced by silver nanoparticles (Geoprincy et al. 2011). The antibacterial assays were performed on human pathogenic bacteria like *Escherichia coli* and *Pseudomonas aeruginosa* by standard disc diffusion method. Luria Bertani (LB) broth/agar medium was used to cultivate bacteria. Basically, nanoparticle has antimicrobial (including antibacterial and antifungal) applications. The silver or gold nanoparticles that are produced extracellular from *Fusarium oxysporum* can be used in several materials like clothes. Such clothing is sterile and used in hospitals to prevent or to minimize infections with pathogenic bacteria like *Staphylococcus aureus.* The average zones of inhibition expressing a profound inhibitory effect was represented as 35 mm in *P. aeruginosa*, 30 mm in *K. pneumonia*, 36 mm in *S. aureus*, 40 mm in *S. typhi*, 38 mm in *S. epidermis*, and 34 mm in *E. coli* (Shirley et al. 2010).

For the concentration of 20 µg, 40 µg, 60 µg, and 80 µg of the nanoparticle, *Staphylococcus aureus* exhibited characteristic inhibitory zones of 14 mm, 16 mm, 18 mm, and 20 mm diameter, respectively, whereas *Enterococcus faecalis* exhibited inhibitory zones of 11 mm, 13 mm, 14 mm, and 17 mm diameter, respectively (Karthick et al. 2011). Nanoparticles were also used in controlled drug delivery, biological detection, optical filters, and sensor design.

3.6.1.1 Disk Diffusion Method

Four major pathogens—*B. cereus, B. subtilis, K. pneumonia*, and *V. cholerae*—were taken for analysis. Nutrient agar media were prepared and poured into Petri dish plates and swabbed with respective microbial inoculum. In each plate, four disks were fixed at equal distance. The first one was loaded with a concentration of 25 µg/ml (A) silver nanoparticle. The second disk was impregnated with a 50 µg/ml (B) concentration of silver nanoparticle. The third disk was impregnated with a 50 µg/ml (D) equimolar concentration of silver nanoparticle and rifamycin, and the fourth disk was loaded with a 50 µg/ml concentration of reference antibiotic rifamycin. Well C was maintained as blank without nanoparticle and rifamycin. All 16 plates were kept at 37°C for incubation overnight. The zone of inhibition was measured after 24 h of incubation.

3.6.1.2 Thin Layer Chromatography

Thin layer chromatography (TLC) is used to separate the individual compounds formulated in antibiotics (crude). The separation of the compound also depends on the type of solvent used. The maximum zone of inhibition (rifamycin) based on the disk diffusion method was used for TLC analysis. For that, a 10 mg/ml concentration of antibiotics in methanol was prepared.

From this solution, 4 µl of the sample was taken and spotted on the silica-coated TLC plates. Then it was kept in slanting position with the solvent to run under capillary pressure. Methanol and chloroform in the ratio of 3:7 was used as a solvent. The spots were then identified both in the UV light, far infrared light in the iodine chamber (Nabi et al. 2006).

The R_f values were calculated. The R_f value is defined as the distance traveled by a given component divided by distance traveled by the solvent front.

3.6.1.3 Bioautography Agar over Layer Method

The developed TLC plates were kept in the sterile Petri plate. The nutrient agar was prepared and poured in a thin layer to cover the entire Petri dish. Twenty-four-hour cultures of *B. cereus* were swabbed on it. The plates were then incubated at 37°C for 24 hours. The zone of inhibition obtained at varying separation points were observed. Similarly chromatogram was developed by loading a 1 µg concentration of nanoparticle impregnated with rifamycin (antibiotic). Thus the characteristic zone of inhibition was obtained by the agar over layer method (Pandey et al. 2004).

3.6.2 Biological Applications of Gold Nanoparticles

Gold nanoparticles have been widely used in many fields, such as chemical and biological sensors, electronics, dyes, conductive coatings, catalysis, fundamental research, and electron microscopy. Almost every chemical process involves catalysis. Nanoparticles can efficiently act as catalysts, since they have a large surface-to-volume-ratio and special binding sites. Gold nanoparticles exhibit high catalytic activity in the oxidation and reduction of hydrocarbons (Zhong and Mathew 2001). Nanoparticles exhibit different physical and chemical properties from their bulk solid materials. When gold nanoparticles are conjugated with saccharide and oligo deoxyribonucleic acid, the combination of organic functionality with dielectric properties of gold nanoparticles resulted in a new material, which can be used as a sensitive colorimeter for the detection of polynucleotides (Zhu et al. 2004).

Gold nanoparticles can also be used as biosensors due to their unique optical properties. Anti-EGFR (epidermal growth factor receptor) gold nanoparticles conjugated with gold nanoparticles can distinguish cancerous and noncancerous cells, which is proven by SPR (surface plasmon resonance) scattering and SPR absorption spectroscopy. Hence, it can be used in cancer diagnostics. Gold nanoparticles play an important role in drug and gene delivery. Gold nanoparticles when irradiated with light in water will create local heating and can be used in photothermal destruction of tumors. Also gold nanoparticles enhance the efficiency of photothermal therapy 20-fold (Guo and Wang 2007). The excellent biocompatibility and unique properties made gold nanoparticles attractive material for biosensors, chemosensors, and electrocatalysts. An electrochemical device with gold nanoparticles will provide new opportunity for gene diagnostics.

3.6.3 Silver Nanoparticles: Biological and Clinical Significance

In general, silver nanoparticles (Ag NPs) are nontoxic to humans, but most effective against bacteria, virus, and other eukaryotic microorganism at low concentrations (Krutyakov et al. 2008). Moreover, several salts of silver and their derivatives are commercially manufactured as antimicrobial agents (Kasthuri et al. 2009). These silver nanoparticles provide significant pharmaceutical, clinical, biological, and immunological applications. They are used to prevent infection, in (burn and traumatic) wound dressings, diabetic ulcers, coating of catheters, dental works, scaffold, and medical devices (Thomas et al. 2007). In addition, they are used as anticancer agents in the treatment of cancer, in the field of tissue engineering, and in drug delivery. The bactericidal properties of silver nanoparticles are due to the release of silver ions from the particles, which is highly effective for antimicrobial activity (Amarendra and Krishna 2010). Also, the potency of the antibacterial effects corresponds to the size of the nanoparticle. The smaller particles have higher antibacterial activities. With respect to the clinical applications of nanoparticles, microorganisms including diatoms, fungi, bacteria, and yeast-producing nanoparticles through biological synthesis were found to be more biocompatible (Guidelli et al. 2011).

3.7 CONCLUSION

It is concluded that plant-mediated synthesis of silver nanoparticles possesses potential antimicrobial applications. The characterization analysis proved that the particle, produced in nano dimensions, would be equally effective as that of antibiotics and other drugs in pharmaceutical applications. The use of silver nanoparticles in drug delivery systems might be the future thrust in the field of medicine. It was concluded that silver nanoparticles can serve as potential drugs with various clinical and pharmacological properties, thereby demonstrating enhanced characteristic anticancer activity, antiapoptotic activity, antioxidant activity, wound healing activity, antimicrobial activity, and in tissue engineering. Hence, it was demonstrated that as a novel therapeutic agent green synthesized silver nanoparticle will be useful in many biomedical applications.

REFERENCES

Aditi, P.K., Ankita, A.S., Pravin, M.H., and Rajendra, S.Z. 2011. Plant mediated synthesis of silver nanoparticles-tapping the unexploited sources. *Journal of Natural Product and Plant Resources* 1, no. 4:100–107.

Ahmad, Z., Pandey, R., Sharma, S., and Khuller, G.K. 2005. Alginate nanoparticles as anti-tuberculosis drug carriers: Formulation development, pharmacokinetics and therapeutic potential. *Indian Journal of Chest Diseases and Allied Sciences* 48:171–176.

Akash Raj, S.R., Divya, S., Sindhu, S., Kasinathan, K., and Arumugam, P. 2014. Studies on synthesis, characterization and application of silver nanoparticles using *Mimosa Pudica* leaves. *International Journal of Pharmacy and Pharmaceutical Sciences* 6, no. 2:0975–1491.

Amarendra, D.D., and Krishna, G. 2010. Biosynthesis of silver and gold nanoparticles using *Chenopodium album* leaf extracts. *Colloids Surface A* 369, no. 3:27–33.

Ankamwar, B., Damle, C., Ahmad, A., and Sastry, M. 2005. Biosynthesis of gold and silver nanoparticles using *Emblics officinalis* fruit extract and their phase transfer and transmetallation in an organic solution. *Journal of Nanoscience and Nanotechnology* 5, no. 10:1665–1671.

Athanassiou, E.K., Grass, R.N., and Stark, W.J. 2006. Large-scale production of carbon-coated copper nanoparticles for sensor applications. *Nanotechnology* 17:1668–1673.

Balamurughan, M.G., Mohanraj, S., Kodhaiyolii, and Pugalenthi, V. 2014. *Ocimum sanctum* leaf extract mediated green synthesis of iron oxide nanoparticles: Spectroscopic and microscopic studies. *Journal of Chemical and Pharmaceutical Sciences* ISSN: 0974–2115.

Bali, R., Razak, N., Lumb, A., and Harris, A.T. 2006. The synthesis of metallic nanoparticles inside live plants. International Conference on Nanoscience and Nanotechnology, ICONN'06, Brisbane, July 3–7.

Bharathi, R.S., Suriya, J., Sekar, V., and Rajasekaran, R. 2012. Biomimetic of silver nanoparticles by *Ulva lactuca* seaweed and evaluation of its antibacterial activity. *International Journal of Pharmacy and Pharmaceutical Sciences* 4, no. 3:139–143.

Borkow, G. 2010. Molecular mechanisms of enhanced wound healing by copper oxide–impregnated dressings. *Wound Repair and Regeneration* 18, no. 2:266–275.

Borkow, G., Zatcoff, R.C., and Gabbay, J. 2009. Reducing the risk of skin pathologies in diabetics by using copper impregnated socks. *Medical Hypotheses* 73, no. 6:883–886.

Chastellain, M., Petri, A., Gupta, A., Rao, K.V., and Hofmann, H. 2004. Super-paramagnetic silica-iron oxide nanocomposites for application in hyper-thermia. *Advanced Engineering Materials* 6, no. 4:235–241.

Chen, L., Zhang, D., Chen, J., Zhou, H., and Wan, H. 2006. The use of CTAB to control the size of copper nanoparticles and the concentration of alkylthiols on their surfaces. *Materials Science and Engineering A* 415:15–161.

Chithrani, D.B., Dunne, M., Stewart, J., Allen, C., and Jaffray, D.A. 2010. Cellular uptake and transport of gold nanoparticles incorporated in a liposomal carrier. *Nanomedicine: Nanotechnology, Biology, and Medicine* 6:161–169.

Cioffi, N., Ditaranto, N., Torsi, L., Picca, R.A., Giglio E.D., and Zambonin, P.G. 2005. Synthesis analytical characterization and bioactivity of Ag and Cu nanoparticles embedded in poly-vinyl-methyl-ketone films. *Analytical and Bioanalytical Chemistry* 382:1912–1918.

Dadgostar, N. 2008. Investigations on colloidal synthesis of copper nanoparticles in a two-phase liquid-liquid system. MSc thesis, University of Waterloo, Ontario, Canada.

Devendra, J., Hemant, K.D., Sumitha, K., and Kothari, S.L. 2009. Synthesis of plant-mediated silver nanoparticles uses papaya fruit extract and evaluation of their anti-microbial activities. *Digest Journal of Nanomaterials and Biostructures* 4, no. 4:723–727.

Feng, Q.L., Wu, J., Chen, G.Q., Cui, F.Z., Kim, T.N., and Kim, J.O. 2000. A mechanistic study of the antibacterial effect of silver ions on *Escherichia coli* and *Staphylococcus aureus*. *Journal of Biomedical Materials Research* 52, no. 4:662–668.

Ganesan, S.V., Kaithavalappil, S., Kannappan, M., Deepa, T., and Vasudevan. 2014. Formulation and in-vitro antimicrobial evaluation of herbosomal gels of extracts of *Quercus infectoria* and *Acorus calamus*. *Asian Journal of Research in Pharmaceutical Sciences and Biotechnology* 2, no. 2:47–54.

Geoprincy, G., Saravanan, P., Nagendra, G.N., and Renganathan, S. 2011. A novel approach of studying the combined antimicrobial effects of silver nanoparticles and antibiotics through agar over layer method and dice diffusion method. *Digest Journal of Nanomaterials and Biostructures* 6, no. 4:1557–1565.

Gong, P., Li, H., He, X., Wang, K., Hu, J., and Tan, W. 2007. Preparation and antibacterial activity of $Fe_3O_4@Ag$ nanoparticles. *Nanotechnology* 18:604–611.

Gu, H., Ho, P.L., Tong, E., Wang, L., and Xu, B. 2003. Presenting vancomycin on nanoparticles to enhance antimicrobial activities. *Nano Letters* 3, no. 9:1261–1263.

Guidelli, E.J., Ramos, M.E., Zaniquelli, D., and Baffa, O. 2011. Green synthesis of colloidal silver nanoparticles using natural rubber latex extracted from *Hevea brasiliensis*. *Spectrochimca Acta A Molecular and Biomolecular Spectroscopy* 82, no. 1:140–145.

Guo, S., and Wang, E. 2007. Synthesis and electrochemical applications of gold nano particles. *Analytica Chimica Acta* 598, no. 2:181–192.

Gupta, A.K., and Gupta, M. 2005. *Biomaterials* 26, no. 18:3995.

Huang, H.H., Yan, F.Q., Kek, Y.M., Chew, C.H., Xu, G.Q., Ji, W., and Tang, S.H. 1997. Synthesis, characterization, and nonlinear optical properties of copper nanoparticles. *Langmuir* 13:172.

Jain, D., Kumar Daima, H., Kachhwaha, S., and Kothari, S.L. 2009. 2 synthesis of plant-mediated silver nanoparticles using papaya fruit extract and evaluation of their anti-microbial activities. *Digest Journal of Nanomaterials and Bio Structures* 4, no. 3:557–563.

Jeong, S.H., Yeo, S.Y., and Yi, S.C. 2005. The effect of filler particle size on the antibacterial properties of compounded polymer/silver fibers. *Journal of Material Science* 40, no. 1:5407–5411.

Jiang, W., and Yatsui, K. 1998. Pulsed wire discharge for nanosize powder synthesis. *IEEE Transactions on Plasma Science* 26:1498–1501.

Kantam, M.L., Jaya, V.S., Lakshmi, M.J., Reddy, B.R., Choudary, B.M., and Bhargava, S.K. 2007. Alumina supported copper nanoparticles for aziridination and cyclopropanation reactions. *Catalysis Communications* 8, no. 12: 1963–1968.

Kapoor, S., Joshi, R., and Mukherjee, T. 2002. Influence of I⁻ anions on the formation and stabilization of copper nanoparticles. *Chemical Physics Letters* 354, no. 56:443.

Karthick, R.N.S., Ganesh, S., and Avimanyu. 2011. Evaluation of anti-bacterial activity of silver nanoparticles synthesized from *Candida glabrata* and *Fusarium oxysporum*. *International Journal of Medicobiological Research* 1, no. 3:131–136.

Kasthuri, J., Kathiravan, K., and Rajendran, N. 2009. Phyllanthin assisted biosynthesis of silver and gold nanoparticles a novel biological approach. *Journal of Nanoparticle Research* 11, no. 1:1075–1085.

Kaviya, S., Santhanalakshmi, J., and Viswanathan, B. 2008. Green synthesis of silver nanoparticles using *Polyalthia longifolia* leaf extract along with D-Sorbitol: Study of antibacterial activity. *Journal of Nanotechnology* 2011, no. 10:1–5, Article ID 152970, 5 pages, doi:10.1155/2011/152970.

Kim, J.S., Kuk, E., Yu, K.N., Kim, J.H., Park, S.J., Lee, H.J., Kim, S.H. et al. 2007. Antimicrobial effects of silver nanoparticles. *Nanomedicine* 3, no. 1:95–101.

Kitchens, C.L., and Roberts, C.B. 2004. Copper nanoparticle synthesis in compressed liquid and supercritical fluid reverse micelle systems. *Industrial and Engineering Chemistry Research* 43:6070–6081.

Krishnaraj, C., Muthukumaran, P., Ramachandran, R., Balakumaran, M.D., and Kalaichelvan, P.T. 2014. *Acalypha indica* Linn: Biogenic synthesis of silver and gold nanoparticles and their cytotoxic effects against MDA-MB-231, human breast cancer cells. *Biotechnology Reports* 4, 42–49.

Krutyakov, Y.A., Kudrynskiy, A., Olenin, A.Y., and Lisichkin, G.V. 2008. Extracellular biosynthesis and antimicrobial activity of silver nanoparticles. *Russian Chemical Reviews* 77:233.

Kuppusamy, P., Mashitah, M., Narasimha, Y., Parine, R., and Govindan, N. 2015. Evaluation of in-vitro antioxidant and antibacterial properties of *Commelina nudiflora* L. extracts prepared by different polar solvents. *Saudi Journal of Biological Science* 22, no. 3:293–301.

Lavanya, M., Veenavardhini, S.V., Gim, G.H., Kathiravan, M.N., and Kim, S.W. 2013. Synthesis, characterization and evaluation of antimicrobial efficacy of silver nanoparticles using *Paederia foetida* L. leaf extract. *International Research Journal of Biological Sciences* 2, no. 3:28–34. ISSN 2278-3202.

Logeswari, P., Silambarasan, S., and Abraham, J. 2015. Synthesis of silver nanoparticles using plants extract and analysis of their antimicrobial property. *Journal of Saudi Chemical Society* 19:311–317.

Lok, C.N., Ho, C.M., Chen, R., He, Q.Y., Yu, W.Y., Sun, H., Tarn, P.K., Chiu, J.F., and Che, C.M. 2007. Silver nanoparticles: Partial oxidation and antibacterial activities. *Journal of Biological Inorganic Chemistry* 12, no. 4:572–534.

Mohzen, Z., Azizah, A.H., Fatima, A.B., Mariana, N.S., Kamyar, S., Fatemeh, J., and Farah, F. 2011. Green synthesis and antibacterial effect of silver nanoparticles using *Vitex negundo* L. *Molecules* 16, no. 8:6667–6676; doi:10.3390/molecules16086667.

Morones, J.R., Elechiguerra, J.L., Camacho, A., and Ramirez, J.T. 2005. The bactericidal effect of silver nanoparticles. *Nanotechnology* 16:2346–2353.

Muraia, K., Watanabea, Y., Saitoa, Y., Nakayamaa, T., Suematsua, H., Jianga, W., Yatsuia, K., Shimb, K.B., and Niiharaa, K. 2007. Preparation of copper nanoparticles with an organic coating by a pulsed wire discharge method. *Journal of Ceramic Processing Research* 8, no. 2:114–118.

Nabi, S.A., Khan, M.A., and Khowaja, S.N. 2006. Thin-layer chromatographic separation of penicillins on stannic arsenate–cellulose layers. *Acta Chromatographica* 16:164–172.

Pandey, B., Ghimire, P., and Agrawal, V.P. 2004. International conference on the Great Himalayas: Climate, health, ecology, management and conservation. Organized by Kathmandu University and the Aquatic Ecosystem Health and Management Society.

Panigrahi, S., Kundu, S., Ghosh, S. K., Nath, S., Praharaj, S., Soumen, B., and Pal, T. 2006. Selective one-pot synthesis of copper nanorods under surfactantless condition. *Polyhydron* 25:1263–1269.

Petri-Fink, A., Chastellain, M., Juillerat-Jeanneret, L., Ferrari, A., and Hofmann, H. 2005. Development of functionalized super paramagnetic iron oxide nanoparticles for interaction with human cancer cells. *Biomaterials* 26, no. 15:2685–2694.

Ponarulselvam, S., Panneerselvam, C., Murugan, K., Aarthi, N., Kalimuthu, K., and Thangamani, S. 2012. Synthesis of silver nanoparticles using leaves of *Catharanthus roseus* Linn. G. Don and their antiplasmodial activities. *Asian Pacific Journal of Tropical Biomedicine* 2, no. 7:574–580.

Raimondi, F., Scherer, G.G., Kotz, R., and Wokaun, A. 2005. Nanoparticles in energy technology: Examples from electrochemistry and catalysis. *Angewandte Chemie, International Edition* 44:2190–2209.

Rathi Sre, Pr., Reka, M., Poovazhagi, R., Arul Kumar, M., and Murugesan, K. 2015. Antibacterial and cytotoxic effect of biologically synthesized silver nanoparticles using aqueous root extract of *Erythrina indica lam. Spectrochimica Acta Part A: Molecular and Biomolecular Spectroscopy* 135:1137–1144.

Raut, R.W., Lakkakula, J.R., Kolekar, N.S., Mendhulkar, V.D., and Kashid, S.B. 2009. Photosynthesis of silver nanoparticle using Gliricidia sepium. *Current Nanoscience* 5:117–122.

Retchkiman-Schabe, P.S., Canizal, G., Becerra-Herrera, R., Zorrilla, C., Liu, H.B., and Ascencio, J.A. 2006. Biosynthesis and characterization of Ti/Ni bimetallic nanoparticles. *Optical Materials* 29:95–99.

Roh, Y., Vali, H., Phelps, T.J., and Moon, J.W. 2006. Extracellular synthesis of magnetite and metal-substituted magnetite nanoparticles. *Journal of Nanoscience Nanotechnology* 11:3517–3520.

Sankar, R., Manikandan, P., Malarvizhi, V., Fathima, T., Shivashangari, K.S., and Ravikumar, V. 2014. Green synthesis of colloidal copper oxide nanoparticles using *Carica papaya* and its application in photocatalytic dye degradation. *Spectrochimica Acta Part A: Molecular and Biomolecular Spectroscopy* 121:746–750.

Saranyaadevi, K., Subha, V., Ernest Ravindran, R.S., and Renganathan, S. 2014. Green synthesis and characterization of silver nanoparticle using leaf extract of *Capparis zylanica*. Vol. 7, suppl 2.

Sastry, M., Ahmad, A., Khan, M.I., and Kumar, R. 2003. Biosynthesis of metal nanoparticles using fungi and actinomycetes. *Current Science* 85:162–170.

Sastry, M., Mayyaa, K.S., and Bandyopadhyay, K. 1997. pH dependent changes in the optical properties of carboxylic acid derivatized silver colloid particles. *Colloids and Surfaces A* 127:221–228.

Shankar, S.S., Ahmad, A., and Sastry, M. 2003. Geranium leaf assisted biosynthesis of silver nanoparticles. *Biotechnology Progress* 19:1627–1631.

Shankar, S.S., Rai, A., Ahmad, A., and Sastry, M. 2004a. Rapid synthesis of Au, Ag and bimetallic Au core-Ag shell nanoparticles using Neem (*Azadirachta indica*) leaf broth. *Journal of Colloid and Interface Science* 275, no. 2:496–502.

Shankar, S.S., Rai, A., Akkamwar, B., Singh, A., Ahmed, A., and Sastry, M. 2004b. Biological synthesis of triangular gold nanoprisms. *Nature Materials* 3:482–488.

Shirley, A.D., Sreedhar, B., and Syed, G.D. 2010. Antimicrobial activity of silver nanoparticles synthesized from novel Streptomyces species. *Digest Journal of Nanomaterials and Biostructures* 5, no. 2:447–451.

Silver, S., Phung, L.T., and Silver, G. 2007. Silver as biocides in burn and wound dressings and bacterial resistance to silver compounds. *Journal of Indian Microbiology Biotechnology* 33, no. 1:627–634.

Slawson, R.M., Trevors, J.T., and Lee, H. 1992. Silver accumulation and resistance in *Pseudomonas stutzeri. Archives of Microbiology* 158:398–404.

Song, R.G., Yamaguchi, M., Nishimura, O., and Suzuki, M. 2007. Investigation of metal nanoparticles produced by laser ablation and their catalytic activity. *Applied Surface Science* 253:3093–3097.

Sougata, G., Sumersing, P., Mehul, A., Rohini, K., Sangeeta, K., Karishma, P., Cameotra, S.S., Jayesh, B., Dilip, D., Amit, J., and Balu, A.C. 2012. Synthesis of silver nanoparticles using *Dioscorea bulbifera* tuber extract and evaluation of its synergistic potential in combination with antimicrobial agent. *International Journal of Nanomedicine* 7:483–496.

Spadaro, J.A., Berger, T.J., Barranco, S.D., Chapin, S.E., and Becker, R.O. 1974. Antibacterial effects of silver electrodes with weak direct current. *Antimicrobial Agents Chemotherapy* 6, no. 1:637–642.

Srivastava, A.A., Kulkarni, A.P., Harpale, P.M., and Zunjarrao, R.S. 2011. Plant mediated synthesis of silver nanoparticles using a bryophyte: *Fissidens minutus* and its antimicrobial activity. *International Journal of Engineering Science and Technology* 3, no. 12:8342–8347.

Suryanarayana, C. 2001. Mechanical alloying and milling. *Progress in Materials Science* 46:1–184.

Suslick, K.S., Hyeon, T., Fang, M., and Cichowlas, A.A. 1996. Sono chemical preparation of nanostructured catalyst. In *Advanced Catalysts and Nanostructured Materials*, edited by W.R. Moser, chap. 8. Academic Press, New York.

Thomas, V., Yallapu, V.V., Sreedhar, B., and Bajpai, S.K. 2007. A versatile strategy to fabricate hydrogel silver nano composites and investigation of their antimicrobial activity. *Journal of Colloid and Interface Science* 315, no. 1:389–395.

Venkatesan, B., Subramanian, V., Tumala, A., and Vellaichamy, E. 2014. *sian Pacific Journal of Tropical Biomedicine* 7S1:S294–300.

Viveka, R., Ramar Thangama, B., Muthuchelianc, K., Gunasekaranb, P., Krishnasamy Kaveri, B., and Kannana, S. 2012. Green biosynthesis of silver nanoparticles from *Annona Squamosa* leaf extract and its in vitro cytotoxic effect on MCF-7 cells. *Process Biochemistry* 47:2405–2410.

Wang, X., Ding, Y., Summers, C.J., and Wang, Z.L. 2004. Large scale synthesis of six-nanometer-wide ZnO nano belts. *Journal of Physical Chemistry B* 108, no. 26:8773–8777.

Zhao, G., and Stevens, S.E. 1998. Multiple parameters or the comprehensive evaluation of the susceptibility of *Escherichia coli* to the silver ion. *Biometals* 11, no. 1:27–32.

Zhong, C.-J., and Mathew, M.M. 2001. Core-shell assembled nanoparticles as catalysts. *Advanced Materials* 13, no. 19:1507–1511.

Zhu, M.-Q., Wang, L.-Q., Exarhos, G.J., and Li, A.D.Q. 2004. Thermosensitive gold nanoparticles. *Journal of the American Chemical Society* 126, no. 9:2656–2657.

CHAPTER 4

Role of Phytoremediation in Maintaining Environmental Sustainability

Sahadevan Renganathan, Bhargavi Gunturu, and Dharmendira Kumar Mahendradas

CONTENTS

Abstract

Bioremediation deals with different methods to resolve various environmental problems by biological means. It is one of the latest developments in waste management strategies, whereby pollutants in contaminated sites are eliminated or neutralized by organisms. Bioremediation techniques include bioaugmentation, bioleaching, bioventing, biosparging, bioslurping, and phytoremediation. Since the Industrial Revolution, water pollution has become a serious threat and cause of death and disease. Nowadays, textile dyes contaminate huge amount of water due to improper management of effluents from industries. Phytoremediation is a burgeoning technology where plants have found enormous applications in the remediation of contaminated soil and water by, for example, degradation, extraction, containment, and immobilization. Being an eco-friendly and cost-effective technique, phytoremediation is extensively used in removing textile dyes, heavy metals, and other pollutants from tainted water. This chapter focuses on various phytoremediation techniques used for the removal of textile dyes from contaminated water.

4.1   INTRODUCTION

Untreated wastewater contains a large number of organic and inorganic constituents, noxious microorganisms, and toxic metal components in addition to nutrients. Hence, it is necessary to treat the wastewater at the source level before releasing in to the environment in order to reduce detrimental effects. The protection of the environment to sustain public health, aquatic life, and other socioeconomic concerns are the ultimate goals of wastewater management (Tchobanoglous and Burton 1991).

Major organic chemical parameters of wastewater are total organic carbon (TOC), biochemical oxygen demand (BOD), chemical oxygen demand (COD), and total oxygen demand (TOD). Acidity and alkalinity, pH, salinity, hardness, ionized metals, and anionic salts such as iron and manganese, sulfates, nitrates, sulfides, chlorides, and phosphates are the major inorganic chemical constituents in wastewater. In addition to these, coliforms, feces, viruses, and other pathogens also exist. So, based on the constituents and their concentrations, wastewater is classified as strong, medium, or weak (Economic and Social Commission for Western Asia [ESCWA] 2003).

Processed water from industries possibly containing dissolved toxic metals has been drained into freshwater. It is recommended by the World Health Organization (WHO) that care should be taken with metals such as chromium, cadmium, iron, zinc, nickel, mercury, cobalt, and copper. These metals have the capability to bioaccumulate in aquatic fauna if taken as food and cause cancer and inhibit cell metabolism in the body. The currently used method of heavy metals removal usually in the form of their hydroxides from waste waters by adding lime or caustic soda is not cost effective and cannot be encourageable (ESCWA 2003).

Copious physical, chemical, and biological methods have been put forward to separate contaminants from wastewater. These procedures have been classified into primary, secondary, and tertiary, depending upon the levels of contaminant removed during the treatment course. In addition, natural systems are also being used to treat wastewater in land-based applications. The physical operations include screening, comminution, flow equalization, sedimentation, flotation, and granular medium filtration. Chemical precipitation, adsorption, disinfection, and dechlorination operations are some of the chemical methods being used. Biological treatment methods include, activated sludge process, aerated lagoon, trickling filters, rotating biological contractors, pond stabilization, anaerobic digestion, and biological nutrient removal (ESCWA 2003).

Conversely, the stringent efforts on waste minimization and water conservation resulted in the generation of highly concentrated toxic residues. As environmental laws become more rigorous, the focus should be on efficient techniques for managing proper disposal of these residues and maintaining the concentration of deadly chemicals to minimum levels in effluent streams. Therefore, research needs to be done to unearth new waste management technologies or to improve already existing ones, which is vital in converting complex molecules into simpler entities in the process of wastewater management (Gogate and Pandit 2004). On the other hand, contamination of soils is brought about by the accumulation of heavy metals and metalloids, derived from growing industrial emissions, metal wastes disposal, mining, leaded gasoline and paints, release of wastewater, disposal of sewage sludge, application of fertilizers and pesticides, residues of coal combustion, and spillage of petrochemicals (Wuana and Okieimen 2011).

Several methods have been proposed to remediate contaminated sites. One of the conventional techniques is excavating the contaminated soil to a landfill or covering. However, excavation, handling, and transportation of hazardous materials to disposal sites are risky. Besides, it is very difficult to find the target landfill sites for the disposal of material and it is an expensive method. Some other techniques such as high-temperature incineration and chemical deposition (for example, base catalyzed dechlorination and UV oxidation) have also been proposed, which are effective in reducing the concentration of contaminants such as polycyclic aromatic hydrocarbons, explosives, pesticides, trinitrotoluene, herbicides, heavy metals, chlorinated compounds, inorganic nitrogen, and petroleum-based compounds. Moreover, due to the lack of public acceptance, complexity, and increased chance of exposure of workers and residents to contaminants, these techniques have not been considered as advantageous (Vidali 2001). Also, these methods have resulted in large volumes of concentrated hazardous sludge, and so complete destruction of pollutants cannot be brought about. These drawbacks can be transcended by adapting a technique that can efficiently destroy the pollutants or convert them into biodegradable substances. One such technique is *bioremediation*. It is a competent method to clean the environment through natural biological activity, by which contaminants can be destroyed. As a cost-effective, eco-friendly, and environmentally friendly technology, bioremediation has received public acceptance and can be carried out with ease at any location (Bhatnagar 2013).

Bioremediation is broadly defined as a set of biological treatment systems to remediate contaminated sites by destroying or reducing the concentration of hazardous waste. Briefly, the process of remediation is brought about by biological entities such as microorganisms or plants, which convert hazardous waste into nonhazardous forms (Dwivedi 2012). Based on the process of removal and transport, bioremediation is classified into in situ and ex situ. In in situ remediation, contaminants are treated at the same site. Ex situ remediation involves complete transfer of contaminated material from the affected site to another place where the treatment is carried out by biological agents. These days, the ex situ mode of treatment is preferred over in situ for remediating contaminated water, soil, and environment (Vidali 2001).

Depending on the material utilized to clean up the contaminated environment, bioremediation is classified into three different methods. Bioremediation by using fungus is known as mycoremediation. Phytoremediation is a type of bioremediation that utilizes plants. Bioremediation brought about by using bacteria is called bacterial bioremediation. The method to be applied for the remediation depends on the area and contaminants that are to be processed.

4.2 BIOREMEDIATION TECHNOLOGIES

Bioremediation technologies include bioaugmentation, biostimulation, bioreactors, land-based remediation, fungal treatment, and bioleaching, in addition to phytoremediation.

4.2.1 Bioaugmentation

Bioaugmentation is the in situ bioremediation technique, in which the degradation of contaminated sites is enhanced by introducing a single strain or consortia of microorganisms with specific catalytic abilities. Briefly, bioaugmentation is the introduction of innate or allochthonous strains or genetically engineered microbes to the contaminated sites in order to hasten the degradation of hazardous material into innocuous forms. Genetically engineered microorganisms have shown enhanced degradative potential toward bioaugmentation of soil polluted with aromatic hydrocarbons. Most of the research work has been established with Gram-negative bacteria belonging to the *Flavobacterium, Achromobacter, Sphingobium, Alcaligens*, and *Pseudomonas* genera. In addition, fungi also have been proven as potential agents for bioaugmentation, including *Absidia, Penicillium, Mucor, Achremonium, Aspergillus*, and *Verticillium*. There is no any particular group or solitary microorganism that can be universally applied for augmentation (Mrozik and Seget 2010).

Biodegradation is a type of bioaugmentation where indigenous or engineered bacteria and fungi can be used for the degradation of organic contaminants in soil or groundwater. The degradation may be aerobic or anaerobic. Biodegradation has been efficient in the degradation of hydrocarbons, nitrogen compounds, heavy metals including lead and mercury, halo compounds, radionucleotides, and nonhalogenated pesticides.

4.2.2 Biostimulation

Biostimulation is an in situ, enhanced bioremediation technique like bioaugmentation. It requires the supply of adequate quantities of water, oxygen, and nutrients into the soil to promote indigenous microbial degradation or cometabolism (Kanissery and Sims 2011). Bioventing, chemical oxidation of soils, and in situ lagoon treatments are the techniques categorized under biostimulation. Bioventing is an in situ remediation, where the activity of native bacteria is accelerated by providing oxygen through injection wells into the soil. Most of the oxidizable constituents can be degraded by using bioventing. It has been proven that bioventing is an efficient technique to degrade petroleum derivatives such as kerosene, diesel, jet fuels, and gasoline (U.S. EPA 1994). Chemical oxidation is a process that involves the use of oxidants or oxidizing agents to transform hazardous contaminants into less toxic oxidized forms. Due to electron transfer, oxidation and reduction of contaminants occur. It is known as an in situ process; remediation of contaminants by using oxidizing agents occurs without transferring the contaminated material. The most commonly used oxidizing agents for remediation are hydrogen peroxide, ozone, chlorine dioxide, hypochlorites, and chlorine. This method is well known for the treatment of sites contaminated with pesticides, solvents, and fuels. Chemical oxidation is an efficient technology for the treatment of groundwater and contaminated soils (U.S. EPA 2012). In situ lagoon treatment is a technique where degradation of mixture of organic and inorganic contaminants is brought about inside a lagoon. The choice of aerobic or anaerobic degradation depends on contaminants. Air sparging, pumping, dredging, and other mechanical slurry mixing techniques increase the bioremediation rates. It has been proven that in situ lagoon treatment can remediate phenols, heavy metals, polychlorinated biphenyls, polycyclic aromatic hydrocarbons, and other aromatic organic solvents. Due to low soil and sludge handling, it can be considered cost effective and advantageous (Cookson 1995).

4.2.3 Bioreactors

Using bioreactors is an ex situ remediation technique. It is a controlled method for treating contaminated soil and groundwater. Remediation can be carried out using either compost-based reactors or slurry-based reactors under optimized conditions of pH, temperature, nutrients, and agitation that facilitate maximum biological activity. In slurry-based reactors, contaminated soil is mixed with water and other additives to form slurry. An optimum environment should be maintained for enhanced growth of microbes in addition to adequate oxygen and nutrient supply (Collina et al. 2005). A compost-based reactor is a controlled in-vessel method in which biodegradable pollutants are converted into innocuous by-products by the action of microorganisms at high temperatures. Elevation in temperature is because of the degradative action of microbes. Sewage sludge can be processed through aerobic composting, whereas anaerobic composting suits hazardous waste treatment. Composting technology is found effective in the treatment of soils contaminated with pentachlorophenol, polycyclic aromatic hydrocarbons, insecticides, landfills, and refinery sludges.

4.2.4 Land-Based Treatments

Land-based treatment is an ex situ remediation method. The excavated soil is treated in piles or specially constructed cells. Land-based treatment methods include composting and land farming. Composting is similar to compost-based reactors. It is also a controlled process to degrade hazardous substances into nontoxic by-products through biological processes by using microorganisms at increased temperatures. The addition of sawdust, animal waste, and other materials to the soil increases the porosity (Cookson 1995; Norris et al. 1994). An efficient method to remediate the upper soil zone is land forming. Contaminants in the form of soil, sludge, or sediments are directly mixed with the upper layers of soil and agitated at regular time intervals to maintain aeration. This has been proposed as a successful technique for maintaining petroleum refinery wastes and other oily sludges.

4.2.5 Fungal Remediation

Utilization of fungi for the treatment of contaminated organic soils is known as fungal remediation. Fungi are ubiquitous and have the ability to colonize in all types of environments such as soil, water, and air, playing a vital role in maintaining a sustained environment. Fungi have been proven as successful agents that can completely degrade lignin and in turn contribute to carbon recycling in nature (Anastasi et al. 2013). Hydrocarbons are one of the major organic pollutants that can be degraded by fungi. For example, white rot fungus (*Phanerochaete chyrsosporium*) produces a group of enzymes called ligninases, which are responsible for its biodegradative property. White rot fungus has the ability to degrade chlorinated aromatic hydrocarbons, polyaromatic hydrocarbons, pesticides, polycyclic aromatics, and even azo dyes.

4.2.6 Bioleaching

Bioleaching is a process of extraction of metals from their ores or minerals by the action of naturally occurring microbes. In bioleaching, the solid metals are transformed into their soluble forms by microorganisms. For example, copper metal is extracted into the aqueous phase as copper sulfate by the oxidation of copper sulfide ore, and the waste residue is discarded through solid wastes (Mishra et al. 2005). Bioleaching is an economically feasible and environmentally friendly remediation technique. Most of the heavy-metal retrieving microorganisms are acidophiles, which can withstand an acidic pH range of 2.0 to 4.0. The inorganic and organic substances secreted by these microbes can dissolve the metals into aqueous media. Commonly used acidophilic bacteria for bioleaching applications include *Acidithiobacillus ferrooxidans*, *Leptospirillum ferrooxidans*, *Acidithiobacillus thiooxidans*, and *Sulfolobus* sp. Also, fungal species such as *Aspergillus niger* and *Penicillium* sp. have shown potential for bioleaching (Mishra and Ree 2014).

4.3 PHYTOREMEDIATION

The term *phytoremediation* is derived from Greek prefix *phyto*, meaning "plant," and Latin word *remedium*, meaning "to correct or restore." Phytoremediation refers to a set of techniques where plants have application in the extraction, containment, degradation, immobilization, or destruction of contaminants that exist in water and soil (U.S. EPA 2000).

Industrial manufacturing works, such as dying of textiles, iron and steel, beverages and food products, metal production, engineering, wood products, electronic ware, plastics, chemicals, paper products, transport equipment, poultry, and medical products, are releasing major quantities of untreated or partly treated tainted water into nearby surface water. In addition, domestic wastewater, fertilizer, and pesticide-mixed water from the agricultural sector are also contributing to the contamination of ground as well as surface water. Phytoremediation methods are economically cheap, environmentally friendly, and advantageous compared to other methods. In recent times, plants have received increased attention in remediating contaminated wastewater, soil, and sediments. Many plant species have been known for their remediative capabilities (Hegazy et al. 2011).

In addition, plants have the ability to accumulate metals. A plant root system is capable of selective uptake of contaminants, which can be accumulated, translocated, or degraded in plant body parts. The processes involved in phytoremediation are phytoextraction or phytoaccumulation, phytostabilization, phytotransformation or degradation, phytovolatilization, and rhizofiltration. Phytoextraction and phytostabilization are known for the removal of inorganic contaminants, whereas

TABLE 4.1   Phytoremediation Methods, Mechanisms, and Removable Contaminants

Mechanism	Process	Media	Removable Contaminants
Phytoextraction	Remediation by extraction and capture	Solids, sediments, sludges	Metals: Ag, Au, Cd, Co, Cr, Cu, Hg, Mn, Mo, Ni, Pb, Zn, Radionuclides: ^{90}Sr, ^{137}Cs, ^{239}Pu, ^{234,238}U
Phytostabilization	Containment	Soils, sediments, sludges	As, Cd, Cr, Cu, Pb, Zn
Phytodegradation	Remediation by destruction	Soils, sediments, sludges, groundwater, surface water	Organic compounds, chlorinated solvents, phenols, pesticides, munitions
Rhizodegradation	Remediation by destruction	Soils, sediments, sludges, groundwater	Organic compounds including polycyclic aromatic hydrocarbons and other solvents

Source: Interstate Technology & Regulatory Council, 2001, Phytotechnology technical and regulatory guidance document, Phytotechnologies Work Team, Technical/ Regulatory Guidelines, http://www.state.nj.us/dep/dsr/bscit/Phytoremediation.pdf.

phytotransformation and rhizofiltration are used for organic contaminant remediation (Tangahu et al. 2011). Table 4.1 gives the overall view of phytoremediation methods, mechanisms, and the type of contaminant remediated (Interstate Technology & Regulatory Council [ITRC] 2001).

4.3.1 Phytoremediation Processes

4.3.1.1 Phytoextraction

The best method to adopt for the removal of soil contamination without disturbing soil's structure and fertility is phytoextraction. It is also known as phytoaccumulation. Phytoextraction is the utilization of plants to remove contaminants from the environment. Plants absorb and precipitate toxic metals from soil into their tissues. Figure 4.1 gives the pictorial representation of phytoextraction.

Previously, phytoextraction was limited to the extraction of heavy metals from soil, but recently, application in other media has been discovered. This method is applicable for the remediation of soils with low contamination levels. Two strategies of phytoextraction have been developed: (1) induced or chelate-aided phytoextraction, where artificial chelates are supplied to hasten the mobility and uptake of metals by the plant; and (2) continuous phytoextraction, where the removal of metal depends on the ability of the plant. Growth rate (which is usually slow), low biomass

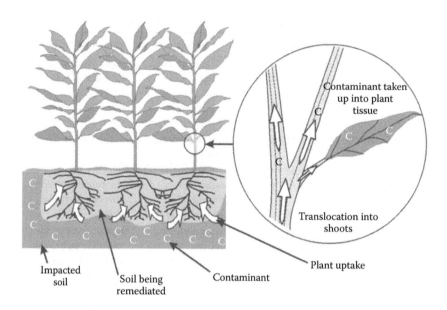

FIGURE 4.1 Phytoextraction. (From Interstate Technology & Regulatory Council, 2001, Phytotechnology technical and regulatory guidance document. Phytotechnologies Work Team, Technical/Regulatory Guidelines, http://www.state.nj.us/dep/dsr/bscit/Phytoremediation.pdf.)

production, a shallow root system, and final disposal of the plant are the major limitations for the use of plant accumulators (Ghosh and Singh 2005). Phytoextraction is mostly applied for the extraction of inorganic metals like chromium, arsenic, mercury, copper, and nickel. Plants with high metal-accumulating capacity are called hyperaccumulator plants (Tangahu et al. 2011). Hyperaccumulators can accumulate and tolerate metal concentrations of 0.1% nickel, lead, copper, and chromium, or 0.1% zinc on dry weight. But most hyperaccumulator plant species seem to uptake only single metal species, though major sites are mixed-metal contaminated. On the other hand, these species produce low biomass levels at high metal concentrations. The measure of phytoextraction success depends on the amount of metal extractable by plant per hectare (Cunningham et al. 1997).

4.3.1.2 Phytostabilization

Phytostabilization is the utilization of plants to stabilize or immobilize contaminants, usually metals, present in the soil and groundwater. The immobilization of contaminants is brought about by accumulation and absorption, adsorption, or precipitation by the root system. It limits the mobility of contaminants by concentrating and clogging them at the rhizosphere (U.S. EPA 1998). Thus the use of metal-tolerant plant species to restrict the mobility of metals reduces the contamination of groundwater and air spreading, which in turn reduces further degradation of the environment. During the process, metals are immobilized as precipitates or complexes by chemical means at root surfaces. Chemical reactions at the root–soil interface make the metals available to the roots for uptake. The possible mechanism of phytostabilization is shown in Figure 4.2.

In addition, through transpiration, metal-polluted water can be treated and can move to the groundwater system in pure form. The selection of plants for phytostabilization depends on their seedlings, ability to acclimatize the new root zone, and their capability to restrict the concentration of metals to only the root system (Neuman and Ford 2006). It is an effective technique for the preservation of surface and groundwater. Reestablishment of vegetative cover over heavily metal-contaminated soils is the major advantage of phytostabilization (U.S. EPA, 2012). However,

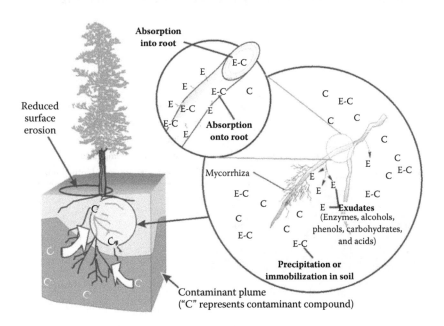

FIGURE 4.2 Phytostabilization. (From Interstate Technology & Regulatory Council, 2001, Phytotechnology technical and regulatory guidance document. Phytotechnologies Work Team, Technical/Regulatory Guidelines, http://www .state.nj.us/dep/dsr/bscit/Phytoremediation.pdf.)

the main impediment is the regular monitoring of plants, as the contaminants or metals remain in the soil (Ghosh and Singh 2005).

4.3.1.3 Phytotransformation

Phytotransformation, also known as phytodegradation, involves the degradation or break down of contaminants by plants. The degradation can occur either by metabolic processes inside the plant or by secondary metabolites such as enzymes produced by the plant to the outside environment. Figure 4.3 shows the mechanism of remediation through phytotransformation.

Complex pollutants are transformed into simpler moieties and are stored inside the plant tissues, and can be utilized for the growth of the plant. Enzymes produced from plants can degrade chlorinated solvents, missile wastes, and even herbicides (U.S. EPA 1998). Plant-derived enzymes for degradation include reductases, oxygenases, and dehalogenases. In addition, rhizodegradation is a type of degradation where microbes present in the root zone break down contaminants into simpler molecules, but it is a late process compared to phytodegradation (Ghosh and Singh 2005).

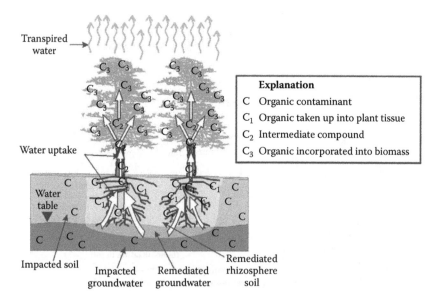

FIGURE 4.3 Phytotransformation. (From Interstate Technology & Regulatory Council, 2001, Phytotechnology technical and regulatory guidance document. Phytotechnologies Work Team, Technical/Regulatory Guidelines, http://www.state.nj.us/dep/dsr/bscit/Phytoremediation.pdf.)

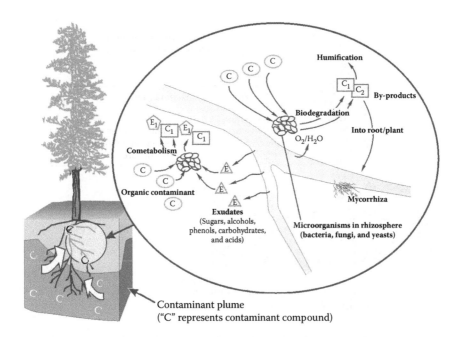

Contaminant plume
("C" represents contaminant compound)

FIGURE 4.4 Rhizodegradation. (From Interstate Technology & Regulatory Council, 2001, Phytotechnology technical and regulatory guidance document. Phytotechnologies Work Team, Technical/Regulatory Guidelines, http://www .state.nj.us/dep/dsr/bscit/Phytoremediation.pdf.)

Figure 4.4 represents the overall mechanism that occurs during rhizodegradation of contaminants.

4.3.1.4 Phytovolatilization

Phytovolatilization is defined as a remediation method in which plants absorb contaminants from soil, transform, and then finally transpire the volatilized contaminants into the environment. This process occurs as plants absorb water, and many organic and inorganic pollutants from soil. Some of them pass along with water to leaves, are volatilized in the leaves, and then transpired into the atmosphere. But through this process, very low concentrations of contaminants are transformed into volatile form. Figure 4.5 shows the mechanism of phytovolatilization through which both organic and inorganic contaminants can be remediated.

Usually the phytovolatilized contaminant is less toxic than its original form. As through phytovolatilization contaminants are released into atmosphere, there is a risk to environment quality and human health. Phytovolatilization is mostly used in the remediation of mercury

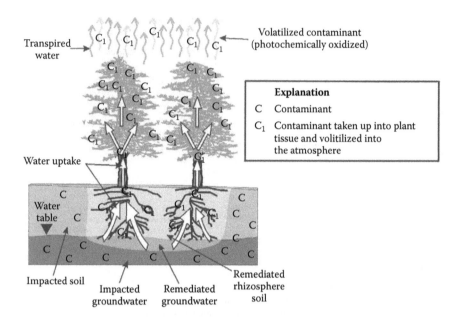

FIGURE 4.5 Phytovolatilization. (From Interstate Technology & Regulatory Council, 2001, Phytotechnology technical and regulatory guidance document. Phytotechnologies Work Team, Technical/Regulatory Guidelines, http://www.state.nj.us/dep/dsr/bscit/Phytoremediation.pdf.)

contaminated sites, where mercuric ions are reduced into less toxic elemental mercury (Ghosh and Singh 2005). Inorganic contaminants solubilized in soil, sediment, water, or sludges, and organic contaminants such as chlorocompounds can be remediated through phytovolatilization along with other phytoremediation techniques (Pivetz 2001).

4.3.1.5 Rhizofiltration

Rhizofiltration is known for the use of plants to absorb, concentrate, and precipitate contaminants present in tainted water. Both aquatic and terrestrial plants have found application in the treatment of contaminated sites. Rhizofiltration is advantageous as it can be used for both ex situ and in situ applications. Plants like tobacco, sunflower, spinach, Indian mustard, and corn have been utilized for the removal of lead (Ghosh and Singh 2005). Rhizofiltration by plant roots depends on their physical and biochemical nature. The efficiency of the process relates to its ability to produce chemicals that hasten the metal uptake. Depending upon the exudates from the roots, some metals get absorbed onto the surface

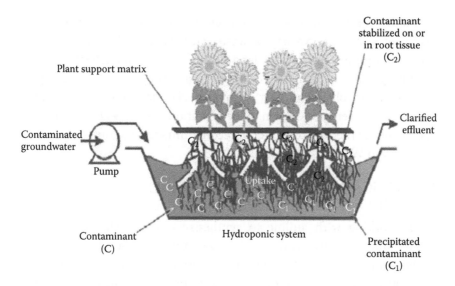

FIGURE 4.6 Rhizofiltration. (From Institute for Clean Energy and Green Environment (IGECE), n.d., Rhizofiltration, http://systemsbiology.usm.edu /BrachyWRKY/WRKY/Rhizofiltration.html.)

of roots. Figure 4.6 refers to the rhizofiltration process using sunflower plants.

Frequent removal of plants is necessary, as they are saturated by the contaminants. Rhizofiltration can be applied for the remediation of industrial and domestic effluents, used water from hydroelectricity, agricultural runoff, mine drainage, radionuclide-contaminated water, stormwater, and other groundwater and surface water bodies. Roots play a vital role in removing toxic metals from solutions. Early on, only aquatic plants were used for rhizofiltration, but recently terrestrial plants are also being widely used to remove toxic metals such as zinc, copper, nickel, chromium, cadmium, lead, and even low amounts of radioactive contaminants from water sources (Rawat et al. 2012).

4.3.1.6 Constructed Wetlands

Constructed wetlands (CWs) are engineered systems designed for the removal of pollutants from waterbodies, and resemble the conditions and habitats of swamp. Their construction is based on normal wetland systems where the treatment occurs by heterotrophic microorganisms and aquatic plants through a combination of naturally occurring physical, chemical, and biological processes. It is an effective, low-energy consuming alternative for the treatment of wastewater compare to other complex treatment systems. A variety of wastewater types, such as agricultural, domestic, mining, stormwater, landfill leachate, and industrial effluents, can be remediated with CWs. Remediation through CWs is a complex process, and involves a number of physical, chemical, and biological processes. The efficiency of CWs in the remediation of wastewater depends on interactions at the soil–root interface, contaminants, and microorganisms (Olejnik and Wojciechowski 2012).

4.4 APPLICATION OF PHYTOREMEDIATION IN TEXTILE DYE REMOVAL

While working with aniline to synthesize quinine, chemist William Perkins unexpectedly synthesized a dye named mauve in 1856. Before the fiction of synthetic dye, coloring materials were derived from natural resources such as roots, bark, leaves, seeds, and shellfish. The Society of Dyers and Colourists (UK) and the American Association of Textile Chemists and Colorists (AATC) classify commercial textile dyes based on generic name and chemical constituents. According to general dye chemistry, textile dyes are classified and grouped into 12 families or groups for ease of understanding: acid dyes, disperse dyes, direct dyes, azodyes, reactive dyes, sulfur dyes, basic dyes, oxidation dyes, vat dyes, chrome or mordant dyes, optical or fluorescent brighteners, and solvent dyes. On the other hand, based on the fibers, dyes are classified into dyes for cellulose fibers, dyes for protein fibers and dyes for synthetic fibers.

Textile industrial effluent is a source of many pollutants ranging from halogen-based organic pesticides to heavy metals conjugated with dyes and the process of dying. High solid content in water reduces the availability of dissolved oxygen. Reduced levels of dissolved oxygen affect the germination and development of seeds and seedlings. Plant growth in the form of chlorophyll, carbohydrate, and protein content indicates the toxicity of textile dye effluent (Ebency et al. 2013). Color that has been released into the water system forms a layer on the surface of water, causing eutrophication, pH fluctuations, reduced photosynthetic activity, increased chemical and biological oxygen demands, and limited dissolved oxygen levels that pose the death of aquatic life (El-Kassas and Mohamed 2014).

Dye removal has received greater attention in recent years due to its toxicity in addition to aesthetic damage. Various physical and chemical methods, including electrocoagulation, activated carbon, photooxidation, ozonation, flocculation, adsorption, and membrane treatment, have been developed to remediate dye from textile effluents. These methods are not cost effective, and they produce concentrated sludge, again raising the problem of safe disposal. Table 4.2 illustrates the advantages and disadvantages associated with currently using physical and chemical methods.

Through conventional techniques, it is very difficult to remove reactive dyes as they can easily pass through the treatment system (Mahmoud 2014). So it is necessary to switch to alternative methods that are cost effective and can remediate large volumes of textile dye effluents (Daneshvar et al. 2005).

TABLE 4.2   Advantages and Disadvantages of Various Physical and Chemical Methods of Dye Removal from Industrial Effluents

Physical/Chemical Methods	Advantages	Disadvantages
Fenton's reagent	Effective decolorization of both soluble and insoluble dyes	Sludge generation
Ozonation	Applied in gaseous state, no alteration of volume	Short half-life (20 min)
Photochemical	No sludge production	Formation of by-products
NaOCl	Initiates and accelerates azo-bond cleavage	Release of aromatic amines
Cucurbituril	Good sorption capacity for various dyes	High cost
Electrochemical destruction	Breakdown compounds are nonhazardous	High cost of electricity
Activated carbon	Good removal of wide variety of dyes	Very expensive
Peat	Good adsorbent due to cellular structure	Specific surface areas for adsorption are lower than activated carbon
Wood chips	Good sorption capacity for acid dyes	Requires long retention times
Silica gel	Effective for basic dye removal	Side reactions prevent commercial application
Membrane filtration	Removes all dye types	Concentrated sludge production
Ion exchange	Regeneration, no adsorbent loss	Not effective for all dyes
Irradiation	Effective oxidation at lab scale	Requires a lot of dissolved O_2
Electrokinetic coagulation	Economically feasible	High sludge production

Source: Robinson, T. et al., *Bioresource Technology* 77, no. 3:247–55, 2001.

The application of biological methods is found effective for the treatment of wastewater contaminated with textile dyes. This biological approach has attracted attention in recent decades due to its efficiency, cost-effectiveness, and environmentally friendly nature. Phytoremediation is one such ecofriendly process involving plants for the remediation of contaminants, and xenobiotics exist in contaminated water or soil. Using several phytoremediation technologies, remediation of contaminated sites can be achieved (Mahajan and Kaushal 2013).

Phytoremediation technologies involving the application of algae and aquatic plants have been found efficient in transcending the environmental problems caused by the accumulation of organic pollutants including dyes and heavy metals in waterbodies (Chekroun and Baghour 2013). Algae are ubiquitous and can grow as autotrophs, heterotrophs, or mixotrophs in any kind of environment. Algae are natural environmental cleaners and they control metal concentration in lakes and oceans. Algae can accumulate and absorb organic pollutants and heavy metals into them, decreasing the concentration of those pollutants in surrounding water. Phototropic algae utilize nutrients present in the water and release oxygen into the environment through photosynthesis by absorbing carbon dioxide, hence, increasing the levels of biological oxygen demand in the water in which they are growing in addition to nutrient removal (Bhatnagar and Kumari 2013). Plants are natural nutrient sinkers; some are able to absorb significant amounts of metals and organic contaminants. Aquatic plants, due to their nutrients and ability to absorb dissolved constituents, make the surrounding environment suitable for the growth of microorganisms that also help in remediation processes. Also, floating varieties supply oxygen to the underneath water layers through the roots from their leaves (Reddy et al. 1989).

Studies by Muthunarayanan et al. (2011) on floating macrophyte including *Eichhornia crassipes* (water hyacinth) on the removal of Red RB and Black B dyes have revealed that the hyacinth is capable of sorbing these dyes up to 95% and 99.5%, respectively. High color removal was observed after 6 days of contact with dyes of 10 ppm concentration each. During the studies, authors observed the increased hexadecanoic acid and decreased phytol levels in the plant. It was also proposed that the color reduction was due to biosorption of dyes onto the roots, shoots, and leaves of plant.

Typha angustifolia Linn (narrow leaved cattail) was studied for the decolorization mechanism of textile dye Reactive Red 141. It was reported that the dye accumulated with silicon and calcium oxalate of plant tissues.

Fourier transform infrared spectroscopic studies revealed that amide and siloxane groups in plants are responsible for the dye uptake. It was suggested that calcium silicate, silica, mono- and dioxalates made the plant tolerable to dye by precipitation in the form of metal-dye complexes (Nilratnisakorn et al. 2008).

Another plant species, *Portulaca grandiflora*, has been used for the decolorization of sulfonated diazo dye Red 5B, effectively removing 92% of the dye in 96 hours from a dye solution concentration of 20 mg/L. Consortium developed with *P. gradiflora* and *Pseudomonas putida* have shown complete decolorization of dye within 72 hours. During decolorization, stimulation of tyrosinase, lignin peroxidase, indophenol reductase, and riboflavin reductase of the plant was observed. The dye was degraded into several nontoxic metabolites, including 7-(benzylamino) naphthalene-2-sulfonicacid, 1-(4-diazenylphenyl)-2-phenyldiazene, 7-aminonaphthalene-2-sulfonic acid, and methylbenzene (Khandare et al. 2013).

The phytoremediation efficiency of *Glandularia pulchella* (Sweet) Tronc was tested against a mixture of structurally different dyes named Navy Blue 2R, Red HE3B, BBR, Rubine GFL, and Scarlet RR and textile effluent. *G. pulchella* pants were subjected to different dye samples of concentration 50 mg/L. Plants were able to decolorize the dyes to various extents through the stimulation of different enzymes specific for dye. With Rubine GFL and Scarlet RR. 100% decolorization was observed after 72 hours. Dye degradation was due to the induction and involvement of laccase, tyrosinase, lignin peroxidase, veratryl alcohol oxidase, and some other reductase enzymes in the metabolism. Infrared spectroscopic analyses of *G. pulchella*-treated effluent have shown different peaks from the original. The probable reason was perhaps after the uptake and degradation, dye metabolites released again into the aqueous system. Complete decolorization of dyes, except BBR, was reported even after 96 hours. The exception of BBR was due to its complex structure. Phytotoxicity studies on treated effluent have not shown any toxic nature of original dyes (Kabra et al. 2012).

A study on tissue cultured *Portulaca grandiflora* Hook (moss-rose) plants against the removal of Navy Blue HE2R, a sulfonated diazo reactive dye, was performed by Khandare et al. (2011a). *P. grandiflora* showed appreciable decolorization for a mixture of dyes (3 mg each) as Navy Blue HE2R (98.20%), Navy Blue RX (94.60%), Rubine GFL (91.45%), Navy Blue 2R (86.75%), Blue GL solo (81.28%), Brown 3REL (74.36%), Red HE7B (56.55%) and Orange HE2R (48.74%). Maximum decolorization was observed for Navy Blue HE2R dye after 47 hours. The degraded dye did not show significant effect on the shoot and root system of plants (*Sorghum vulgare, Phaseolus mungo*) grown as compared to untreated dye, which indicates that the degraded dye metabolites by *P. grandiflora* were nontoxic for plant growth (Khandare et al. 2011a).

Decolorization of various dyes including Methyl orange, Red HE7B, Malachite green, Reactive Red 2, Red HE8B, Golden yellow HER, Direct red 5B, Patent blue, and Brilliant Blue R (BBR) by using in vitro cultured *Typhonium flagelliforme* plants was been performed by Kagalkar et al. (2010). It was observed that many plant enzymes, such as laccase, peroxidase, 2,6-dichlorophenol indophenol reductase, and tyrosinase, were induced during the degradation of BBR dye, indicating the involvement of enzymes in dye metabolism. The degradation resulted in significant decrease in BOD and COD values of dye mixtures, and textile effluent indicates the decolorization and detoxification of dyes by plant.

Color removal studies were performed with individual *Aster amellus* Linn, *Glandularia pulchella* (Sweet) Tronc, and their consortium-AG. Plants and their consortium-AG were exposed to aqueous solutions of a variety of dyes, including Remazol Orange 3R, Red HE3B, Red HE8B, Blue GLL, and Methyl Orange, of concentration 20 gm/L each. Consortium-AG has shown greater potential toward decolorization of dyes compared to its individual counter parts, but in various extents. Complete decolorization of Red HE3B and Remazol Orange 3R was noticed in 60 and 36 hours, respectively, whereas, degradation of the dyes Blue GLL, Methyl orange, and Red HE8B was 85.2%, 72.58%, and 56.27%, respectively, after 96 hours. Complete decolorization of Remazol Orange 3R dye by individual plants *A. amellus* and *G. pulchella* was observed at 72 and 96 hours, respectively. At high dye concentrations, the potential of dye removal by consortium-AG was reduced due to the hindrance of enzyme activity of plant by the dye. Veratryl alcohol oxidase and NADH-DCIP reductases were induced by *A. amellus* in response to dye, whereas *G. pulchella* has shown induction in laccase and tyrosinase. On the other hand, consortium-AG showed induction in enzymes as individuals but to different extents, in addition to the stimulation of lignin peroxidase. *A. amellus* degraded the dye into benzene, acetamide, 3-diazenylnaphthalene-2-sulfonic acid, and naphthalene moieties, while *G. pulchella* degraded the dye into naphthalen-1-ol, acetamide, 3-diazenyl-4-hydroxynaphthalene and (ethylsulfonyl) benzene. Consortium-AG degraded the dye into N-(naphthalene-2yl) acetamide and 2-(phenylsulfonyl) ethanol. Phytotoxicity results have shown that the metabolites are not toxic for the growth of other plants (Kabra et al. 2011).

Experiments were conducted on degradation of two types of azodyes—Tetrazine (monoazodye) and Ponceau (diazodye)—using different green algal strains (*Chlorella vulgaris, Chlorella ellipsodea, Chlorella kessleri, Scenedsmus bijugatus, Scenedesmus bijuga,* and *Scenedesmus obliquus*) and cyanobacteria (including *Anabaena laxa, Anabaena subcylindrica, Oscillatoria angusta,* and *Nostoc muscourm*). Dye removal was tested with initial dye concentrations of 5, 10, and 20 ppm, and monitored for 3 and 6 days. Initial dye concentrations showed significant dye removal by the strains. The dye removal efficiency decreased from 5 ppm to 20 ppm for both green algae and cyanobacteria. Rapid decolorization of dyes was observed after 3 days of incubation; afterward, a decrease in the color removal with time was observed. This was explained as strong attraction forces between algae and dye molecules resulting in rapid diffusion of dye onto algal cells in the initial stages of contact.

The extent of Tatrazine dye removal was 29% to 68% by green algae, whereas it was 57% to 70% after 3 and 6 days of incubation. The removal of Ponceau was observed in the range of 18% to 62% and 23% to 59% by green algae and cyanobacteria, respectively, after 6 days of incubation. Maximum absorption of Ponceau dye was shown by *Scenedsmus bijugatus* (55% to 62%) and *Nostoc muscourm* (55% to 59%) at 5 ppm dye concentration after 3 and 6 days of incubation. Studies on azo reductase activity revealed that presence of azodyes in algal cultures induced the azodye reductase activity (Omar 2008).

Plants of *Brassica juncea, Sorghum vulgare,* and *Phaseolous mungo* were evaluated for their decolorization efficiency of textile dyes obtained from textile effluent. The three species have shown dye removal up to 79%, 57%, and 53%, respectively, after incubation for 48 hours. *Sorghum vulgare* and *Phaseolous mungo* have shown considerable decrease in the shoot and root length when grown in effluent medium. But, *B. juncea* showed significant enhanced growth in original textile effluent. As a response to decolorization of dye, *S. vulgare* and *B. juncea* exhibited induced levels of NADH–dichlorophenol indophenols reductase enzyme by 209% and 194%, respectively, whereas induced extracellular riboflavin reductase by 223% was observed with *P. mungo*. In addition, stimulation of intracellular laccase by 266% was observed in case of *B. juncea*. Induction of various enzymes when grown in textile effluent indicates that the respective enzymes took part in the dye metabolism. In the same study, *B. juncea* was exposed to Reactive red 2 dye; metabolites of the degraded dye were found as 2-amino-4, 6-dichlorotriazine and napthalenesufamide (Ghodake et al. 2009).

Khandare et al. (2011b) studied the color removal efficiency of *Aster amellus* Linn using a sulfonated azodye Ramezol Red (RR). A maximum of 96% color removal was observed. During the process of decolorization, a number of plant enzymes, including tyrosinase, lignin peroxidase, riboflavin reductase, and veratryl alcohol oxidases, were induced, indicating their participation in dye metabolism. Four metabolites of degraded dye have been identified: naphthalene-2-sulfonate, 3-(1,3,5-triazin-2-ylamino)benzenesulfonate, 4-amino-5-hydroxynaphthalene-2, 7-disulfonate, and 2-[(3-diazenylphenyl)sulfonyl] ethanesulfonate. Incubation of *A. amellus* with textile effluent and mixture of dyes caused the decolorization up to 47% and 62%, respectively. BOD was reported as 75% and 48% for textile effluent and mixture of dyes, respectively. COD was reported as 60% and 75% for textile effluent and mixture of dyes, respectively. Reduction in total organic carbon (TOC) was observed after treatment with *A. amellus* for 60 hours. From the results it was proposed that *A. amellus* has the ability to clean the textile effluents (Khandare et al. 2011b).

The free-floating aquatic plant *Lemna minor* was tested for the decolorization of two triphenylmethane dyes: Crystal violet (CV) and Malachite green (MG). Dye removal of 90% of MG and 86% of CV was observed with 5 gm of total plant. The decolorization efficiency of plants was highest at lower dye concentrations. Plants showed 96% and 98% decolorization efficiency toward CV and MG dyes, respectively, at the lower concentration of 40 mg/L. It was reported that the plants are capable of withstanding higher concentration of dyes, that is, (190 mg/L for CV and 300 mg/L for MG. pH of the dyes showed significant effect on dye removal; maximum decolorization was observed at pH 7 for CV and at pH 4 for MG. The optimal temperatures dyes uptake was observed as 40°C for MG (efficiency 98%) and room temperature for CV (efficiency 92%). *L. minor* showed higher uptake of MG dye compared to CV (Torok et al. 2015).

Ten aquatic macrophytes—*Hydrilla verticillata* (submerged), *Eichhornia crassipes, Ceratophyllum demersum, Azolla pinnata, Lemna aequinoctialis, Cyperus alopecuroides, Spirodela polyrhiza* (free floating), *Phragmites karka, Polygonum barbatum,* and *Typha angustata* (emergent)—have been screened for their efficiency to tolerate dyes present in textile wastewater.

Both acidic (4.0) and basic pH (10.0) were found intolerable by *Eichhornia*; optimal growth was observed at pH 7. *Azolla* died in both acidic (4.0) and alkine conditions. At pH below 3 and above 9, the protoplasm of the root zone was severely affected. The damage was due to acidic azo compounds and alkaline silicate waters. Among all the tested species, only *Phragmites* showed pH tolerance and grew well in acidic, basic, and neutral waste-waters. The sensitive species *Ceratophyllum, Azolla, Lemna*, and *Spiroleda* were proposed as markers to assess the toxicity of textile waters (Sharma et al. 2005).

4.5 CONCLUSION

Phytoremediation is a groundbreaking and developing technology that uses plants to remediate soil, sediment, groundwater, and air contaminants. Plants utilize several built-in metabolic processes and metabolites including enzymes to treat a contaminated site. The remediation may be in the form of isolation, degradation, transformation, or accumulation. Phytoremediation can be applied to diverse contaminants, including crude oils, explosives, metals, chlorinated solvents, and other organic and inorganic pollutants. Plants use several mechanisms to treat contamination—inside, outside, or throughout the plant system. Phytoextraction is an affordable method to treat inorganic pollutants, whereas phytodegradation and rhyzodegradation are suitable to treat organic contaminants. Phytovolatilization is applicable to any type of contaminant, whether organic or inorganic in origin.

The complex aromatic structure of dyes is responsible for their resistance to fading, intense color, adherence to fabrics, and water solubility, making them recalcitrant toward degradation. The physical and chemical methods used for removing the dyes require of excessive chemicals and production of large volumes of concentrated sludge, which are obstacles to applying these methods in all situations. The enzyme-synthesizing capacity of plants make them environmentally friendly and cost-effective alternatives for the treatment of dyes. Enzymes can degrade the complex molecular structure of dyes into simpler and innocuous metabolites, thus reducing the risk of concentrated sludge disposal. It is an effective method to be followed to transform polluted sites into productive ones. So far, only a few studies have been reported on the use of plants for textile dye removal. Hence, immense research work needs to be done to elicit new plant-based bioremediation technologies for application toward the remediation of polluted soils and water.

ACKNOWLEDGMENT

The authors would like to acknowledge the Department of Science and Technology (DST) for providing fellowship and a research grant under the Women Scientist Programme (WOS-B).

REFERENCES

Anastasi, A., Tigini, V., and Varese, G.C. 2013. The bioremediation potential of different ecophysiological groups of fungi. In *Fungi as Bioremediators*, edited by E.M. Goltapeh, Y.R. Danesh, and A. Varma, 29–49. Cambridge: Cambridge University Press.

Bhatnagar, S., and Kumari, R. 2013. Bioremediation: A sustainable tool for environmental management—A review. *Annual Review & Research in Biology* 3, no. 4:974–93.

Chekroun, K.B., and Baghour, M. 2013. The role of algae in phytoremediation of heavy metals: A review. *Journal of Materials and Environmental Science* 4:873–80.

Collina, E., Bestetti, G., and Gennaro, P.D., Franzetti, A., Gugliersi, F., Lasagni, M., and Pitea, D. 2005. Naphthalene biodegradation kinetics in an aerobic slurry-phase bioreactor. *Environment International* 31:167–71.

Cookson, J.T. Jr. 1995. *Bioremediation Engineering Design and Application*. New York: McGraw-Hill.

Cunningham, S.D., Shann, J.R., Crowley, D.E., and Anderson, T.A. 1997. Phytoremediation of contaminated water and soil. In *Phytoremediation of Soil and Water Contaminants* (ACS Symposium Series 664), edited by E.L. Kruger, T.A. Anderson, and J.R. Coats. Washington, DC: American Chemical Society.

Daneshvar, N., Ayazloo, M, Khataee, A.R., and Pourhassan, M. 2005. Biodegradation of the textile dye malachite green by *Microalgae cosmarium* sp. 4th National Biotechnology Congress, Islamic Republic of Iran, August.

Dwivedi, S. 2012. Bioremediation of heavy metal by algae: Current and future perspective. *Journal of Advanced Laboratory Research in Biology* 3, no. 3.

Ebency, C.I.L., Rajan, S., Murugesan, A.G., Rajesh, R., and Elayarajah, B. 2013. Biodegradation of textile azo dyes and its bioremediation potential using seed germination efficiency. *International Journal of Current Microbiology and Applied Sciences* 2:496–505.

Economic and Social Commission for Western Asia (ESCWA). 2003. Wastewater treatment technologies: A general review. http://www.igemportal.org/Resim/Wastewater%20Treatment%20Technologies_%20A%20general%20rewiev.pdf.

El-Kassas, H.Y., and Mohamed, L.A. 2014. Bioremediation of the textile waste effluent by Chlorella vulgaris. *Egyptian Journal of Aquatic Research* 40, no. 3:301–8.

Ghodake, G.S., Talke, A.A, Jadhav, P., and Govindwar, S.P. 2009. Potential of *Brassica juncea* in order to treat textile-effluent-contaminated sites. *International Journal of Phytoremediation* 11, no. 4:297–312.

Ghosh. M., and Singh, S.P. 2005. A review on phytoremediation of heavy metals and utilization of its byproducts. *Applied Ecology and Environmental Research* 3:1–18.

Gogate, P.R., and Pandit, A.B. 2004. A review of imperative technologies for wastewater treatment I: Oxidation technologies at ambient conditions. *Advances in Environmental Research* 8:501–51.

Hegazy, A.K., Abdel-Ghani, N.T., and El-Chaghabye, G.A. 2011. Phytoremediation of industrial wastewater potentiality by *Typha domingensis*. *International Journal of Environmental Science and Technology* 8:639–48.

Institute for Clean Energy and Green Environment (IGECE). n.d. Rhizofiltration. http://systemsbiology.usm.edu/BrachyWRKY/WRKY/Rhizofiltration.html.

Interstate Technology & Regulatory Council (ITRC). 2001. Phytotechnology technical and regulatory guidance document. Phytotechnologies Work Team, Technical/Regulatory Guidelines. http://www.state.nj.us/dep/dsr/bscit /Phytoremediation.pdf.

Kabra, A.N., Khandare, R.V., Waghmode, T.R., and Govindwar, S.P. 2012. Phytoremediation of textile effluent and mixture of structurally different dyes by *Glandularia pulchella* (Sweet) Tronc. *Chemosphere* 87, no. 3:265–72.

Kabra, A.N., Khandare, R.V., Waghmode, T.R., and Govindwar, S.P. 2011. Differential fate of metabolism of a sulfonated azo dye Remazol Orange 3R by plants *Aster amellus* Linn., *Glandularia pulchella* (Sweet) Tronc. and their consortium. *Journal of Hazardous Materials* 190:424–31.

Kagalkar, A.N., Jagtap, U.B., Jadhav, J.P., Govindwar, S.P., and Bapat, V.A. 2010. Studies on phytoremediation potentiality of *Typhonium flagelliforme* for the degradation of Brilliant Blue R. *Planta* 232, no. 1:271–85.

Kanissery, R.G., and Sims, G.K. 2011. Biostimulation for the enhanced degradation of herbicides in soil. *Applied and Environmental Soil Science* 2011:1–10.

Khandare, R.V., Kabra, A.N., Awate, A.V., and Govindwar, S.P. 2013. Synergistic degradation of diazo dye Direct Red 5B by *Portulaca grandiflora* and *Pseudomonas putida*. *International Journal of Environmental Science and Technology* 10:1039–50.

Khandare, R.V., Kabra, A.N., Kurade, M.B., and Govindwar, S.P. 2011a. Phytoremediation potential of *Portulaca grandiflora* Hook. (Moss-Rose) in degrading a sulfonated diazo reactive dye Navy Blue HE2R (Reactive Blue 172). *Bioresource Technology* 102, no. 12:6774–7.

Khandare, R.V., Kabra, A.N., Tamboli, D.P., and Govindwar, S.P. 2011b. The role of *Aster amellus* Linn. in the degradation of a sulfonated azo dye Remazol Red: A phytoremediation strategy. *Chemosphere* 82, no. 8:1147–54.

Mahajan, P., and Kaushal, J. 2013. Degradation of Congo red dye in aqueous solution by using phytoremediation potential of *Chara Vulgaris*, *Chitkara Chemistry Review* 1:67–75.

Mahmoud, M.S. 2014. Decolorization of certain reactive dye from aqueous solution using Baker's Yeast (*Saccharomyces cerevisiae*) strain. *Housing and Building National Research Center Journal*.

Mishra, D., Kim, D.J, Ahn, J.G, and Rhee, Y.H. 2005. Bioleaching: A microbial process of metal recovery: A review. *Metals and Materials International* 11, no. 3:249–56.

Mishra, M., and Rhee, Y.H. 2014. Microbial leaching of metals from solid industrial wastes. *Journal of Microbiology* 52, no. 1:1–7.

Mrozik, A., and Seget, Z.P. 2010. Bioaugmentation as a strategy for cleaning up of soils contaminated with aromatic compounds. *Microbiological Research* 165, no. 5: 363–75.

Muthunarayanan, V., Santhiya, M., Swabna, V., and Geetha, A. 2011. Phytodegradation of textile dyes by water hyacinth (*Eichhornia Crassipes*) from aqueous dye solutions. *International Journal of Environmental Sciences* 1, no. 7:1702–17.

Neuman, D., and Ford, K.L. 2006. Phytostabilization as a remediation alternative at mining sites. U.S. Bureau of Land Management Papers, paper 21. http://digitalcommons.unl.edu/usblmpub/21.

Nilratnisakorn, S., Thiravetyan, P., and Nakbanpote, W. 2008. Synthetic reactive dye wastewater treatment by narrow-leaved cattail: Studied by XRD and FTIR, *Asian Journal on Energy and Environment* 9, no. 3–4:231–52.

Norris, R.D., Hinchee, R.E. et al. 1994. *Handbook of Bioremediation*. Boca Raton: CRC Press/Lewis Publishers.

Olejnik, D., and Wojciechowski, K. 2012. The conception of constructed wetland for dyes removal in water solutions. *CHEMIK* 66:611–14.

Omar, H.H. 2008. Algal decolorization and degradation of Monoazo and Diazo dyes. *Pakistan Journal of Biological Sciences* 11, no. 10:1310–16.

Pivetz, B.E. 2001. Phytoremediation of contaminated soil and ground water at hazardous sites (Ground water issue). EPA/540/S-01/500.

Rawat, K., Fulekar, M.H., and Pathak, B. 2012. Rhizofiltration: A green technology for remediation of heavy metals. *International Journal of Innovations in Bio-Sciences* 2:193–99.

Reddy, K.R., D'Angelo, E.M. and DeBusk, T.A. 1989. Oxygen transport through aquatic macrophytes: The role in wastewater treatment. *Journal of Environmental Quality* 19:261–67.

Robinson, T., McMullan, G., Marchant, R., and Nigam, P. 2001. Remediation of dyes in textile effluent: A critical review on current treatment technologies with a proposed alternative. *Bioresource Technology* 77, no. 3:247–55.

Sharma, K.P., Sharma, K., Kumar, S., Sharma, S., Grover, R., Soni, P., Bhardwaj, S.M., Chaturvedi, R.K., and Sharma, S. 2005. Response of selected aquatic macrophytes towards textile waste waters. *Indian Journal of Biotechnology* 4:538–45.

Tangahu, B.V., Abdullah, S.R.S., Basri, H., Idris, M., Anuar, N., and Mukhlisin, M. 2011. A review on heavy metals (As, Pb, and Hg) uptake by plants through phytoremediation. *International Journal of Chemical Engineering* 2011:1–31.

Tchobanoglous, G., and F. L. Burton. 1991. *Wastewater Engineering: Treatment, Disposal and Reuse*. New York: McGraw-Hill.

Torok, A., Buta, E., Indolean, C., Tonk. S, Silaghi-Dumitrescu, L., and Majdik, C. 2015. Biological removal of triphenylmethane dyes from aqueous solution by *Lemna minor. Acta Chimica Solvenica* 62, no. 1:1–10.

United States Environmental Protection Agency (U.S. EPA). 1994. Bioventing. In *How to Evaluate Alternative Cleanup Technologies for Underground Storage Tank Sites: A Guide for Corrective Action Plan Reviewers*, chap. 3. http://www .epa.gov/oust/pubs/tum_ch3.pdf.

United States Environmental Protection Agency (U.S. EPA). 1998. A citizen's guide to phytoremediation. Office of Solid Waste and Emergency Response (5102G). EPA 542-F-98-011. http://www.clu-in.org/download/remed/phyto2.pdf.

United States Environmental Protection Agency (U.S. EPA). 2000. Introduction to phytoremediation. National Risk Management Research Laboratory, EPA/600/R-99/107. http://clu-in.org/download/remed/introphyto.pdf.

United States Environmental Protection Agency (U.S. EPA). 2012. A citizen's guide to in situ chemical oxidation. Office of Solid Waste and Emergency Response, EPA 542-F-12-011. https://cluin.org/download/Citizens/a_citizens_guide _to_in_situ_chemical_oxidation.pdf.

Vidali, M. 2001. Bioremediation—An overview. *Pure and Applied Chemistry* 73: 1163–72.

Wuana, R.A., and Okieimen, F.E. 2011. Heavy metals in contaminated soils: A review of sources, chemistry, risks and best available strategies for remediation. *International Scholarly Research Network* 20:1–20.

CHAPTER 5

Algal Biosorption of Heavy Metals

M. Jerold, C. Vigneshwaran,
A. Surendhar, B.G. Prakash Kumar,
and Velmurugan Sivasubramanian

CONTENTS

Abstract

Metal-bearing effluent can be removed by conventional treatment such as chemical precipitation, electrochemical cells, reverse osmosis, and ion exchange; however, each treatment method has limitations. Sorption, particularly biosorption, has become one of the alternative treatments to conventional treatments of wastewaters and industrial effluents. The treatment of industrial wastewater from aqueous solutions employing biosorption technology is an advanced and novel technology. Basically, sorption is the process of assimilation of particles from one phase to another phase. The particles move from the bulk liquid and get amalgamated in the solid surface. The particles bound to the solid surface by physical (Van der Waal's forces) or chemical interaction (chemical bonds). Biosorption has been chosen as an alternative remedial solution because of its high metal-uptake capacity, greater surface area with reactive sorbents, and, above all, low cost. Many biomaterials available in nature have been employed as biosorbents for the desired pollutant removal. Algae are of special interest for the development of new biosorbent material due to their high sorption capacity and ready availability in practically unlimited quantity. Particularly, macroalgae are found to have greater metal uptake capacity. Seaweed (macroalgae) collected from the ocean has shown impressive biosorption of metals. Brown algae, especially, contain high amounts of alginate, which are well protected within brown algae's cellular structures, and copious carboxylic groups capable of capturing cations present in solutions. This chapter discusses the significance of the algal resource in the removal of metals from waste streams and provides a brief overview of marine brown algae, its properties, and their potential applications for biosorption.

5.1 INTRODUCTION

Today, heavy metal pollution is one of the major problems in the environment. Various industries contribute to environmental pollution by the release of various hazardous materials. Such toxic materials are released including the mining and smelting industry, battery industry, energy and fuel production, fertilizer and chemical industry and application, iron and steel industry, electroplating, electrolysis, leatherworking, photography, electric appliance manufacturing, and nuclear power plants (Wang and Chen, 2009). Particularly, in the industrialized city, the following categories are important concerns (Volesky, 2007):

- Accumulated acid mine drainage (AAMD), coupled with mining source

- Waste metal solution from the electroplating industry (enlarged polluting industry)

- Thermal power plants (throughput of coal and other hazardous materials)

- Nuclear power plants (usage of toxic chemicals, such as uranium and polonium)

The heavy metals are grouped into three main categories: toxic metals, radionuclides, and precious metals (Wang and Chen, 2006, 2009). Mercury, chromium, lead, copper, nickel, cadmium, and cobalt are listed under the category of toxic metals. The metals such as uranium, thorium, radium, and polonium are included under radionuclides. Precious metals include platinum, palladium, gold, and silver. There are three methods for the removal of metal ion from aqueous solution: physical, chemical, and biological technologies. Conventional methods have been practiced for many years for the removal of metal ions from various aqueous solutions (such as chemical precipitation, chemical and electrocoagulation, filtration, ion exchange, electrochemical treatment, membrane technologies, adsorption on activated carbon, zeolite, and evaporation). However, its operation is restricted nowadays due to various limitations and drawbacks. The membrane technologies and activated carbon adsorption process cannot be adopted for large-scale operations because they are extremely expensive and only a low concentration of heavy metals can be treated. The various other advantages and disadvantages of the conventional metal removal technologies are summarized by Volesky (2001). In order to improve the competitiveness of industrial operation processes, it is essential to develop and implement a cost-effective process for removal/recovery of metals. The various drawbacks such as high cost, less efficiency, release of secondary pollutants, etc. have led to the evolution of novel and sophisticated separation technologies (Volesky and Naja, 2007).

Biohydrometallurgy is a modern trend based on the application of microbial interaction for the extraction of metals from specific raw materials. The biotechnological approach has opened up various processes for the recovery of metals. Therefore, integration of such promising technology in the recovery of metal from primary material (such as ores and concentrates) and secondary waste materials (such as mining, metallurgical, and power plant wastes) would certainly improve the process of metal recovery. The most promising biotechnological approach covers all the cutting-edge areas of biohydrometallurgy, which include bioleaching, bioprecipitation, bioflotation, bioflocculation, biooxidation, biosorption, bioreduction, bioaccumulation, and biosensors. These techniques are not only cost effective but also environmentally friendly. Biosorption is one such significant and widely applied biological process, whereby the sorption of metals takes place on the surface of the biomass. Microbial biomass provides a natural metal sink, by the process of biosorption where various toxic hazardous materials are adsorbed by the biomass. The biosorbents derived from microbial cells are effective for the removal of several metal ions from the solution, except the alkali metal ions such as Na^+ and K^+. However, this can be passively removed by living or dead organisms (Gadd, 1993).

Biosorption is an entirely different form of bioaccumulation where sorption takes place on the surface of the biomass, which was pioneered by Volesky and his team in 1981 at McGill University (Montreal) (Tsezos and Volesky, 1981). Earlier, most research was carried out with live organisms (Lesmana et al., 2009). However, it was noticed that dead biomass possessed a greater metal removal capacity (Volesky, 1990), therefore, most researchers turned their attention towards biosorption (Asthana et al., 1995; Bossrez et al., 1997; Fourest and Roux, 1992; Holan et al., 1993; Niu et al., 1993; Selatnia et al., 2004; Yetis et al., 2000; Zhou, 1999). The research and development in biosorption have exploited various inactive and dead biomasses for the removal of toxic metal pollutants from wastewater. Hence, biomass derived from agro waste (Kuyucak, 1990), algal biomass, aquatic ferns, and dead microbial cells are used as biosorbents. Out of which macroalgae are reported to have an excellent biosorptive behavior (Kuyucak and Volesky, 1990) due to the presence of alginate in their cell walls.

Algae in nature are broadly classified into microalgae and macroalgae, of which macroalgae is reported to have an imperative biosorption property. These macrophytes flourish largely in shallow coastal areas and are also present in the many areas of the oceanic parts of world. Generally macroalage (seaweed) are of three types based on their color: red, green, and brown, of which brown algae belonging to the class Phaeophyta is found to be an excellent biosorbent for the removal of metal ion from the aqueous solution. Various researchers have reported the inherent potential of brown algae (Hamdy, 2000; Kuyucak and Volesky, 1990; Ofer et al., 2003; Zhou et al., 1998). Thus, this chapter will describe the involvement of algae in the biosorption of heavy metals. Also, the behavior of brown algae in the sequestration of heavy metals is summarized.

5.2 SALIENT FEATURES OF BROWN ALGAE

5.2.1 A Comparison with Other Algae

Algae are an extensive and divergent group present in nature that possess chlorophyll and carry out oxygenic photosynthesis. Blue-green algae are also oxygenic phototrophs but they belong to a separate class of eubacteria (true bacteria). Basically algae are microscopic in morphological shape. Hence, they are also included under the group of microorganisms. Bold and Wynne (1985) classified algae into the following divisions: Cyanophyta, Prochlorophyta, Phaeophyta, Chlorophyta, Charophyta, Euglenophyta, Chrysophyta, Pyrrhophyta, Cryptophyta, and Rhodophyta. In our context we mainly discuss Phaeophyta (brown algae). The storage products distinguish the Phaeophyta from the other two groups Chlorophyta (green algae) and Rhodophyta (red algae). Laminaran is the main storage product in the case of Phaeophyta, whereas floridean starch is produced and stored in the Rhodophyta groups. Motility is a salient feature of the organism. Rhodophyta lacks the flagellar system; however, it is present in Chlorophyta and Phaeophyta. The cell wall chemistry plays an important role in the biosorptiom mechanism(s); however, the electrostatic attraction and complexation also contribute to the biosorption process. The other groups of algae are not suitable for the sorption, because Cryptophyta lacks cell walls (Lee, 1989) and Pyrrhophyta (dinoflagellates) can be "naked" or protected by cellulosic "thecal" plates (Bold and Wynne, 1985; Lee, 1989). Last, Chrysophyta can be either "naked" or even have an enveloped cell wall (Lee, 1989). The forms of algae commonly used in the sorption process are Phaeophyta, Rhodophyta, and Chlorophyta. These possess cellulose made of fibrillar skeleton material. However, the cellulose can be modified with xylan in Chlorophyta and Rhodophyta. Perhaps, in some cases, mannan is present in the Chlorophyta. In Phaephyta, the cell wall is embedded with alginate or alginic acid with trace amounts of sulfated glycans, whereas Rhodaphyta contains a large quantity of sulfated galactans. The extra polysaccharide accessories present in the cell wall are important factors for the sorption of metals. Hence, these classes of algae are potentially excellent for the biosorption of heavy metals.

5.2.2 Brown Macroalgae

Brown algae are classified into 265 genera and more than 1500 species (Bold and Wynne, 1985). The presence of carotenoid fucoxanthin in their chloroplast yields a brown color for these classes of algae. In addition the brown color is also due to the presence of pheophycean tannins. These algal species are found mainly in the littoral zone of the marine environment. These species are scattered in the brackish environment and known to exhibit as salt fenland fauna (Lee, 1989). The brown algae class Phaeophyta is divided into 13 orders (Bold and Wynne, 1985); however, only two orders are important for biosorption, namely, Laminariales and Fucales. These two orders are richly available in the marine environment. Among these two orders, Fucales is a huge and diversified order, with a range of morphological diversity (Bold and Wynne, 1985). One of the best genus of Fucales tested for biosorption is *Sargassum*. Figure 5.1 shows the cell wall structure in brown algae. A semispeculative model of brown algae with regard to the structure of the cell wall has been proposed (Kloareg et al., 1986). As discussed earlier, cellulose forms the strong backbone structural network in which biopolymers such as xylofucoglucans, xylofucoglycuronans, alginates, and homofucans are linked. The alginic content of various *Sargassum* species is summarized in Table 5.1. The biosorption performance is also due to the presence of two common moieties: sulfate esters in the cellular polysaccharides and the presence

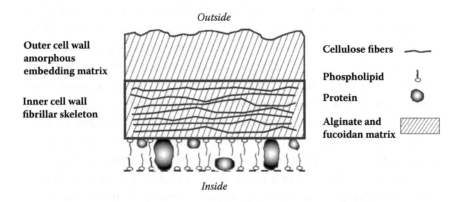

FIGURE 5.1 Cell wall structure in brown algae. (From Schiewer, S., Volesky, B., 2000, Biosorption by marine algae, in *Remediation*, edited by J.J. Valdes, 139–169, Dordrecht, Netherlands: Kluwer Academic Publishers.)

TABLE 5.1   Alginic Acid Contents of *Sargassum*

Sargassum Species	Alginic Acid (Percent of Dry Weight)
Sargassum longifolium	17%
Sargassum wightii	30%
Sargassum tenerium	35%
Sargassum fluitans	45%
Sargassum oligocystum	≈45%

Source: Adapted from Davis, T.A. et al., *Water Research* 37:4311–4330, 2003.

TABLE 5.2   Binding Groups in Brown Algae

Binding Chemical Group	Ligand Atom	Biopolymer
Carboxyl	Oxygen	Alginic acid
Thiol	Sulfur	Amino acids
Sulfonate	Sulfur	Sulfate polysaccharides fucoidan
Amine	Nitrogen	Amino acids peptidoglycan
Amide	Nitrogen	Amino acids

Source: Adapted from Ofer, R. et al., *Biotechnology and Bioengineering* 87: 451–458, 2004.

of polyuronides (Holan et al., 1993). Hence, the presence of functional groups in the uronic acids and the availability of sulfated moieties serve as excellent ligands for the sorption of metal ions from aqueous solution (Crist et al., 1991). Table 5.2 lists the major binding groups present in the brown algae.

5.3   HI-TECH CRAMS OF ALGAL BIOSORPTION

5.3.1   Mechanisms of Heavy Metal Uptake by Algae

Apparently, there is no complete evidence for the mechanism behind the biosorption of metal. However, it is believed that metal biosorption is due to the formation of ions on the surface of the biomass. Biosorption of metals is a two-step process. First, the metals ions bind and, second, the metal ions accumulate on the binding sites (Ahluwalia and Goyal, 2003). The metal uptake capacity of algal biomass depends on the availability of polysaccharide contents such as alginates and fucoidans on the cell surface (Herrero et al., 2006). Perhaps, marine brown alga has rich content of extracellular polysaccharide. Hence, they exhibit a prosperous metal sorption compared to other algal species (Herrero et al., 2006).

5.3.2 Key Functional Groups in the Algal Cell Wall

A typical brown algal biomass consists of abundant carboxylic groups that are principally acidic functional groups. Esterification of carboxylic sites occurs during the reduction of cadmium and lead by *Sargassum* biomass (Fourest and Volesky, 1996). Fourier transform infrared (FTIR) analysis reveals that formation of complex with the carboxylate groups of alginate moieties (Fourest and Volesky, 1996) as explained by Rees and coworkers (Rees et al., 1982; Thom et al., 1982) in the "egg-box" model. Sulfonic acid is the second most abundant functional group present in brown algae. However, it plays only a secondary role in the biosorption process and shows its action only at the acidic pH level. Likewise, hydroxyl groups are present in all polysaccharides, which are less abundant in brown algae and comes into existence only at pH >10.

5.3.3 Ion Exchange

Ion exchange is the principal mechanism for algal biosorption (Herrero et al., 2006; Kaewsarn, 2002; Ofer et al., 2004). It is known that alginate plays an important role in algal biosorption, therefore, it is appropriate that ion exchange takes place between the binding of metals and alginate moiety (Myklestad, 1968). In the biosorption of cobalt by *Ascophyllum nodosum*, there was an enhanced release of ions, including Ca^{2+}, K^+, Mg^{2+}, and Na^+, from the algae in the cobalt-bearing solution compared to the cobalt-free solution (Kuyucak and Volesky, 1989a,b,c). Hence, it is proved and concluded that ion exchange was the prevailing mechanism behind the biosorption. Treated biomass has been shown to have high sorption. The treatment can be made by any one of the chemical modifications. First, the protonation of biomass with strong acid will displace the light metals occupying the binding sites such as carboxylic, sulfonic, and others. Second, treating the biomass at high concentration of the given metal solution possibly most of the sites are occupied with potassium or calcium. Raw *Sargassum*, when allowed contact with heavy metal ion solution, releases light metal ions from the biomass because, generally, the raw biomass contains light metal ions such as K^+, Na^+, Ca^{2+}, and Mg^{2+}, which bind to the acid functional group of alga.

5.3.4 Complexation

Complex formation is also a kind of mechanism for metal biosorption. The metal complexation in brown algae is well addressed by the binding of metal ions with alginate. Haug (1961) reported that during the binding of metal-ion to the alginic acid extracted from *Laminaria digitata* protons are released into the solution, which get diminished in the order $Pb^{2+} >$ $Cu^{2+} > Cd^{2+} > Ba^{2+} > Sr^{2+} > Ca^{2+} > Co^{2+} > Ni^{2+} > Mn^{2+} > Mg^{2+}$. Therefore, the metal sequestration is a complexation or coordination process of centralized heavy metals to the alginate, a multidendate ligand. Further, the egg-box model explains the steric interaction of ligand and metal ion supported by x-ray diffraction (Mackie et al., 1983) and nuclear magnetic resonance spectroscopic analyses (Steginsky et al., 1992).

5.3.5 Alginate and Its Role in Selectivity

A description of ion exchange and its involvement in the biosorption process was given in the preceding section. Thus, we will briefly cover the importance of the macromolecular structure alginate in the metal selectivity process of biosorption. Alginate is the common term applied to a polysaccharide family containing acid residues 1,4-linked α-D-mannuronic (M) and R-L-guluronic (G) in blockwise fashion, as shown in Figure 5.2. The affinity of alginate toward the divalent metal ion is determined by the availability of M and G residues in the alginate macromolecular structures (Haug, 1967).

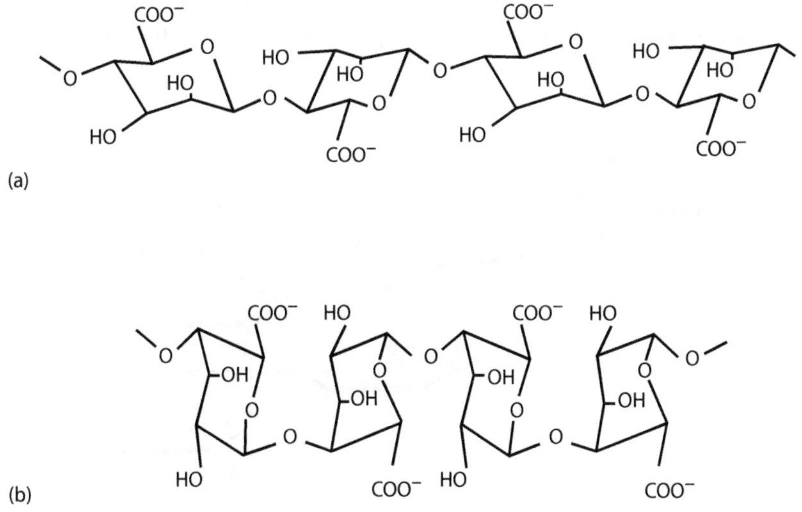

FIGURE 5.2   Main sections of alginic acid: (a) poly(D-mannuronosyl) segment and (b) poly(L-gluronosyl) segment.

5.4 BIOSORPTION BY SEAWEED

The large surface area of seaweeds is an advantage for the biosorption process. They contain the polyfunctional groups on the metal-binding site for both cationic and anionic complexes, as shown in Figure 5.3. In the cell wall of algal biomass, a good number of cationic binding sites are present, which include carboxyl, amine, imidazole, phosphate, sulfate, sulfhydryl, and hydroxyl. In addition, some functional groups are identified in the cell proteins and sugar molecules. The ligands of seaweeds form an ionic interaction with metal ions in the solutions, which is the virtual mechanism behind the binding (Yun et al., 2001). All seaweeds collected from the oceanic regions are shown to have an impressive metal sorption property (Kuyucak and Volesky, 1990). Table 5.3 shows the application of various macroalgae in the biosorption of metals, and Table 5.4 gives a gist of various operating parameters for a higher sorption. Brown seaweeds are highly suited for the binding of metal ions because of their rich content of polysaccharides in the cell wall (Percival and McDowell, 1967). Figure 5.4 shows the surface morphological changes during Cr (VI) biosorption.

5.5 CONCLUSION

The chapter has summarized the admirable achievements of algal biosorption that have been gathered over the past two decades, especially the significance of marine brown algal biosorption. It is significant to realize that

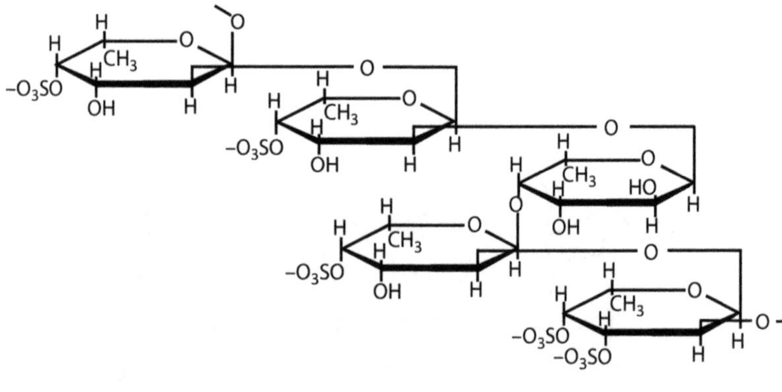

FIGURE 5.3 Structure of fucoidan. (From Davis, T.A. et al., *Water Research* 37: 4311–4330, 2003.)

TABLE 5.3   Macroalgae Used for the Biosorption of Metals

Macro Algae	Sorbate Used	Reference
Ascophyllum sp.	Pb, Cd	Volesky and Holan, 1995
Cladophora crispata	Cd, Pb, Cu, Ag	Gin et al., 2002
Cladophora fascicularis	Pb	Deng et al., 2007
Fucus ceranoides	Cd	Herrero et al., 2006
Fucus serratus	Cd	Herrero et al., 2006
Fucus spiralis	Cu	Murphy et al., 2007
Gracilaria fischeri	Cd, Cu	Chaisuksant, 2003
Gracilaria sp.	Pb, Cu, Cd, Zn, Ni	Sheng et al., 2004
Jania rubrens	Pb	Hamdy, 2000
Laminaria digitata	Cd, Zn, Pb, Cu	Sandau et al., 1996
Laurencia obtusa	Cr, Co, Ni, Cu, Cd	Hamdy, 2000
Palmaria palmata	Cu	Murphy et al., 2007
Petalonia fascia	Cu, Ni	Schiewer and Wong, 2000
Porphyra columbina	Cd	Basso et al., 2002
Sargassum asperifolium	Pb	Hamdy, 2000
Sargassum hemiphyllum	Cu, Ni	Schiewer and Wong, 2000
Sargassum hystrix	Pb	Jalali et al., 2002
Sargassum natans	Pb	Jalali et al., 2002
Sargassum vulgaris	Cd, Ni	Ofer et al., 2003
Sargassum kjellmanianum	Cd, Cu	Zhou et al., 1998
Turbinaria conoides	Pb	Senthilkumar et al., 2007
Ulva fascia	Cu, Ni	Schiewer and Wong, 2000
Ulva lactuca	Pb	Hamdy, 2000

20 years prior, knowledge on the mechanism of biosorption was vague. Today, scientists and engineers have identified the various technological phenomena behind the process of biosorption. And another development is the exploitation of various bioresources for the remediation of heavy metals. Brown algae have contributed to solving the problems that have arisen due to heavy metal pollution. The various traits of brown algae and their influence over of heavy metal sequestration have been explained. Thus, it is important to recognize the key aspects of algae in metal sorption. Finally, it should be noted that algae are not only applied for biosorption of heavy metals but also for the treatment of various rare earth metals. Hence, macroalgae are considered giant biosorbents for the removal of metals.

TABLE 5.4 Comparison of Metal Uptake Capacity of Different Marine Algae

Adsorbent (Seaweed)	Color	Adsorbate	Adsorption Capacity (mg/g)	pH	Temperature	Kinetic Model	Isotherm	Reference
Turbinaria conoides	Brown	Lead	439.40	4.5	30°C	–	Langmuir	Senthilkumar et al., 2007
Turbinaria ornata	Brown	Copper	147.06	6.0	–	–	Langmuir	Vijayaraghavan et al., 2004
Sargassum wightii	Brown	Copper (II)	115.00	4.5	–	–	Langmuir	Vijayaraghavan and Prabu, 2006
Sargassum polcysstum	Brown	Total chromium	69.4	2.0	–	Pseudo second order	–	Senthilkumar et al., 2010
Ulva reticulate	Green	Zinc (II)	135.5	5.5	30°C	Pseudo second order	Langmuir	Senthilkumar et al., 2006
Sargassum sp.	Brown	Cadmium	0.90 mmol/g	4.5	–	–	–	Davis and Volesky, 2000
Ulva reticulate	Green	Copper (II)	74.63	5.5	–	–	Freundlich	Vijayaraghavan et al., 2004
Fucus vesiculosus	Olive-brown	Copper (II)	1.85 mmol/g	–	–	–	Langmuir	Ahmady-Asbchin and Mohammadi, 2011
Sargassum myriocystum	Brown	Lead	179.5	5.0	25°C	–	–	Jeba et al., 2014
Pithophora varia	Green	Chromium (III)	60.6	5.0	20°C	–	Langmuir	Michalak et al., 2007

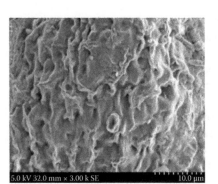
(a)

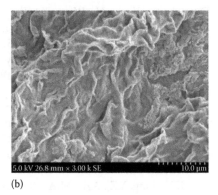
(b)

FIGURE 5.4   SEM image of *Sargassum*: (a) before Cr (VI) sorption, (b) after Cr (VI) sorption.

REFERENCES

Ahluwalia, S.S., Goyal, D. 2003. Removal of lead from aqueous solution by different fungi. *Indian Journal of Microbiology* 43, no. 4:237–241.

Ahmady-Asbchin, S., Mohammadi, M. 2011. Biosorption of copper ions by marine brown alga *Fucus vesiculosus*. *Journal of Biological and Environmental Sciences* 5, no. 15:121–127.

Asthana, R.K., Chatterjee, S., Singh, S.P. 1995. Investigations on nickel biosorption and its remobilization. *Process Biochemistry* 30:729–734.

Basso, M.C., Cerrella, E.G., Cukierman, A.L. 2002. Empleo de algas marinas para la biosorcion de metales pesados de aguas contaminadas asides. *Avances en Energías Renovables y Medio Ambiente* 6:69–74.

Bold, H.C., Wynne, M.J. 1985. *Introduction to the Algae*. Englewood Cliffs, NJ: Prentice-Hall.

Bossrez, S., Remacle, J., Coyette, J. 1997. Adsorption of nickel on *Enterococcus hirae* cell walls. *Journal of Chemical Technology and Biotechnology* 70:45–50.

Chaisuksant, Y. 2003. Biosorption of cadmium (II) and copper (II) by pretreated biomass of marine alga *Gracilaria fisheri*. *Environmental Technology* 24:1501–1508.

Crist, R.H., Martin, J.R., Crist, D.R. 1991. Interaction of metals and protons with algae. Equilibrium constants and ionic mechanisms for heavy metal removal as sulfides and hydroxides. In *Mineral Bioprocessing*, edited by R.W. Smith and M. Misra, 275–287. Washington, DC: Minerals, Metals and Materials Society.

Davis, T.A., Volesky, B., Mucci, A. 2003. A review of the biochemistry of heavy metal biosorption by brown algae. *Water Research* 37:4311–4330.

Davis, T.A., Volesky, B., Vieira, R.H.S.F. 2000. *Sargassum* seaweed as biosorbent for heavy metals. *Water Research* 34, no. 17:4270–4278.

Deng, L., Su, Y., Su, H., Wang, X., Zhu, X. 2007. Sorption and desorption of lead (II) from wastewater by green algae *Cladophora fascicularis. Journal of Hazardous Materials* 143:220–225.

Fourest, E., Roux, J.C. 1992. Heavy-metal biosorption by fungal mycelial byproducts—Mechanisms and influence of pH. *Applied Microbiology and Biotechnology* 37:399–403.

Fourest, E., Volesky, B. 1996. Contribution of sulfonate groups and alginate to heavy metal biosorption by the dry biomass of *Sargassum fluitans. Environmental Science and Technology* 30, no. 1:277–282.

Gadd, G.M. 1993. Interactions of fungi with toxic metals. *New Phytologist* 124:25–60.

Gin, K.Y., Tang, Y.Z., Aziz, M.A. 2002. Derivation and application of a new model for heavy metal biosorption by algae. *Water Research* 36:1313–1323.

Hamdy, A.A. 2000. Removal of Pb^{2+} by biomass of marine algae. *Current Microbiology* 41:239–245.

Haug, A. 1961. The affinity of some divalent metals to different types of alginates. *Acta Chemica Scandinavica* 15:1794–1799.

Herrero, R., Cordero, B., Lodeiro, P., Rey-Castro, C., Vicente, M.E.S.D. 2006. Interactions of cadmium (II) and protons with dead biomass of marine algae *Fucus* sp. *Marine Chemistry* 99:106–116.

Holan, Z.R., Volesky, B., Prasetyo, I. 1993. Biosorption of cadmium by biomass of marine algae. *Biotechnology and Bioengineering* 41:819–825.

Jalali, R., Ghafourian, H., Asef, Y., Davarpanah, S.J., Sepehr, S. 2002. Removal and recovery of lead using nonliving biomass of marine algae. *Journal of Hazardous Materials* 92:253–262.

Jeba Sweetly, D., Sangeetha, K., Suganthi, B. 2014. Biosorption of heavy metal lead from aqueous solution by non-living biomass of *Sargassum myriocystum. International Journal of Application or Innovation in Engineering & Management* 3, no. 4:39–40.

Kaewsarn, P. 2002. Biosorption of copper(II) from aqueous solutions by pre-treated biomass of marine algae Padina sp. *Chemosphere* 4:1081–1085.

Kloareg, B., Demarty, M., Mabeau, S. 1986. Polyanionic characteristics of purified sulfated homofucans from brown algae. *International Journal of Biological Macromolecules* 8:380–386.

Kuyucak, N. 1990. Feasibility of biosorbents application. In *Biosorption of Heavy Metals*, edited by B. Volesky, 371–378. Boca Raton: CRC Press.

Kuyucak, N., Volesky, B. 1989a. Accumulation of cobalt by marine alga. *Biotechnology and Bioengineering* 33, no. 7:809–814.

Kuyucak, N., Volesky, B. 1989b. Desorption of cobalt-laden algal biosorbent. *Biotechnology and Bioengineering* 33, no. 7:815–822.

Kuyucak, N., Volesky, B. 1989c. The mechanism of gold biosorption. *Biorecovery* 1:219–235.

Kuyucak, N., Volesky, B. 1990. Biosorption by algal biomass. In *Biosorption of Heavy Metals*, edited by B. Volesky, 173–198. Boca Raton: CRC Press.

Lee, R.E. 1989. *Phycology*. Cambridge, UK: Cambridge University Press.

Lesmana, S.O., Febriana, N., Soetaredjo, V., Sunarso, J., Ismadji, S. 2009. Studies on potential applications of biomass for the separation of heavy metals from water and wastewater. *Biochemical Engineering Journal* 44:19–41.

Mackie, W., Perez, S., Rizzo, R., Taravel, F., Vignon, M. 1983. Aspects of the conformation of polyguluronate in the solid state and in solution. *International Journal of Biological Macromolecules* 5:329–341.

Michalak, I., Zielinska, A., Chojnacka, K., Matula, J. 2007. Biosorption of Cr (III) by microalgae and macroalgae: Equilibrium of the process. *American Journal of Agricultural and Biological Sciences* 2, no. 4:284–290.

Murphy, V., Hughes, H., McLoughlin, P. 2007. Cu (II) binding by dried biomass of red, green and brown macroalgae. *Water Research* 41:731–740.

Myklestad, S. 1968. Ion-exchange properties of brown algae. Determination of rate mechanism for calcium–hydrogen ion exchange for particles from *Laminaria hyperborea* and *Laminaria digitata*. *Journal of Applied Chemistry* 18:30–36.

Niu, H., Xu, X.S., Wang, J.H., Volesky, B. 1993. Removal of lead from aqueous-solutions by *Penicillium* biomass. *Biotechnology and Bioengineering* 42:785–787.

Ofer, R., Yerachmiel, A., Yannai, S. 2003. Marine macroalgae as biosorbents for cadmium and nickel in water. *Water Environmental Research* 75:246–253.

Ofer, R., Yerachmiel, A., Yannai, S. 2004. Mechanism of biosorption of different heavy metals by brown marine macroalgae. *Biotechnology and Bioengineering* 87:451–458.

Percival, E.G.V., McDowell, R.H. 1967. *Chemistry and Enzymology of Marine Algal Polysaccharides*. London: Academic Press.

Rees, D.A., Morris, E.R., Thom, D., Madden, J.K. 1982. Shapes and interactions of carbohydrate chains. In *The Polysaccharides*, vol. 1 (Molecular Biology: An International Series of Monographs and Textbooks), edited by G.O. Aspinall, 195–290. New York: Academic Press.

Sandau, E., Sandau, P., Pulz, O., Zimmermann, M. 1996. Heavy metal sorption by marine algae and algal by-products. *Acta Biotechnologica* 16:103–119.

Schiewer, S., Volesky, B. 2000. Biosorption by marine algae. In *Remediation*, edited by J.J. Valdes, 139–169. Dordrecht, Netherlands: Kluwer Academic Publishers.

Schiewer, S., Wong, M.H. 2000. Ionic strength effects in biosorption of metals by marine algae. *Chemosphere* 41:271–282.

Selatnia, A., Boukazoula, A., Kechid, N., Bakhti, M.Z., Chergui, A., Kerchich, Y. 2004. Biosorption of lead (II) from aqueous solution by a bacterial dead *Streptomyces rimosus* biomass. *Biochemical Engineering Journal* 19:127–135.

Senthilkumar, R., Vijayaraghavan, K., Jegan, J., Velan, M. 2010. Batch and column removal of total chromium from aqueous solution using *Sargassum polycystum*. *Environmental Progress and Sustainable Energy* 29, no. 3:334–341.

Senthilkumar, R., Vijayaraghavan, K., Thilakavathi, M., Iyer, P.V.R., Velan, M. 2007. Application of sea weeds for the removal of lead from aqueous solution. *Biochemical Engineering Journal* 33:211–216.

Senthilkumar, R., Vijayaraghavan, K., Thilakavathi, M., Iyer, P.V.R., Velan, M. 2006. Seaweeds for the remediation of wastewaters contaminated with zinc (II) ions. *Journal of Hazardous Materials* 136, no. 3:791–799.

Sheng, P.X., Tan, L.H., Chen, J.P., Ting, Y.P. 2004. Biosorption performance of two brown marine algae for removal of chromium and cadmium. *Journal of Dispersion Science and Technology* 25:681–688.

Steginsky, C.A., Beale, J.M., Floss, H.G., Mayar, R.M. 1992. Structural determination of alginic acid and the effects of calcium binding as determined by high-field NMR. *Carbohydrate Research* 225, no. 1:11–26.

Thom, D., Grant, G.T., Morris, E.R., Rees, D.A. 1982. Characterisation of cation binding and gelation of polyuronates by circular dichroism. *Carbohydrate Research* 100:29–42.

Tsezos, M., Volesky, B. 1981. Biosorption of uranium and thorium. *Biotechnology and Bioengineering* 23:583–604.

Vijayaraghavan, K., Jegan, J., Palanivelu, K., Velan, M. 2004. Copper removal from aqueous solution by marine green alga *Ulva reticulata*. *Electronic Journal of Biotechnology* 7, no. 1:20–30.

Vijayaraghavan, K., Prabu, D. 2006. Potential of *Sargassum wightii* biomass from copper (II) removal from aqueous solutions: Application of different mathematical models to batch and continuous biosorption data. *Journal of Hazardous Materials* 137, no. 1:558–564.

Volesky, B. 1990. *Biosorption of heavy metals*. Boca Raton: CRC Press.

Volesky, B. 2001. Detoxification of metal-bearing effluents: Biosorption for the next century. *Hydrometallurgy* 59:203–216.

Volesky, B. 2007. Biosorption and me. *Water Research* 41:4017–4029.

Volesky, B., Holan, Z.R. 1995. Biosorption of heavy metals. *Biotechnology Progress* 11:235–250.

Volesky, B., Naja, G. 2007. Biosorption technology: Starting up an enterprise. *International Journal of Technology Transfer and Commercialisation* 6:196–211.

Wang, J.L., Chen, C. 2006. Biosorption of heavy metals by *Saccharomyces cerevisiae*: A review. *Biotechnology Advances* 24:427–451.

Wang, J.L., Chen, C. 2009. Biosorbents for heavy metals removal and their future a review. *Biotechnology Advances* 27:195–226.

Yetis, U., Dolek, A., Dilek, F.B., Ozcengiz, G. 2000. The removal of Pb (II) by *Phanerochaete chrysosporium*. *Water Research* 34:4090–4100.

Yun, Y.S., Parck, D., Park, J.M., Volesky, B. 2001. Biosorption of trivalent chromium on the brown seaweed biomass. *Environmental Sciences and Technology* 35:4353–4358.

Zhou, J.L. 1999. Zn biosorption by *Rhizopus arrhizus* and other fungi. *Applied Microbiology and Biotechnology* 51:686–693.

Zhou, J.L., Huang, P.L., Lin, R.G. 1998. Sorption and desorption of Cu and Cd by macroalgae and microalgae. *Environmental Pollution* 10:67–75.

CHAPTER 6

Current Developments in Mass Production of Microalgae for Industrial Applications

S. Ramachandran, P. Sankar Ganesh,
Velmurugan Sivasubramanian,
and B.G. Prakash Kumar

CONTENTS

Abstract

Microalgae are widely mass-produced for applications such as nutritional supplements, protein sources, food colorants and pigments, pharmaceuticals, animal feed, biodiesel production, CO_2 sequestration, and bioremediation. The most common methods of mass cultivation of microalgal biomass are open pond, closed bioreactor systems, and photobioreactors. Photobioreactors (PBRs) are a preferred mode of production due to their inherent advantages such as high yield, minimal contamination, better control on environmental parameters, continuous production capability throughout the year, and ability to use indoor controlled environments with natural or artificial lighting or outdoor cultivation using sunlight. This chapter focuses on current developments in the photobioreactors, as they hold promise for economic production and offer a continuous supply of biomass for biotechnological applications. PBRs are increasingly used in different modes like flat bed, tubular, and helical based on the type of microalgae being cultivated and parameters chosen for mass cultivation. Though the economics of biomass produced using PBR systems are considered to be high, the current developments in PBR-based technologies offer lower price than current estimates to produce phototrophic microalgal biomass, which can change the scenario for large-scale utilization of microalgae, such as biodiesel production. This chapter outlines the recent developments in the design, use of light source, harvesting of light source for indoor applications, use of low-cost materials for PBRs, control of environmental conditions, automation in operation and control mechanisms, developments in harvesting methods, strain improvements on some of the most used microalgal strains, potential for new opportunities for finding suitable microalgal strains, extraction of biotechnologically important compounds, and the use of sustainable resources for microalgal biomass production.

6.1 INTRODUCTION

Microalgae are photosynthetic organisms mostly found in freshwater, hypersaline, and marine environments. Microscopic algae (Richmond and Tomaselli, 2007) include prokaryotic cyanobacteria and eukaryotic algae. The advantages of microalgae are, in addition to fixing atmospheric carbon dioxide, adapting to diverse environments and having high growth rates with the ability to double in biomass content within 3.5 to 24 h (Raheem et al., 2015). Microalgae have been widely studied due to their intense biological roles in both natural and manmade ecosystems. During the past decades, microalgae biomasses have been exclusively used in the health food market (Chacón-Lee and González-Mariño, 2010). Today there are numerous commercial applications of microalgae, out of which some of the important ones are

- Enhancing the nutritional value of human food and animal feed with their unique chemical composition (for example, polyunsaturated fatty acids [PUFAs])

- Playing a crucial role in aquaculture, especially in the developmental stages of the cultured organisms

- Providing vital ingredient in cosmetics

- Source of useful molecules such as algal pigments, which are widely used as food coloring agents and natural dyes

- Potential feedstock for biodiesel production

Cyanobacteria produce substances critical to biological functions and have found a niche in healthcare product preparations due to the high amount of PUFAs, antioxidants, vitamins, amino acids, and pigments (Christaki et al., 2011). Consuming edible microalgae has eye-related benefits such as reducing the risk of cataracts and age-related macular degeneration (Yu et al., 2012). Microalgae are widely cultivated all over the world for a variety of applications. A very brief introduction is provided here about the importance of microalgae and its mass production. The Chinese, who used *Nostoc* sp. to survive during famine, provide the first recorded human use of microalgae on a commercial scale almost 2000 years ago. But in modern terms, commercial culture of microalgae began only after 1950s at a global scale. Now, several cyanobacteria are commercially grown for food supplements including *Chlorella*, *Hematococcus pluvialis*, *Dunaliella salina*, and *Spirulina* sp. Microalgae enjoy a favored position as the purveyors of human nutritional supplements. Since 1950s, many research groups have acknowledged microalgae as a very important and plentiful source of quality protein and is used for children affected by malnutrition in underdeveloped countries. Of all the algae cultivated in culture systems around the world, over 75% of the annual biomass production is mostly used in the manufacture of supplement tablets, syrups, and capsules of microalgae and with desired composition. Yet the scope remains broad for the food industry, as microalgae could be incorporated in food products in several ways that are yet to be realized. Microalgae are a good source of carbohydrates (starch, cellulose, sugars, and other polysaccharides). These carbohydrates are produced in easily digestible forms, contributing to their wide use in different foods for patients of all ages (Chacón-Lee and González-Mariño, 2010). As for the lipid complement, the average lipid content in microalgae varies between 1% and 40%, and can be as high as 85% of dry weight if grown under optimal conditions. The most important lipids are the essential polyunsaturated fatty acids such as linoleic, eicosapentaenoic (EPA), and docosahexaenoic (DHA). The PUFA obtained from algal biomass have a great advantage over those obtained conventionally from fish oil in that they are uncontaminated by native fish lipids and present an easier mode of production as compared to the aquatic choice, especially in the case of low taste intensity and off-odor problems (Chacón-Lee and González-Mariño, 2010). These are also used in different forms such as processed powder in capsules and added to food as one of the ingredients in pasta, beverages, and noodles (Raheem et al., 2015).

Biofuels from microalgae are considered to be more advantageous over plant-based resources due to their inherent characteristics such as high productivity, minimal land requirements, and use of wastewater for cultivation. This would widen the scope of microalgal products and they would become economically competitive (Spolaore et al., 2006). Biofuels derived from microalgae include biodiesel, bioethanol, biohydrogen, and syngas (Raheem et al., 2015). Optimization strategies that can increase the microalgal lipid content for biodiesel production remains attractive to scientists and venture capitalists. Many governments have initiated research and subsidize sustainable alternative fuels derived from microalgae in order to reduce dependence of fossil fuels.

Microalgae are cultured as feeds in commercial aquaculture. Over 50 species of algae are used worldwide for this purpose and this number varied with the variety of the local seafood being commercially produced. They are also used as feed for cattle, pigs, and poultry in order to rear healthy animals. Microalgae may be incorporated in feeds at different percentages, depending on the final feed product. During the past decades, microalgae biomass has been significantly used in healthcare products such as antiaging, sunscreen lotions as UV rays protectant, and is the source of vibrant food colors. The most important microalgae mass produced for biotechnological applications and their products are presented in Table 6.1.

This chapter outlines the most common microalgal species being mass produced on a commercial scale. Economical methods for mass production are discussed with emphasis on current developments in mass production of industrially important microalgal species.

TABLE 6.1　Most Important Microalgae Mass Produced for Biotechnological Applications

Application and Product	Microalgae	References
Human nutrition	Spirulina, Chlorella, Dunaliella salina, Aphanizomenon flos-aquae	Lorenz and Cysewski, 2000; Hejazi and Wijffels, 2004; Pulz and Gross, 2004;
Animal nutrition	Spirulina	Ratledge, 2004; Spolaore et al., 2006
Cosmetics	Spirulina, Chlorella, Dunaliella salina	
Phycobiliproteins	Spirulina	
Aquaculture	Chlorella, Haematococcus pluvialis	
Beta-carotene	Dunaliella salina	
Astaxanthin	Haematococcus pluvialis	
DHA oil	Crypthecodinium cohnii, Shizochytrium	

6.2 CULTIVATION METHODS

Cultured microalgae have several advantages, as they can grow 20 to 30 times faster than traditional food crops and do not compete for arable land. Table 6.2 presents the comparative lipid yield from microalgae and commercially important oil crops. In addition, the secondary metabolites produced from microalgae are diverse and difficult to produce both technically and commercially. Thus, developing optimal microalgal production techniques that have commercial applications and economic viability are necessary. For large-scale cultivation, both indoor and outdoor microalgae cultivation are practiced, and each method has its merits and demerits. In outdoor cultivation, some of the microalgae have light sensitivity at intense solar radiation (Pruvost et al., 2015). Such microalgae of importance can be cultivated indoors using artificial sources of illumination.

Microalgae can be cultured either as monoculture or part of a mixed culture that competes with other organisms. Figures 6.1 to 6.3 present the most common methods of microalgal cultivation with commercial success for biomass, pigments, protein, and animal feed production purposes. The site for installation requires careful analysis of environmental parameters especially temperature, water, sunlight, and humidity. In the case of open pond systems, site preference is of particular concern, as animals and birds are potential sources of contamination.

TABLE 6.2 Lipid Productivity from Microalgae and Commercial Crops

Microalgae/Commercial Crops	Lipid Productivity (>100 mg/L/Day)	Reference
Nannochloris sp. UTEX LB1999	15.6–109.3	Takagi et al., 2000
Chlorella protothecoides	1881.3–1840	Cheng et al., 2009
	1209.6–3701	Xiong et al., 2008
	932	Xu et al., 2006
	732.7–932	Li et al., 2007
Chlorella sp. Phototrophica	121.3–178.8	Chiu et al., 2008
Nannochloropsis oculata NCTU-3	84.0–142	Chiu et al., 2009
Neochloris oleoabundans UTEX 1185	38.0–133.0	Li et al., 2008
Chlamydomonas sp. JSC4	223	Ho et al., 2014
Chlorella sorokiniana CY1	140.8	Chen et al., 2013
Nannochloropsis sp. F&M-M24	110.1	Bondioli et al., 2012
Isochrysis zhangjiangensis	136.2	Feng et al., 2011
Chlorella sp. GD	155	Kuo et al., 2015

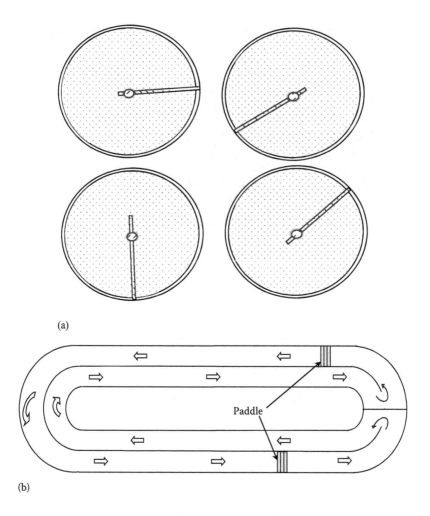

(a)

(b)

FIGURE 6.1   Most common methods of microalgal cultivation: (a) open pond, (b) open raceway pond.   (*Continued*)

Monocultures are of great research interest and yield better commercial profit, hence are most commonly practiced. Each type of microalgae needs an optimum mix of light, oxygen, nutrients, and temperature conditions to thrive. The most common method of mass cultivation of microalgae are open pond and in photobioreactors (PBRs). Table 6.3 presents major microalgal strains produced using different methods for commercial applications.

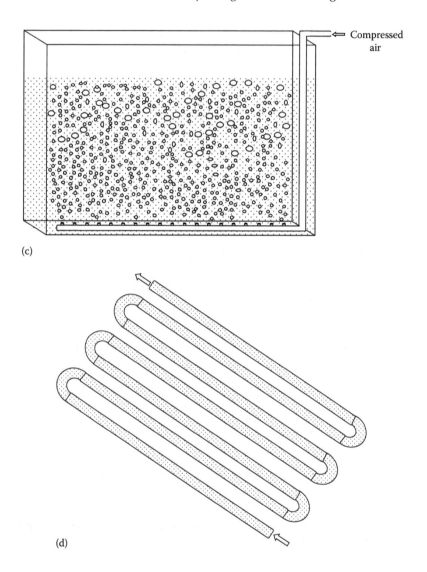

(c)

(d)

FIGURE 6.1 (CONTINUED) Most common methods of microalgal cultivation: (c) flat plate PBR, (d) tubular horizontal. *(Continued)*

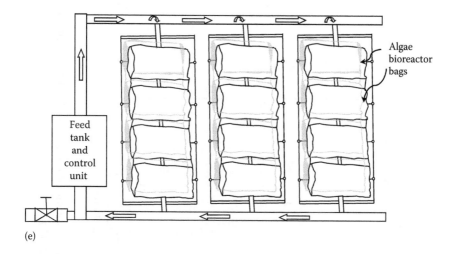

(e)

FIGURE 6.1 (CONTINUED)   Most common methods of microalgal cultivation: (e) vertical poly bag systems.

6.2.1 Open Pond System

The open pond method has been adopted to meet the demand for microalgal biomass since the 1950s. The major advantages of culturing microalgae in open pond system are

- Easy to build and operate
- Cheaper than closed bioreactors
- High production capacity due to large area of cultivation
- Continuous harvesting is possible

Major drawbacks of the open pond system are the lack of control over temperature and illumination. In addition, open systems are highly vulnerable to contamination by bacteria, algae, and other organisms (thus not suitable for monocultures). The growth of microalgae in open ponds is heavily dependent on the environment, and biomass production capacity fluctuates with the climatic conditions and length of the day–night cycle. *S. plantensis*, *H. pluvialis*, *Chlorella*, and *Dunaliella salina* are some of the microalgal species cultivated in open pond systems for mass production. To minimize contamination, some improvements are made to traditional systems. A combination of closed and open systems in series, and surrounding the pond with a transparent or translucent barrier to turn the

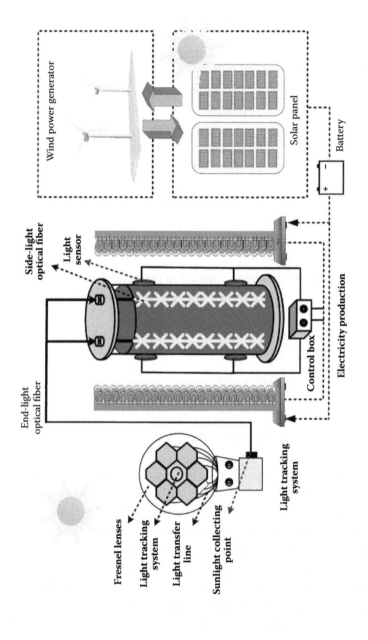

FIGURE 6.2 Photobioreactor for indoor cultivation system with solar-energy-based LED illumination and sunlight transmitted through optical fibers to PBRs. (From Chen C-Y et al., *Bioresource Technology* 102:71–81, 2011.)

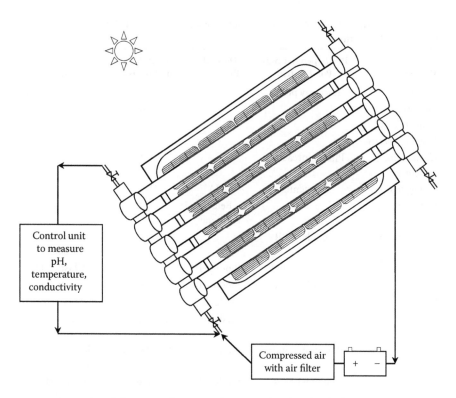

FIGURE 6.3   Modification of a vertical system that has both the advantages of a vertical system and a tubular reactor for outdoor and indoor applications.

pond into a greenhouse are successful designs. Volaris™ (Figure 6.4) as developed by Heliae, in Arizona, is one advanced system that allows continuous biomass productivity throughout the year. This design allows more variety of species to be grown, yet allows monoculture conditions to be maintained. It extends the growing season, as the climate is not allowed to interfere. Open pond systems for the production of biofuels are of specific interest as they are the most economical method, where large volume cultivation for lipid extraction and microbial contamination are not significant factors. Sapphire Energy Inc., have developed commercial-scale open pond production facilities in Las Cruces and in Columbus, New Mexico and is aiming to produce Omega oils, high-value protein and biofuel from microalgae.

6.2.2 Photobioreactors

Photobioreactors (PBRs) are among the preferred modes of microalgae production due to their inherent characteristics such as high yield, minimal

TABLE 6.3   Strains Cultivated, Type of Photobioreactor, Cost, and Derived Products from Mass Production

Microalgae	PBR/Open Raceway Pond System	Derived Products	Cost ($/kg of Biomass)	Reference
Spirulina	Raceway ponds	–	10	Darzins et al., 2010
Chlorella vulgaris,			20	
Haematococcus			100	
H. pluvialis, C. vulgaris	Photobioreactors		100	
C. vulgaris	Fermenter	DHA	100	
D. salina		Beta-carotene	>50	
P. tricornutum	Tubular PBR		34	Grima et al., 2003
–	Raceway pond	Biodiesel	3.8	Chisti, 2007
–	Tubular PBR	Biodiesel	2.95	Chisti, 2007
S. almeriensis	Tubular PBR	Biofuels	34	Shen, 2009
–	Raceway ponds	Biodiesel	2.33–3.33 $/gal	Gallagher, 2011
–	Raceway ponds	Biodiesel	8.75 $gal/L	Kovacevic and Wesseler, 2011
–	Polymer bags	Biodiesel	300–600	Grima et al., 2003

contamination, better control on environmental parameters, continuous production capability, and ability to use indoor artificial illumination or natural sunlight. The closed-system photobioreactor holds promise for economic production in terms of high-quality biomass and high yield. PBRs are increasingly used in different modes like flat-bed (Huang, 2014), tubular (Ashokkumar et al., 2015), and helical (Soletto et al., 2008), and also in several types of disposable systems (Ojo et al., 2014). Though the cost of biomass produced using PBR systems is considered to be higher, current developments in PBR-based technologies have reduced the production cost of phototrophic microalgal biomass. Such developments can change the scenario of microalgae for large-scale cultivation using PBRs. Current research on PBRs focus on minimizing the construction cost of the reactors; controlling environmental conditions; illumination systems; and using low-cost nutrients, specifically the macronutrients such as carbon and nitrogen for biomass production. Table 6.3 shows the types of PBR, cost per kilogram of biomass produced, nutrients used, strains

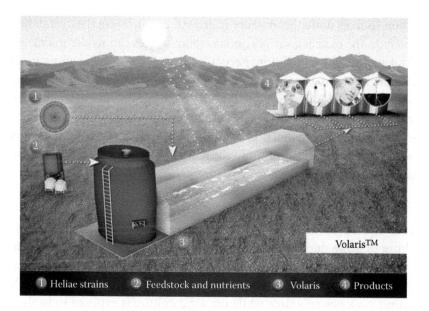

FIGURE 6.4 Volaris™ Commercial Production Platform: Heliae-Volaris platform uses sunlight and waste carbon feedstocks, which significantly reduce capital costs. The versatile system reduces most sources of contamination resulting in increased product quality.

cultivated, and products derived through mass production. Carbon capture and sequestration through biological processes are among the economical means for reducing CO_2 from atmosphere and capturing flue gases from industrial emissions. Microalgae have shown to be more efficient in CO_2 sequestration than the physicochemical scrubbing process. Microalgal species can capture up to 5% to 15% of CO_2 from CO_2-rich industrial flue gases, whereas plants can take up only 0.03% to 0.06% (Raheem et al., 2015). This equals up to 10 to 50 times higher than terrestrial plants. This high rate of carbon sequestration can be exploited for reducing CO_2 using photobioreactors, which can be a solution for point-source reduction of carbon emissions.

6.3 CURRENT DEVELOPMENTS IN PHOTOBIOREACTOR DESIGN

Efforts to reduce the cost of biodiesel production from microalgal species include cost-effective methods for biomass production, efficient harvesting, and oil extraction methods. Though developments using organic carbon sources under heterotrophic growth can increase the overall lipid content in the biomass, such organic carbon sources can render microalgae with bacterial contamination. Particularly in the open pond systems, the addition of organic carbon leads to uncontrolled microbial contamination. Thus, for biodiesel production, the phototrophic mode is most preferred for lipid production (Chen et al., 2011). In addition, the phototrophic mode can be optimized for scale-up in both open pond systems and closed photobioreactor systems. Some of the more modern bioreactor designs are discussed next.

6.3.1 Airlift Method

Airlift and bubble-column bioreactors have gained widespread use in the culture of algae due to advantages such as constant agitation and better air dispersal that aid the growth of organisms. This method is used in outdoor cultivation of microalgae, where the air is agitated through the column via a power source while the algal culture is exposed to sunlight for continuous growth (Richmond, 1986).

6.3.2 Closed Reactors

The biggest advantage of culturing microalgae within a closed system is that it provides control over the physical, chemical, and biological environment of the culture. Factors such as temperature gradients, winds, and contaminants that make open ponds unsafe are avoided in closed reactors (Tredici and Materassi, 1992). This kind of bioreactor is supported by photosynthetic reactions using dissolved carbon dioxide and sunlight. The carbon dioxide is mixed into the closed reactor to allow it to reach the algae. The bioreactor is made out of transparent material such that sunlight can penetrate.

Photobioreactors give the perfect example of a closed system where relevant factors such as temperature, oxygen demand, pH, and prevention of microbial contamination can be controlled. They can operate continuously by resupplying nutrients and controlling all elements to prevent collapse due to waste buildup. Proper removal of waste and addition of nutrients allow reactors to operate for long periods. In such a continuous mode, algae can be made to grow constantly in the "log phase," and stationary and death phases are avoided. Batch operation is also possible wherein a stipulated amount of nutrients is supplied to the algae for a time and then restocked for the next batch to grow in.

Major types of PBRs that are commonly used for microalgal cultivation include:

- Horizontal PBRs

- Vertical PBRs

- Flat plate PBRs

6.3.2.1 Horizontal Photobioreactors

In horizontal PBRs tubes laid on the ground form a network of loops. Mixing of microalgal suspended culture occurs through a pump that raises the culture vertically at timed intervals into a photobioreactor. Pulsed mixing at intervals produces better results than the use of continuous mixing. Better temperature control also increases production of microalgae. For example, *Spirulina* could be cultivated intensively at a wide temperature range and an extended cultivation period, even over summer months in a closed system (Tredici and Materassi, 1992).

6.3.2.2 Vertical Systems

Vertical reactors use vertical polyethylene sleeves hung from an iron frame. Alternatively, glass tubes can also be used. The two main types of vertical photobioreactors are the flow-through vertical alveolar panel (VAP) and the bubble column VAP (Tredici and Materassi, 1992).

Advantages

- High mass transfer

- Good mixing with low shear stress

- Low energy consumption

- Reduced photoinhibition

- High potential for scalability

- Easy to sterilize

- Readily tempered

- Good for immobilization of algae and photooxidation

Limitations

- Small illumination surface area
- Deoxygenation troubles
- Requires sophisticated materials
- Shear stress to algal cultures
- Decrease of illumination surface area upon scale-up

6.3.2.3 Flat Plate Reactors

Flat plate reactors (FPRs) are built using narrow panels and are placed horizontally to maximize sunlight input to the system. The design of FPRs helps to increase the surface-area-to-volume ratio such that sunlight is efficiently used. Originally it was thought that there might not be sufficient mixing and this system was considered an expensive alternative, but now with the incorporation of gas exchange units, FPRs have been widely used for production of *Spirulina* and other unicellular microalgae that can grow under outdoor cultivation conditions. FPRs overcome issues of circulation and provide an advantage of an open gas transfer unit that reduces oxygen build up (Carvalho et al., 2006), for example, in production of *Nannochloropsis* sp. and *H. pluvialis* for their high levels of astaxanthin. Generally, flat plate photobioreactors are made of transparent materials like acrylic sheets for maximum utilization of solar light energy and to reduce capital investment. Field studies have reported that with flat plate photobioreactors, high photosynthetic efficiencies can be achieved.

Advantages

- Large illumination surface area, suitable for outdoor cultures
- Good for immobilization of algae, good light path
- Good biomass productivities
- Relatively minimum cost
- Easy to clean up
- Readily tempered
- Low oxygen buildup

Limitations

- Scale-up require many compartments and support materials
- Difficulty in controlling culture temperature
- Some degree of wall growth and possibility of hydrodynamic stress to some algal strains

6.4 FERMENTOR-TYPE REACTORS

Fermentor-type reactors (FTRs) are bioreactors where fermentation is carried out. FTRs have now been used to culture algae as well, but they are still in a nascent stage and have several disadvantages like bad surface-area-to-volume ratio and less than satisfactory efficiency in sunlight penetration (Richmond, 1986; Carvalho et al., 2006). FTRs have been created with a combination of artificial light and automatic control over other culture factors that can help produce algae. But these systems are still in the laboratory stage and right now only see major use in producing metabolites for the pharmaceutical industry (Carvalho et al., 2006).

6.5 TUBES IN DESERT

A new method for microalgal cultivation is the "tubes in desert" as developed by the Biodesign Institute at Arizona State University to produce engine fuel in a sustainable way. The microalgae are grown in transparent tubes that are exposed to sunlight and are provided with white reflectors placed beneath the tubes to enhance light penetration into the tubes flowing with algal cultures in circulation. It is vital to place the flexible tubes in sunlight-rich environs like the desert, hence the name tubes in the desert. This project has future plans of constructing several acres of tubes in the desert to optimize bacteria biofuel production.

6.6 PARAMETERS FOR MICROALGAL GROWTH

Algae are mostly grown in a bioreactor or photobioreactor. Algae bioreactors can be used to produce biofuels, biomass to produce biologically important molecules, or to reduce pollutants such as NO_2 and CO_2 in industry emissions. There are pertinent concerns bioreactors and the major areas that influence algal productivity are under rapid development using current technology in terms of materials, energy efficiency, and alternative sustainable resources.

6.6.1 Lighting

With regard to lighting, the light intensity, spectral quality, and photoperiod are critical to cell growth. As algae are photosynthetic, they convert light energy to chemical energy and thus require precise light intensity, because at a certain level, saturation of light occurs and algae dissipate the excess energy as heat. The spatial dilution of light (distributing solar radiation on a greater photosynthetic surface area) can be used to lessen light saturation. The problem of mutual shading of cells in the culture is alleviated by spatial dilution, which results in lower concentrations of accessory pigments and increased growth rate. With increasing depths and high cell concentrations, the light intensity must be correspondingly increased to pierce through the thick suspension. About 10,000 lux are used for large cultures in industry. With regard to the light source, either fluorescent tubes or natural sunlight can be used.

In algae, chlorophyll absorption peaks between 650 and 700 nm (red region), whereas the carotenoids strongly absorb between 450 and 500 nm, which is the blue region. Thus fluorescent tubes emitting either in the blue or the red light spectrum are preferred as accords to the cultured algae. Light and dark cycles strongly influence the growth of algae in culture. The time length of the dark reactions in photosynthesis serves as the rate-limiting step for photosynthesis and growth. The light regime in a photobioreactor must be modulated to provide an optimal dark period, which depends on the photon flux density of the previous light period and the fluid residence time in zones of different irradiance. The flashing light effect, alternating dark and light seconds between 40 microseconds and a second, can be used to increase algal growth under high light intensity, while also increasing the instantaneous photonic flux. A photoperiod of 18 hours light and 6 hours dark is widely used for algal cultivation. This kind of light regime paves the way for photoacclimation (physiological responses of cells to rapid changes in light intensity). A photobioreactor's design principle is to maximize the surface-area-to-volume ratio for efficient light harvesting in photosynthesis.

6.6.2 Mixing

Mixing is crucial to algal growth, as it increases the frequency of cell exposure to light and dark volumes of the reactor and mass transfer between the nutrients and cells. Transfer of mass in and out of the cell is an important factor that determines growth. The overall mass transfer coefficient in a reactor is used to compare reactors. The coefficient depends on the agitation rate, type of sparger, surfactants/antifoam agents, and the temperature. Continuous monitoring is needed because high levels of mixing will result in cell death due to shear (Kunjapur and Eldridge, 2010).

6.6.3 Water Consumption

Algae are generally divided based on whether they grow ideally in freshwater or saltwater. The optimum salinity range was 20 to 40 ppt for *Nannochloropsis*, 20 to 35 ppt for *Tetraselmis*, and 25 to 35 ppt for *Isochrysis* (Kunjapur and Eldridge, 2010). Instead of using pristine water for algal growth, treated wastewater can be used as it is generally high in nutrients used by the cell (high concentrations of nitrogen and phosphorus). Instead of releasing such wastewaters into water sources and causing unwanted algae growth and eutrophication, they can be used for biomass production.

6.6.4 CO_2 Consumption

CO_2 can be supplied to the culture via a gas permeable membrane by diffusion. It is crucial to ensure availability of CO_2 while preventing CO_2 inhibition at high gaseous pCO_2. CO_2 concentrations from 1% to 5% (by volume) allow maximum growth. A pCO_2 value of at least 0.15 kPa is needed to prevent kinetic CO_2 uptake limitation. In addition, a ratio of about 1.7 to 1.8 g CO_2/g dry biomass is required (Kunjapur and Eldridge, 2010). Instead of using CO_2 gas commercially produced for this purpose, flue gas from industrial emissions that contains up to 13% CO_2 can be used. It also reduces greenhouse gas emissions while reducing the cost of algal biofuel production.

6.6.5 O_2 Removal

It is undesirable to have a high concentration of oxygen around algae cells because photooxidative damage can occur from the combination of harsh sunlight and high oxygen concentration. Generally, oxygen concentrations should be maintained below 400% of air saturation value (Bajpai et al., 2014). Thus, tube length is limited in horizontal tubular reactors because of the constraint on the concentration of dissolved oxygen. This makes it difficult for reactor scale-up. As oxygen does not amass much in open ponds, they are better than closed reactors in this aspect. Modifications are introduced to limit oxygen build by centralized agitation and recirculation systems, lowering the temperatures and pumping air into the tubular photobioreactors.

6.6.6 Nutrient Supply

Major nutrients that control growth of algae are carbon, nitrogen and phosphorus, along with an array of other nutrients including hydrogen, oxygen, sulfur, calcium, magnesium, sodium, potassium, and chlorine. Minor nutrients include iron, boron, manganese, copper, molybdenum, vanadium, cobalt, nickel, silicon, and selenium. Nutrient stress can cause metabolic changes within the biomass. For example, phosphorus deficiency (–50%) caused an increase in the chlorophyll *a* level in *Picochlorum* sp. (El-Kassas, 2013). Starvation and low concentrations of nutrients (0.005 mM P, 0.2 mM N, $1.1*10^{-5}$ mM/S) increased carotenoids/chlorophyll ratios, specifically for sulfur limitation in *Nannochloropsis gaditana* (Rukminasari, 2013).

6.6.7 Temperature and pH

Optimal growth is limited to a narrow range specific to each strain of algae though they are quite diverse in nature. For example, Abu-Rezq et al. (1999) found that the optimum temperature range for *Nannochloropsis* was 19°C to 21°C and *Isochrysis* was 24°C to 26°C. A thermal growth model may be used to describe microalgal growth. Temperature concerns are especially rampant in outdoor conditions. Low temperature conditions with consistent light intensity result in carboxylase activity going up and overproduction of energy supply plus an increase in chlorophyll activity in the case of *Skeletonemacostatum*. The pH of the medium is linked to the concentration of CO_2 and pH rises steadily in the medium as CO_2 is consumed during flow downstream (Suh and Lee, 2008).

6.7 ADVANCES IN MICROALGAE CULTURE

6.7.1 Light Harvesting Complexes (LHCs)

Genetic engineering is a factor that has tremendous potential and has already achieved success in the laboratory. For example, Mussgnug et al. (2007) described experiments that altered the LHCs in the chlorophyll. The purpose of LHCs is to capture solar energy and regulate the flow of excitation signals to the photosynthetic machinery. In addition to this they disperse light energy as heat or fluorescence when irradiation exceeds photosynthetic capacity. This second trait in algal bioreactors reduces efficiency and therefore efforts are being made to curb it. To resolve this issue, the authors used RNAi technology to create a mutant of *C. reinhardtii* (referred to as Stm3LR3) that significantly downregulated the amount of LHC I and LHC II complexes. Their experiments, which were successful, also showed that the reduction was permanent, something that had not previously been reported in literature.

The strain of Stm3LR3 resulted in the

- Decrease in dissipation of captured light energy

- Increase in photosynthetic quantum yield

- Reduction of sensitivity of the system to photoinhibition

6.7.2 Diatom Genomics

Diatoms are photosynthetic microalgae that can store energy in the form of chrysolaminarin or as lipids. Diatoms are most common types of phytoplankton (unicellular). So far this group of organisms has not been fully utilized due to the lack of genetic tools. *Phaeodactylum tricornutum* is a marine diatom that has been modified using both meganucleases and transcription activator-like effector nucleases. The whole genome sequence of the species has distinct features. It has the different combinations of specific metabolism and a number of diatom genes, which have specific application in biotechnology. In a specific genetic manipulation, a 45-fold increase in triacylglycerol (TAG) lipid content was achieved through the disruption of the UDP-glucose pyrophosphorylase gene. This strain will assure the mass production of biofuels as well as high value molecules (Daboussi et al., 2014; Russo et al., 2015). For example, the lipids can be extracted from the cells by a milking process without destroying the cells. The various techniques of milking the cells are spontaneous oozing, mechanical pressure, pulsed electric field and centrifugation. Understanding of cell stress is very essential to carrying out these techniques (Vinayak et al., 2015).

6.7.3 Radiation Breeding

Radiation breeding is another technique to grow algae using gamma rays with doses up to 800 Gy, with different exposure times. Choi et al. (2014) developed a high lipid-accumulating mutant strain (*S. dimorphus*) and found a 25% increase in lipid content and a faster growth rate compared to the wild type in the laboratory-scale culture.

6.7.4 Nitrogen Starvation

Light transfer plays a vital role in triglyceride fatty-acid content and productivity from microalgae during nitrogen starvation. Nitrogen starvation of microalgae produces more triglyceride fatty-acid for biofuels. Triglyceride fatty-acid productivity correlated with the mean volumetric rate of energy absorption and reached a maximum of 13 μmolh$v$/gs. Thus, triglyceride fatty-acid synthesis was limited by the photon absorption rate in the photobioreactor (Kandilian, 2014). Along with nitrogen starvation, starchless mutants liberate photosynthetic capacity more toward the production of superior triacylglycerol. Especially, *Scenedesmus obliquus* (slm1) yielded more than 51% of the triacylglycerol (Breuer et al., 2014).

6.7.5 Freshwater Microalga

Most freshwater microalgae isolated for biomass production are tested for their performance in low quality water, as water is the most precious resource and adds cost to the productivity. Because treated wastewater is abundant, microalgal growth characteristics are studied in wastewater medium. There are several reports of algal strains adapting to wastewater conditions. Six newly isolated microalgae strains were tested for growth, fatty acid methyl ester profiles, and biodiesel properties. It has been found that *Scenedesmus* SDEC-8 species produces with favorable fatty acids (FAs) of 73.43%. This is industrially pertinent as the selection of the right strains is of fundamental importance to the success of the algae-based oil industry (Song et al., 2014).

6.7.6 Growth Stimulant Lighting

There are different techniques to stimulate growth, pigment, and other metabolite production in microalgal biomass being employed. Light as a stimulant is one of the most successful in microalgae in addition to salinity. *C. vulgaris* grow best in the blue LED illumination when compared to white fluorescent light (Atta et al., 2013). The increased lipid content productivity of *C. vulgaris* under blue light illumination at different photoperiods (12 h each of light and dark) using LED are encouraging for PBR applications. The effect of flashing light on the growth of *Chlorella* sp. as well as flow rate using PBR was carried out by Fathurrahman et al. (2013). This study was compared to bright culture conditions. The results showed that both parameters were insignificant on growth of *Chlorella* sp.

6.8 CONCLUSION

Microalgae and their applications as sources of renewable energy have a long history, which is not covered in this chapter. The improvements and current highlights with respect to commercial, successful, renewable, and environmentally friendly microalgae energy sources for humanity have been discussed. Algae are sources of nutritional supplement for humans, animals and fisheries, biofuels, for fertilizers, wastewater treatments, high-value chemicals, and carbon dioxide sequestration. From the survey, microalgae are a promising feedstock for the production of superior quantities of TAGs and FAs with new technologies for the production of biofuels and bioproducts. The major challenges would be bleaching, evaporation, temperature, lag period during hot and cold seasons, nutrients, stress, and other environmental factors for outdoor as well as indoor cultivations for the production of biomass and lipids. The production costs will very much depend on the systems considered. The various approaches detailed in this chapter could improve commercialization of microalgal biomass production, lipids, and pigments.

REFERENCES

Abu-Rezq TS, Al-Musallam L, Al-Shimmari J, and Dias P. 1999. Optimum production conditions for different high-quality marine algae. *Hydrobiologia* 403:97–107.

Ashokkumar V, Zainal Salam F, Tiwari ON, Chinnasamy S, Mohammed S, and Ani FN. 2015. An integrated approach for biodiesel and bioethanol production *from Scenedesmus bijugatus* cultivated in a vertical tubular photobioreactor. *Energy Conversion and Management* 101:778–786.

Atta M, Idris A, Bukhari A, and Wahidin S. 2013. Intensity of blue LED light: A potential stimulus for biomass and lipid content in fresh water microalgae Chlorella vulgaris. *Bioresource Technology* 148:373–378.

Bajpai R, Prokop A, and Zappi M, eds. 2014. *Algal Biorefineries. Volume 1: Cultivation of Cells and Products*, 160–161. Springer, Dordrecht.

Bondioli P, Della Bella L, Rivolta G, Chini Zittelli G, Bassi N, Rodolfi L, Casini D, Prussi M, Chiaramonti D, and Tredici MR. 2012. Oil production by the marine microalgae *Nannochloropsis* sp. F&M-M24 and *Tetraselmis suecica* F&M-M33. *Bioresource Technology* 114:567–572.

Breuer G, Jaeger L, Artus VPG, Martens DE, Springer J, Draaisma RB, Eggink G, Wijffels RH, and Lamers PP. 2014. Superior triacylglycerol (TAG) accumulation in starchless mutants of *Scenedesmus obliquus*: (II) evaluation of TAG yield and productivity in controlled photobioreactors. *Biotechnology for Biofuels* 12, no. 7:70.

Carvalho AP, Meireles LA, and Malcata FX. 2006. Microalgal reactors: A review of enclosed system designs and performances. *Biotechnology Progress* 22, no. 6:1490–1506.

Chacón-Lee TL, and González-Mariño GE. 2010. Microalgae for healthy foods—Possibilities and challenges. *Comprehensive Reviews in Food Science and Food Safety* 9 no. 6:655–675.

Chen C-Y, Chang J-S, Chang H-Y, Chen T-Y, Wu J-H, and Lee W-L. 2013. Enhancing microalgal oil/lipid production from *Chlorella sorokiniana* CY1 using deep-sea water supplemented cultivation medium. *Biochemical Engineering Journal* 77:74–81.

Chen C-Y, Yeh K-L, Aisyah R, Lee DJ, and Chang JS. 2011. Cultivation, photobioreactor design and harvesting of microalgae for biodiesel production: A critical review. *Bioresource Technology* 102:71–81.

Cheng Y, Zhou, WG, Gao CF, Lan K, Gao Y, and Wu Q. 2009. Biodiesel production from Jerusalem artichoke (*Helianthus Tuberosus* L.) tuber by heterotrophic microalgae *Chlorella protothecoides*. *Journal of Chemical Technology and Biotechnology* 84:777–781.

Chisti Y. 2007. Biodiesel from microalgae. *Biotechnology Advances* 25:294–306.

Chiu SY, Kao CY, Chen CH, Kuan TC, Ong SC, and Lin CS. 2008. Reduction of CO_2 by a high-density culture of *Chlorella* sp. in a semicontinuous photobioreactor. *Bioresource Technology* 99:3389–3396.

Chiu SY, Kao CY, Tsai MT, Ong SC, Chen CH, and Lin CS. 2009. Lipid accumulation and CO_2 utilization of *Nannochloropsis oculata* in response to CO_2 aeration. *Bioresource Technology* 100:833–838.

Choi JI, Yoon M, Joe M, Park H, Le SG, Han SJ, and Lee PC. 2014. Development of microalga Scenedesmus dimorphus mutant with higher lipid content by radiation breeding. *Bioprocess and Biosystems Engineering* 37, no. 12:2437–2444.

Christaki E, Florou-Paneri P, and Bonos E. 2011. Microalgae: A novel ingredient in nutrition. *International Journal of Food Sciences and Nutrition* 62, no. 8:794–799.

Daboussi F, Leduc S, Mare´chal A, Dubois G, Guyot V, Perez-Michaut C, Amato A et al. 2014. Genome engineering empowers the diatom *Phaeodactylum tricornutum* for biotechnology. *Nature Communications* 1–7.

Darzins A, Pienkos P, and Edye L. 2010. Current status and potential for algal biofuels production: A report to IEA Bioenergy Task 39. National Renewable Energy Laboratory, Food and Agriculture Organization of the United Nations.

El-Kassas HY. 2013. Growth and fatty acid profile of the marine microalga *Picochlorum* sp. grown under nutrient stress conditions. *Egyptian Journal of Aquatic Research* 39:233–239.

Fathurrahman L, Hajar AH, Sakinah DW, Nurhazwani Z, and Ahmad J. 2013. Flashing light as growth stimulant in cultivation of green microalgae, *Chlorella sp.* utilizing airlift photobioreactor. *Pakistan Journal of Biological Sciences* 16, no. 22:1517–1523.

Feng D, Chen Z, Xue S, and Zhang W. 2011. Increased lipid production of the marine oleaginous microalgae *Isochrysis zhangjiangensis* (Chrysophyta) by nitrogen supplement. *Bioresource Technology* 102:6710–6716.

Gallagher BJ. 2011. The economics of producing biodiesel from algae. *Renewable Energy* 36:158–162.

Grima EM, Belarbi E-H, Acién Fernández FG, Medina AR, and Chisti Y. 2003. Recovery of microalgal biomass and metabolites: Process options and economics. *Biotechnology Advances* 20:491–515.

Hejazi MA, and Wijffels RH. 2004. Milking of microalgae. *Trends in Biotechnology* 22, no. 4:189–194.

Ho SH, Nakanishi A, Ye X, Chang JS, Hara K, Hasunuma T, and Kondo A. 2014. Optimizing biodiesel production in marine *Chlamydomonas* sp. JSC 4 through metabolic profiling and an innovative salinity-gradient strategy. *Biotechnology for Biofuels* 7:97.

Huang J, Li Y, Wan M, Yan Y, Feng F, Qu X, Wang J et al. 2014. Novel flat-plate photobioreactors for microalgae cultivation with special mixers to promote mixing along the light gradient. *Bioresource Technology* 159:8–16.

Kandilian R, Pruvost J, Legrand J, and Pilon L. 2014. Influence of light absorption rate by *Nannochloropsis oculata* on triglyceride production during nitrogen starvation. *Bioresource Technology* 163:308–319.

Kovacevic V, and Wesseler J. 2010. Cost-effectiveness analysis of algae energy production in the EU. *Energy Policy* 38:5749–5757.

Kunjapur AM, and Eldridge RB. 2010. Photobioreactor design for commercial biofuel production from microalgae. *Industrial and Engineering Chemistry Research* 49:3516–3526.

Kuo C-M, Chen T-Y, Lin T-H, Lai JT, Chang JS, and Lin CS. 2015. Cultivation of *Chlorella* sp. GD using piggery wastewater for biomass and lipid production. *Bioresource Technology* 194:326–333.

Li, XF, Xu H, and Wu Q. 2007. Large-scale biodiesel production from microalga *Chlorella protothecoides* through heterotropic cultivation in bioreactors. *Biotechnology Bioengineering* 98:764–771.

Li Y, Horsman M, Wang B, Wu N, and Lan CQ. 2008. Effects of nitrogen sources on cell growth and lipid accumulation of green alga Neochloris oleoabundans. *Applied Microbiology and Biotechnology* 81:629–636.

Lorenz RT, and Cysewski GR. 2000. Commercial potential for *Haematococcus* microalgae as a natural source of astaxanthin. *Trends in Biotechnology* 18, no. 4:160–7.

Mussgnug JH, Thomas-Hall S, Rupprecht J, Foo A, Klassen V, McDowall A, Schenk PM, Kruse O, and Hankamer B. 2007. Engineering photosynthetic light capture: Impacts on improved solar energy to biomass conversion. *Plant Biotechnology Journal* 5 no. 6:802–814.

Ojo EO, Auta H, Baganz F, and Lye GJ. 2014. Engineering characterisation of a shaken, single-use photobioreactor for early stage microalgae cultivation using *Chlorella sorokiniana*. *Bioresource Technology* 173:367–375.

Pruvost J, Cornet JF, Borgne FL, Goetz V, and Legrand J. 2015. Theoretical investigation of microalgae culture in the light changing conditions of solar photobioreactor production and comparison with cyanobacteria. *Algal Research* 10:87–99.

Pulz O, and Gross W. 2004. Valuable products from biotechnology of microalgae. *Applied Microbiology and Biotechnology* 65, no. 6:635–648.

Raheem A, Wan Azlina WAKG, Taufiq Yap YH, and Danquah MK. 2015. Thermochemical conversion of microalgal biomass for biofuel production. *Renewable and Sustainable Energy Reviews* 49:990–999.

Ratledge C. 2004. Fatty acid biosynthesis in microorganisms being used for single cell oil production. *Biochimie* 86, no. 11:807–815.

Richmond A, and Tomaselli L. 2007. *Handbook of Microalgal Culture: Biotechnology and Applied Phycology*, 1–19. John Wiley & Sons, Oxford, UK.

Richmond A. 1986. *Handbook of Microalgal Mass Culture*. Boca Raton: CRC Press.

Rukminasari N. 2013. Effect of nutrient depletion and temperature stressed on growth and lipid accumulation in marine-green algae *Nannochloropsis* sp. *American Journal of Research Communication* 2, no. 5:82–87.

Russo MT, Annunziata R, Sanges R, Ferrante MR, and Falciatore A. 2015. The upstream regulatory sequence of the light harvesting complex Lhcf2 gene of the marine diatom *Phaeodactylum tricornutum* enhances transcription in an orientation- and distance-independent fashion. *Marine Genomics* 24:69–79.

Shen Y, Yuan W, Pe ZJ, Wu Q, and Mao E. 2009. Microalgae mass production methods. *Trans. ASABE* 52:1275–1287.

Soletto D, Binaghi L, Ferrari L, Lodi A, Carvalho JCM, Zilli M, and Concwerti A. 2008. Effects of carbon dioxide feeding rate and light intensity on the fed-batch pulse-feeding cultivation of *Spirulina platensis* in helical photobioreactor. *Biochemical Engineering Journal* 39:369–375.

Song M, Pei H, Hu W, Zhang S, Ma G, Han L, and Ji Y. 2014. Identification and characterization of a freshwater microalga Scenedesmus SDEC-8 for nutrient removal and biodiesel production. *Bioresource Technology* 162:129–135.

Spolaore P, Joannis-Cassan C, Duran E, and Isambert A. 2006. Commercial applications of microalgae. *Journal of Bioscience and Bioengineering* 101, no. 2:87–96.

Suh IS, and Lee, CG. 2003. Photobioreactor engineering: Design and performance. *Biotechnology and Bioprocess Engineering* 8, no. 6:313–321.

Takagi M, Watanabe K, Yamaberi K, and Yoshida T. 2000. Limited feeding of potassium nitrate for intracellular lipid and triglyceride accumulation of *Nannochloris* sp. UTEX LB1999. *Applied Microbiology and Biotechnology* 54:112–117.

Tredici M, and Materassi R. 1992. From open ponds to vertical alveolar panels: The Italian experience in the development of reactors for the mass cultivation of phototrophic microorganisms. *Journal of Applied Phycology* 4, no. 3:221–231.

Vinayak V, Manoylov KM, Gateau H, Blanckaert V, Herault J, Pencreach G, Marchand J, Gordon R, and Schoefs B. 2015. Diatom milking: A review and new approaches. *Marine Drugs* 13, no. 5:2629–2665.

Xiong W, Li XF, Xiang JY, and Wu QY. 2008. High-density fermentation of microalga *Chlorella protothecoides* in bioreactor for microbio-diesel production. *Applied Microbiology and Biotechnology* 78:29–36.

Xu H, Miao XL, and Wu QY. 2006. High quality biodiesel production from a microalga *Chlorella protothecoides* by heterotrophic growth in fermenters. *Journal of Biotechnology* 126:499–507.

Yu B, Wang J, Suter PM, Russel RM, Grusak MA, Wang Y, Wang Z, Yin S, and Tang G. 2012. Spirulina is an effective dietary source of zeaxanthin to humans. *British Journal of Nutrition* 108, no. 4:611–619.

CHAPTER 7

Integrated Biological System for the Treatment of Sulfate-Rich Wastewater

K. Vasantharaj, Raja Sivashankar, C. Vigneshwaran, and Velmurugan Sivasubramanian

CONTENTS

Abstract

Sulfate, a major pollutant, increases the salinity of receiving water bodies, which in turn, reduces the availability of potable and usable water. In addition to brackish and saline water, various sulfate-laden wastewaters are also produced from industrial processes, including pulp and paper, fermentation, pharmaceutical production, food production, tannery operations, petrochemical manufacturing, and mining. In anaerobic biological sulfate reduction, sulfate-reducing bacteria (SRB) utilize the sulfate as an electron acceptor, reducing it to sulfide, which is corrosive to metals and is toxic to living species. A sulfide removal step is essential to proper treatment of sulfate-laden wastewaters. The biological sulfide oxidation process is accomplished by the respiratory activity of various chemolithotrophic sulfide-oxidizing bacteria (SOB), using a reduced sulfur compound as the electron donor, oxygen as the electron acceptor, and CO_2 as the carbon source. Integration of the sulfate-to-sulfide conversion step by SRB and sulfide-to-elemental sulfur conversion by SOB into one single reactor is of practical interest. This chapter illustrates the suitable biological system coupling sulfate reduction and sulfide oxidation with maximum recovery of elemental sulfur.

7.1 INTRODUCTION

The natural and anthropogenic source of sulfur contributes to environmental pollution. The natural source of sulfur includes a stable form of reduced pyrite (FeS_2) and oxidized gypsum ($CaSO_4$) in sediment. Seawater also contains sulfate ions generated photochemically from volcanic SO_2 and H_2S (Farquhar and Wing 2003). Seawater contains about 2700 mg of sulfate per liter (Hitchcock 1975). The presence of sulfate increases the salinity of receiving waterbodies, thereby reducing the availability of potable and usable water (Pulles et al. 1995). According to GEMS/WATER, a global network of water monitoring stations, typical sulfate levels in freshwater are in the vicinity of 20 mg/liter and range from 0 to 630 mg/liter in rivers, from 2 to 250 mg/liter in lakes, and from 0 to 230 mg/liter in groundwater (United Nations Environment Programme [UNEP] 1990). The average daily intake of sulfate from drinking water, air, and food is approximately 500 mg, with food being the major source. However, in areas with drinking-water supplies containing high levels of sulfate, drinking water may constitute the principal source of intake. Cathartic effects are commonly reported by people consuming drinking water containing sulfate in concentrations exceeding 600 mg/liter (United States Department of Health, Education and Welfare 1962), although it is also reported that humans can adapt to higher concentrations with time (United

States Environmental Protection Agency 1985). Dehydration has also been reported as a common side effect following the ingestion of large amounts of magnesium or sodium sulfate (Fingl 1980).

At present, the generation of wastewaters is rapidly increased with an increase in global population and industrial activities (Cenni et al. 2001). Acid mine drainage (AMD) rich in sulfate and heavy metals is one of the major sulfate-rich wastewater generated from mining industries. AMD is generated on subsequent weathering of waste rock, tailings, and surfaces exposed on mining. The AMD generated causes the acidification and contamination of the groundwater with heavy metals. When discharged untreated, it poses a threat to the quality of freshwater resources and consequently the well-being of humans and the environment at large. Its management thus requires attention (Naicker et al. 2003). Industrial activities such as flue-gas scrubbing, galvanic processes, battery, paint, and chemical manufacturing discharge effluents with similar characteristics also generates a large volume of sulfate- and metal-laden wastewaters (Jong and Parry 2003).

Hydrogen sulfide (H_2S) is a highly toxic, corrosive, and flammable gas with an unpleasant odor (Visser et al. 1997). H_2S is toxic to aquatic animal life in very low concentrations. The threshold limit value for freshwater or saltwater fish is 0.5 ppm (Environmental Protection Service 1984). H_2S is emitted into the atmosphere through volcanic activities and evaporation from oceanic waters. H_2S is emitted into the atmosphere as dissolved sulfide in wastewaters and H_2S in waste gases (Brimblecombe et al. 1989). H_2S is a major environmental contaminant from many industrial effluents, including petroleum refining (Henshaw 1990), pulp and paper manufacturing (Wani et al. 1999), and food processing (Chung et al. 2001). Sulfide-rich wastewaters can be treated using physicochemical methods, such as Claus, Alkanolamine, Lo-Cat, and Holmes-Stretford (Iliuta and Larachi 2003), and biological processes. Since the physicochemical methods are costly and energy intensive, biological sulfide oxidation is mostly preferred for the treatment of sulfide-rich wastewaters.

Biological sulfate reduction results in sulfide, which is toxic and corrosive. Hence, biological sulfate reduction coupled with biological sulfide oxidation is found to be suitable for effectively treating sulfate-rich wastewaters. However, coupling both processes in a single reactor is of practical interest. This chapter highlights the integrated biological systems used for treating sulfate-rich wastewaters.

7.2 THE SULFUR CYCLE

Sulfur is one of the most abundant elements on Earth. Sulfur, a multivalent nonmetal, occurs in different oxidation states such as −2 (sulfide and reduced organic sulfur), 0 (elemental sulfur), +2 (thiosulfite), +4 (sulfite), and +6 (sulfate), of which sulfate is the most significant in nature (Muyzer and Stams 2008). The transformation of sulfur from one state to another occurs by means of chemical or biological agents. The biological sulfur transformation carried out by microorganisms comprises many oxidation-reduction reactions as shown in Figure 7.1. Dissimilatory sulfate reduction is a form of anaerobic respiration that uses sulfate as the terminal electron acceptor. It is done by a group of strict anaerobic microorganisms, both bacteria and archaea, often termed sulfate-reducing bacteria (SRB). The dissimilatory sulfur reduction is a process that uses elemental sulfur as the electron acceptor. Assimilatory sulfate reduction is a process where the reduced sulfide is assimilated in biomass, proteins, amino acids, and cofactors by plants, fungi, and microorganisms (Sanchez-Andrea et al. 2014). Phototrophic or chemolithothrophic bacteria use reduced sulfur compounds such as sulfide as electron donors and convert these compounds to elemental sulfur or sulfate, and are called sulfide-oxidizing bacteria (SOB) (Robertson and Kuenen 2006). In sulfur

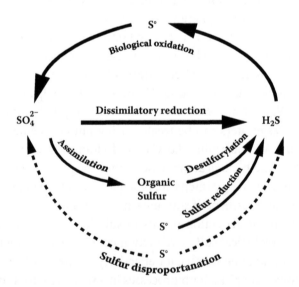

FIGURE 7.1 Biological sulfur transformations. (From Tang, K. et al., *Biochemical Engineering Journal*, 44:73–94, 2009.)

disproportionation, elemental sulfur or thiosulfate functions both as electron donor and electron acceptor, resulting in simultaneous formation of sulfate and sulfide. Sulfur disproportionation is an energy-generating process that is carried out by some species of sulfate-reducing bacteria and other highly specialized bacteria (Bak and Pfennig 1987).

7.3 SULFUR BACTERIA

7.3.1 Sulfate-Reducing Bacteria (SRB)

Sulfate-reducing bacteria (SRB) are important members of microbial communities, with economic, environmental, and biotechnological interest. They can exist in a variety of environments such as soils; sediments; and domestic, industrial, and mining wastewaters (Y. Chang et al. 2001). SRB are a group of obligate anaerobes that are able to survive in the anoxic environments containing organic substances and sulfate. SRB utilize organic substances or hydrogen as electron donors and reduces sulfate to sulfide in accordance with Equation 7.1 (Postgate 1984). Some species of SRB may utilize thiosulfate and sulfite as electron acceptors. Sulfur-reducing bacteria are the other group of obligate anaerobes that involves reduction of sulfur to sulfide but not sulfate to sulfide. Examples include *Desulfuromonas*, *Desulfurella*, *Sulfurospirrilium*, and *Campylobacter* (Rabus et al. 2006).

$$SO_4^{2-} + 8e^- + 4H_2O \rightarrow S^{2-} + 8OH^- \tag{7.1}$$

7.3.2 Sulfide-Oxidizing Bacteria (SOB)

Sulfide-oxidizing bacteria are classified into photoautotrophs and chemolithotrophs. CO_2 is used as a terminal electron acceptor in the case of photoautotrophs, whereas oxygen (aerobic species) or nitrate and nitrite (anaerobic species) in case of chemolithotrophs (Pagella and De Faveri 2000). Hydrogen sulfide acts as an electron donor in both cases. Photoautotrophs include green sulfur bacteria such as *Chlorobium* and purple sulfur bacteria such as *Allochromatium* (Madigan et al. 2006). The reactions carried out by photoautotrophic bacteria is given in Equation 7.2, which is referred to as the van Niel reaction (Janssen et al. 1999; Madigan and Martinko 2006).

$$2H_2S + CO_2 \xrightarrow{\text{light}} 2S^0 + CH_2O(\text{carbohydrate}) + H_2O \tag{7.2}$$

Chemolithotrophs include colorless sulfur bacteria such as *Thiobacillus, Acidithiobacillus, Achromatium, Beggiatoa, Thiothrix, Thioplaca, Thiomicrospira, Thiosphaera,* and *Thermothrix* (Robertson and Kuenen 2006). The following equations show the reactions carried out by chemolithotrophs in oxidations of sulfide (Equations 7.3 and 7.4), sulfur (Equation 7.5), and thiosulfate (Equation 7.6) (Madigan and Martinko 2006). A list of sulfate reducing bacteria and sulfide oxidizing bacteria is shown in Table 7.1.

$$H_2S + \frac{1}{2}O_2 \rightarrow S^0 + H_2O \tag{7.3}$$

$$H_2S + 2O_2 \rightarrow SO_4^{2-} + 2H^+ \tag{7.4}$$

$$S^0 + \frac{3}{2}O_2 + H_2O \rightarrow SO_4^{2-} + 2H^+ \tag{7.5}$$

$$S_2O_3^{2-} + H_2O + 2O_2 \rightarrow 2SO_4^{2-} + 2H^+ \tag{7.6}$$

7.4 BIOLOGICAL SULFATE REDUCTION

Biological sulfate reduction (BSR) has a number of advantages over other processes and hence has been cited as a suitable method for the treatment of sulfate and metal-rich waters originating from the mining industry (I.S. Chang et al. 2000). Biological sulfate reduction is a process under anaerobic conditions where sulfate is converted into sulfide by sulfate reducing bacteria. Sulfide and bicarbonates are the end products of this process of which the dissolved metals are precipitated by the sulfide and the bicarbonate neutralizes acidity (Drury 1999). The mechanism of biological sulfate reduction by sulfate-reducing bacteria is shown in Figure 7.2. BSR occurs within the cell cytoplasm through a sequence of reactions catalyzed by three enzymes, namely, ATP sulfurylase, APS reductase, and sulfite reductase (Hansen et al. 1994), through an intermediate called sulfite (Shen and Buick 2004).

In this process, sulfate-reducing bacteria obtain energy by oxidizing organic compounds (carbon source) and utilize sulfate as a terminal electron acceptor (Drury 1999). In most cases the organic carbon source

TABLE 7.1   Sulfate-Reducing and Sulfide-Oxidizing Bacteria

Type		Name	Reference
	Sulfate-Reducing Bacteria		
Acetate oxidizers		*Desulfobacter*	Postgate 1984;
		Desulfobacterium	Rabus et al. 2006;
		Desulfococcus	Widdel 1988;
		Desulfonema	Colleran et al.
		Desulfosarcina	1995
		Desulfoarculus	
		Desulfoacinum	
		Desulforhabdus	
		Desulfomonile	
		Desulfotomaculum	
		acetoxidans	
		Desulfotomaculum	
		sapomandens	
		Desulfovibrio baarsii	
Nonacetate oxidizers		*Desulfovibrio*	Madigan and
		Desulfomicrobium	Martinko 2006;
		Desulfobotulus	Colleran et al.
		Desulfofustis	1995
		Desulfotomaculum	
		Desulfomonile	
		Desulfobacula	
		Archaeoglobus	
		Desulfobulbus	
		Desulforhopalus	
		Thermodesulfobacterium	
	Sulfide-Oxidizing Bacteria		
Photoautotrophs	Green sulfur	*Chlorobium*	Madigan and
	bacteria	*Prosthecochloris*	Martinko 2006
		Pelodictyon	
		Ancalochloris	
		Chloroherpeton	
	Purple sulfur	*Chromatium*	Friedrich et al.
	bacteria	*Thioalkalicoccus*	2001
		Thiorhodococcus	
		Thiocapsa	
		Thiocystis	
		Thiococcus	
		Thiospirillum	
		Thiodictyon	
		Thiopedia	

(Continued)

TABLE 7.1 (CONTINUED)   Sulfate-Reducing and Sulfide-Oxidizing Bacteria

Type		Name	Reference
Sulfide-Oxidizing Bacteria			
Chemolithotrophs	Colorless sulfur bacteria	*Thiobacillus* *Acidithiobacillus* *Achromatium* *Beggiatoa* *Thiothrix* *Thioplaca* *Thiomicrospira* *Thiosphaera* *Thermothrix*	Robertson and Kuenen 2006

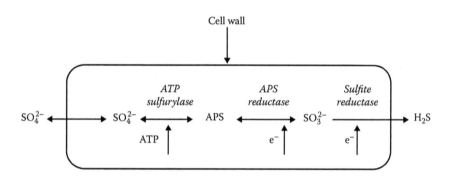

FIGURE 7.2   Pathway of biological sulfate reduction. (From Shen, Y., Buick, R., *Earth Science Reviews* 64:243–272, 2004.)

can also act as an electron donor (Postgate 1984). The electron donors may include molecular hydrogen, organic acids (such as formate, acetate, propionate, and butyrate), alcohols (such as methanol and ethanol), and aromatic compounds (Postgate 1984; Hansen 1988; Widdel 1988; Widdel and Hansen 1992; Colleran et al. 1995). Kaksonen (2004) and Oyekola (2008) reported that supplying a suitable carbon source and electron donor is significant in the biological sulfate reduction process. They also recommended lactate as a potential carbon source and electron donor. Sulfide toxicity has been reported to decrease when lactate was used as a substrate (Kuo and Shu 2004). Also Kaksonen et al. (2004) reported that using lactate increases the sulfate reduction and alkalinity generation. Organic waste such as straw and hay (Bechard et al. 1994), sawdust (Wakao et al. 1979), peat (Eger and Lapakko 1988), spent mushroom compost (Dvorak et al. 1992), and whey (Christensen et al. 1996) were also reported as electron

TABLE 7.2   Organic Carbon Sources Used for Biological Sulfate Reduction

Carbon Source	Sulfate Removal	Reference
Tannery effluent	60%–80%	Boshoff et al. 2004a
Microalgal biomass	–	Boshoff et al. 2004b
Sweet meat waste (filtered)	70%	Das et al. 2013
Marine waste extract	83%	Dev et al. 2015
Chicken manure	79.04%	Zhang and Wang 2014
Dairy manure	64.78%	Zhang and Wang 2014
Saw dust	50.27%	Zhang and Wang 2014
Wine wastes	>90%	Costa et al. 2009
Waste activated sludge without pretreatment	73.6%	Sheng et al. 2010
Waste activated sludge, alkaline pretreatment	27.6%	Sheng et al. 2010
Waste activated sludge, ultrasonic pretreatment	80.7%	Sheng et al. 2010
Waste activated sludge, thermal pretreatment	88.4%	Sheng et al. 2010
Rice husk: pig farm wastewater treatment sludge: coconut husk chips (60:20:20 by volume)	59%	Kijjanapanich et al. 2014

donors. Hence the choice of the carbon source used for the process plays a significant role in sulfate reduction. The various organic carbon sources used for biological sulfate reduction are shown in Table 7.2.

The initial sulfate concentration affects activity of sulfate-reducing bacteria and thereby the sulfate reduction process (White and Gadd 1996; Dries et al. 1998). The kinetics of anaerobic reduction of sulfate in continuous bioreactors was studied by Moosa et al. (2002). The sulfate reduction rate was found to increase from 0.007 kg m^{-3} h^{-1} to 0.17 kg m^{-3} h^{-1} with an increase in initial sulfate concentration in the range 1.0 kg m^{-3} to 10.0 kg m^{-3}. It was also reported that the initial sulfate concentration has no significant effect on the maximum specific growth rate (μ_m) and decay coefficient (k_d). The saturation constant (K_s) was found to be an increasing function of initial sulfate concentration. Hence the growth of SRB and the kinetics of anaerobic sulfate reduction were greatly affected by initial sulfate concentration.

7.5   BIOLOGICAL SULFIDE OXIDATION

Biological sulfide oxidation is an oxidation process carried out by photo-autotrophic or chemolithotrophic sulfide-oxidizing bacteria where sulfide is used as an electron donor. Here sulfide is oxidized into elemental sulfur or sulfate (Pagella and De Faveri 2000). Janssen et al. (1999) studied the biological oxidation of hydrogen sulfide from wastewater and waste gases by aerobic *Thiobacillus*-like bacteria. The colloidal and interfacial aspects of biologically produced sulfur particles were analyzed. The colloidal properties of a biological sulfur sol differ considerably from those of the well-studied standard LaMer sulfur sol.

The biological conversion of hydrogen sulfide to elemental sulfur by the green sulfur bacterium *Chlorobium thiosulfatophilum* in a fixed-film continuous flow photoreactor was reported by Henshaw and Zhu (2000). At sulfide loadings of 111–286 mg/h L, 100% of the influent sulfide was removed, where 92% to 95% of the influent sulfide was converted to elemental sulfur. In this study, the maximum sustainable sulfide loading rate of 286 mg/h L was reported. This loading rate was found to be higher than the previously reported value of 104 mg/h L. This loading rate was achieved at a hydraulic retention time of 0.5 h, and 100% of the influent sulfide was removed of which 92% was converted to elemental sulfur. The high efficiency of this reactor was attributed to the high reactor bacteria concentration and light intensity (light/volume).

Mojarrad Moghanloo et al. (2010) investigated the hydrogen sulfide removal in mineral media using *Thiobacillus thioparus* TK-1 in a biofilm airlift suspension reactor to evaluate the relationship between biofilm formation and changes in inlet sulfide loading rates. At a hydraulic residence time of 3.3 h, the maximum sulfide oxidation rate was found to be 6.7 mol S^{2-} m^{-3} h^{-1} in the mineral medium. The removal efficiency of sulfide was as high as 100% at a sulfide loading rate of 4.8 mol S^{2-} m^{-3} h^{-1}. Since the oxygen transfer is higher in case of an airlift reactor, all the inlet sulfide was completely oxidized into sulfate not elemental sulfur.

Under oxygen-limited condition (0.2–1.0 mg L^{-1}), the airlift reactor was operated to oxidize sulfide biologically into elemental sulfur by Lohwacharin and Annachhatre (2010). It was observed that gradual increase in volumetric sulfide loading rate could increase the production of elemental sulfur. When the sulfide loading rate was 2.2 kg S m^{-3} d^{-1}, 50% of influent sulfide was converted to elemental sulfur. At a maximum volumetric sulfide loading rate of 4.0 kg S m^{-3} d^{-1}, about 90% of influent sulfide was converted into elemental sulfur. The elemental sulfur produced was aggregated effectively by the addition of polyaluminum chloride as coagulant.

In addition to the dissolved oxygen concentration and sulfide loading rate, the carbon source for the growth of sulfide-oxidizing bacteria also plays a role in biological oxidation of sulfide to elemental sulfur. Wang et al. (2016) treated enriched sulfide and nitrogen-rich wastewaters using waste activated sludge (WAS) fermentation liquid as the carbon and energy source for the growth and activity of sulfide-oxidizing bacteria and nitrate-reducing bacteria under microaerobic condition. Waste activated sludge fermentation liquid was found to be more effective than glucose and an economic for treatment of enriched sulfide and nitrogen-rich wastewaters. This could be applied in further biological flue gas desulfurization and denitrification processes. A maximum generation efficiency of S^0 was achieved at a sulfide loading rate of 8.52 kg S m^{-3} d^{-1} and WAS of 25.13 kg COD m^{-3} d^{-1}.

7.6 COUPLED SULFATE REDUCTION AND SULFIDE OXIDATION

Biological sulfate reduction results in sulfide that is corrosive to metals and toxic to living species. For the effective treatment of sulfate-rich wastewaters, a sulfide removal step is essential. Biological sulfate removal through a reactor coupled with the sulfide oxidation process is the most desired (Celis-Garcia et al. 2007). Integration of the sulfate-to-sulfide conversion step by SRB and sulfide-to-elemental sulfur conversion by SOB into one single reactor is of practical interest. Okabe et al. (2005) reported that SRB and SOB can coexist in hydrothermal vents, microbial mats, marine sediments, and wastewater biofilms as a response to high organic input and low dissolved oxygen (DO) concentration. The level of dissolved oxygen has been proposed as an effective process parameter to regulate the activities of SRB and SOB (Okabe et al. 2005). Hence the elemental sulfur recovery can be enhanced in an integrated sulfate-reducing and sulfide-oxidizing reactor under microaerobic condition.

Elemental sulfur has its own advantages that it is noncorrosive and easy to handle and transport. Steam stripping followed by the Claus process or Holmes–Stretford process is the typical method used in petroleum refineries to recover elemental sulfur. Due to the need of replacing poisoned catalysts, contaminated reactor liquids, and corroded reaction vessels, these methods were found to be costly to operate. These processes were also found to be energy intensive. Elemental sulfur is commercially used in sulfuric acid production. Both elemental sulfur and sulfuric acid are used in chemical processing and fertilizer production (Cork et al. 1986). The advantages of elemental sulfur recovery through the biological method include high removal even at low sulfide concentrations and high elemental sulfur recovery (Henshaw and Zhu 2000). The integrated process of sulfate reduction and sulfide oxidation in a single reactor under microaerobic conditions for the treatment of sulfate-rich waste waters with elemental sulfur recovery is schematically shown in Figure 7.3.

An ethanol-fed, sulfate-reducing, four-compartment anaerobic baffled reactor was developed by Bekmezci et al. (2011) to treat synthetic acid mine drainage rich in sulfate (3.0–3.5 g L^{-1}) and various metals (Co, Cu, Fe, Mn, Ni, and Zn). The key parameters chosen were chemical oxygen demand (COD)/sulfate ratios, hydraulic retention times (HRT), pH, and metal concentrations. The sulfate reduction efficiency was found to be 88% with a feed sulfate concentration of 3.5 g L^{-1}, COD/sulfate mass ratio of 0.737, pH of 3.0, and HRT of 2 days without aeration in the fourth compartment. The alkalinity generation neutralizes synthetic acid mine drainage and hence the pH increased up to 7.0–8.0. Metal removal efficiencies were found to be greater than 99%, except for Mn. The proposed anaerobic baffled reactor provided advantages over commonly used reactor configurations, such as up-flow sludge blanket reactors (UASBs) and fluidized bed reactors (FBRs), in recovering valuable metals without interrupting the process. Aeration was provided to the fourth compartment where sulfide produced from sulfate is converted into elemental sulfur. The aeration rate and HRT were adjusted to oxidize the produced sulfide to elemental sulfur and not to sulfates, and observed that 32% to 74% of produced sulfide was oxidized to elemental sulfur. Hence, sulfate reduction, metal removal, alkalinity generation, and excess sulfide oxidation were achieved in an integrated anaerobic baffled reactor.

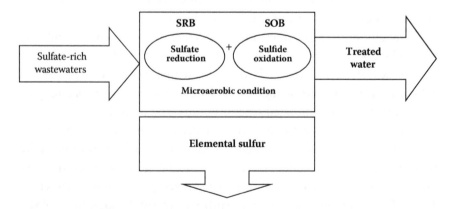

FIGURE 7.3 Elemental sulfur recovery in an integrated system.

Sulfate reduction and sulfide oxidation in a single reactor without any compartments is desired, but the problem of lower elemental sulfur conversion needs to manipulate the dissolved oxygen concentration. The activity of sulfide-oxidizing bacteria is achieved by supplying oxygen whose concentration should not inhibit the activity of sulfate-reducing bacteria. Hence, an integrated reactor coupling sulfate reduction and sulfide oxidation under microaerobic condition (DO = 0.10–0.12 mg L^{-1}) was operated by Xu et al. (2012). At the given dissolved oxygen concentration the removal efficiency for sulfate and recovery of elemental sulfur both peaked at 81.5% and >70%, respectively. When the dissolved oxygen concentration started to increase, sulfide-oxidizing bacteria competed with the chemical oxidation of sulfide with molecular oxygen. Similarly when DO exceed 30 mg L^{-1}, the integrated system failed to operate because of the inhibition of sulfate-reducing bacteria.

The absence of dissolved oxygen also promotes the conversion of sulfide to elemental sulfur. This is based on the fact that the oxidation of sulfides can take place by other intermediates such as sulfites in absence of dissolved oxygen. Since acid mine drainage and other sulfate-laden wastewaters are acidic and contain high metal concentrations, the immobilization of cells proves to be advantageous. The application of these two facts was employed to treat sulfate-rich wastewaters in packed bed reactors and reported by Brahmacharimayum and Ghosh (2014). Elemental sulfur was found to be formed in this process in the absence of dissolved oxygen. The factors chosen were HRT, COD/SO_4^{2-} ratio, and SO_4^{2-} concentration. The results showed that a maximum SO_4^{2-} reduction of 97% could be achieved in the packed bed reactor when operated at an HRT of 24 h with an SO_4^{2-} concentration of 1500 mg/L and COD/SO_4^{2-} ratio of 0.67.

7.7 CONCLUSION

This chapter illustrated the integrated biological system for treating sulfate wastewaters coupled with sulfide oxidation to recover elemental sulfur. The generation of wastewater is rapidly increasing with an increase in the global population and industrial activities. Acid mine drainage from mining industries and sour water from petroleum industries are the major sources of sulfate and sulfide-rich wastewaters, respectively. The physicochemical methods are employed in these industries to treat the wastewaters and recover elemental sulfur. But these methods are found to be costly and energy intensive. An alternate biological removal process proved to be a suitable method of treating sulfate-laden wastewaters even when their concentrations are very low. The biological sulfate removal process results in sulfide formation, which is toxic and corrosive. Hence sulfate removal is coupled with the sulfide oxidation process. The integration of biological sulfate reduction and sulfide oxidation in a single reactor is of practical interest, so as to reduce the bioreactor configuration and thereby reduce their installation costs. Parameters like dissolved oxygen and sulfide loading rate have been studied in the integrated system to maximize elemental sulfur recovery. Elemental sulfur has commercial application such as fertilizer production. Hence, the integrated system is found to be an economic way of treating sulfate-rich wastewaters and a better solution for the mining and petroleum industries.

REFERENCES
Bak, F., Pfennig, N. 1987. Chemolithotrophic growth of *Desulfovibrio sulfodismutans* sp. nov. by disproportionation of inorganic sulfur compounds. *Archives of Microbiology* 147:184–189.

Bechard, G., Yamazaki, H., Gould, W.D., Bedard, P. 1994. Use of cellulosic substrates for the microbial treatment of acid mine drainage. *Journal of Environmental Quality* 23:111–116.

Bekmezci, O.K., Ucar, D., Kaksonen, A.H., Sahinkaya, E. 2011. Sulfidogenic biotreatment of synthetic acid mine drainage and sulfide oxidation in anaerobic baffled reactor. *Journal of Hazardous Materials* 189:670–676.

Boshoff, G., Duncan, J., Rose, P.D. 2004a. Tannery effluent as a carbon source for biological sulphate reduction. *Water Research* 38:2651–2658.

Boshoff, G., Duncan, J., Rose, P.D. 2004b. The use of micro-algal biomass as a carbon source for biological sulphate reducing systems. *Water Research* 38:2659–2666.

Brahmacharimayum, B., Ghosh, P.K. 2014. Sulfate bioreduction and elemental sulfur formation in a packed bed reactor. *Journal of Environmental Chemical Engineering* 2:1287–1293.

Brimblecombe, P., Hammer, C., Rohde, H., Ryaboshapko, A., Boutron, C. 1989. Human influence on the sulphur cycle. In *Evolution of the Global Biogeochemical Sulphur Cycle* (Scope 39), edited by P. Brimblecombe, A. Yu Lein, 77. New York: Wiley.

Celis-Garcia, L.B., Razo-Flores, E., Monroy, O. 2007. Performance of a downflow fluidized-bed reactor under sulfate reduction conditions using volatile fatty acids as electron donors. *Biotechnology and Bioengineering* 97: 771–780.

Cenni, R., Janisch, B., Spliethoff, H., Hein, K.R.G. 2001. Legislative and environmental issues on the use of ash from coal and municipal sewage sludge co-firing as construction material. *Waste Management* 21:17–31.

Chang, I.S., Shin, P.K., Kim, B.H. 2000. Biological treatment of acid mine drainage under sulphate-reducing conditions with solid waste materials as substrate. *Water Research* 34:1269–1277.

Chang, Y., Peacock, A.D., Long, P., Stephen, J.R., McKinley, J.P., Macnaughton, S.J., Hussain, A.K.M., Saxton, A.M., White, D.C. 2001. Diversity and characterization of sulfate-reducing bacteria in groundwater at a uranium mill tailings site. *Applied and Environmental Microbiology* 67:3149–3160.

Christensen, B., Laake, M., Lien, T. 1996. Treatment of acid mind water by sulfate-reducing bacteria: Results from a bench scale experiment. *Water Research* 30:1617–1624.

Chung, Y.C., Huang, C., Tseng, C.P. 2001. Biological elimination of H_2S and NH_3 from waste gases by biofilter packed with immobilized heterotrophic bacteria. *Chemosphere* 43:1043–1050.

Colleran, E., Finnegan, S., Lens, P. 1995. Anaerobic treatment of sulphate-containing waste streams. *Antonie van Leeuwenhoek* 67:29–46.

Cork, D.J., Jerger, D.E., Maka, A. 1986. Biocatalytic production of sulfur from process waste streams. *Biotechnology and Bioengineering* 16:149–162.

Costa, M.C., Santos, E.S., Barros, R.J., Pires, C., Martins, M. 2009. Wine wastes as carbon source for biological treatment of acid mine drainage. *Chemosphere* 75:831–836.

Das, B.K., Gauri, S.S., Bhattacharya, J. 2013. Sweetmeat waste fractions as suitable organic carbon source for biological sulfate reduction. *International Biodeterioration & Biodegradation* 82:215–223.

Dev, S., Roy, S., Das, D., Bhattacharya, J. 2015. Improvement of biological sulfate reduction by supplementation of nitrogen rich extract prepared from organic marine wastes. *International Biodeterioration & Biodegradation* 104:264–273.

Dries, J., De Smul, A., Goethals, L., Grootaerd, H., Verstraete, W. 1998. High rate biological treatment of sulphate-rich wastewater in an acetate-fed EGSB reactor. *Biodegradation* 9:103–111.

Drury, W.J. 1999. Treatment of acid mine drainage with anaerobic solid substrate reactors. *Water and Environmental Research* 71:1244–1250.

Dvorak, D.H., Hedin, R.S., Edenborn H.M., McIntire, P.E. 1992. Treatment of metal-contaminated water using bacterial sulfate reduction: Results from pilot-scale reactors. *Biotechnology and Bioengineering* 40:609–616.

Eger, P., Lapakko, K. 1988. Nickel and copper removal from mine drainage by a natural wetland. In *Mine Drainage and Surface Mine Reclamation*, IC 9183, 1:301–309. Bureau of Mines, US Department of the Interior, Pittsburg, Pennsylvania.

Environmental Protection Service (EPS). 1984. *Hydrogen Sulphide: Technical Information for Problem Spills*. Ottawa, Canada: Environmental Protection Service, Environment Canada.

Farquhar, J., Wing, B.A. 2003. Multiple sulfur isotopes and the evolution of the atmosphere. *Earth and Planetary Science Letters* 213:1–13.

Fingl, E. 1980. Laxatives and cathartics. In *Pharmacological Basis of Therapeutics*, edited by L.S. Goodmand, A.G. Gilman. New York: MacMillan.

Friedrich, C.G., Rother, D., Bardischewsky, F., Quentmeier, A., Fischer, J. 2001. Oxidation of reduced inorganic sulfur compounds by bacteria: Emergence of a common mechanism? *Applied Environmental Microbiology* 67:2873–2882.

Hansen, T.A. 1988. Physiology of sulphate-reducing bacteria. *Microbiological Sciences* 5:81–84.

Hansen, T.A. 1994. Metabolism of sulfate-reducing prokaryotes. *Antonie Van Leeuwenhoek* 66:165–185.

Henshaw, P.F. 1990. Biological removal of hydrogen sulfide from refinery wastewater and conversion to elemental sulfur. MASc thesis, University of Windsor, ON, Canada.

Henshaw, P.F., Zhu, W. 2000. Biological conversion of hydrogen sulphide to elemental sulphur in a fixed-film continuous flow photo-reactor. *Water Research* 35, no. 15:3605–3610.

Hitchcock, D.R. 1975. Biogenic contributions to atmospheric sulphate levels. In *Proceedings of the 2nd National Conference on Complete Water Re-use*. Chicago: American Institute of Chemical Engineers.

Iliuta, I., Larachi, F. 2003. Concept of bifunctional redox iron-chelate process for H_2S removal in pulp and paper atmospheric emissions. *Chemical Engineering Science* 58:5305–5314.

Janssen, A.J.H., Lettinga, G., de Keizer, A. 1999. Removal of hydrogen sulphide from wastewater and waste gases by biological conversion to elemental sulphur: Colloidal and interfacial aspects of biologically produced sulphur particles. *Colloids and Surfaces A: Physicochemical and Engineering Aspects* 151:389–397.

Jong, T., Parry, D.L. 2003. Removal of sulfate and heavy metals by sulfate reducing bacteria in short-term bench scale upflow anaerobic packed bed reactor runs. *Water Research* 37:3379–3389.

Kaksonen, A. 2004. The performance, kinetics and microbiology of sulfidogenic fluidized bed reactors treating acidic metal- and sulphate-containing wastewater. DTech thesis, Tampere University of Technology, Tampere, Finland.

Kaksonen, A.H., Plumb, J.J., Franzmann, P.D., Puhakka, J.A. 2004. Simple organic electron donors support diverse sulfate-reducing communities in fluidized-bed reactors treating acidic metal- and sulfate-containing wastewater. *FEMS Microbiology Ecology* 47:279–289.

Kijjanapanich, P., Annachhatre, A.P., Esposito, G., Lens, P.N.L. 2014. Use of organic substrates as electron donors for biological sulfate reduction in gypsiferous mine soils from Nakhon Si Thammarat (Thailand). *Chemosphere* 101:1–7.

Kuo, W.-C., Shu, T.-Y. 2004. Biological pre-treatment of wastewater containing sulfate using anaerobic immobilized cells. *Journal of Hazardous Materials* 113:147–155.

Lohwacharin, J., Annachhatre, A.P. 2010. Biological sulfide oxidation in an airlift bioreactor. *Bioresource Technology* 101:2114–2120.

Madigan, M.T., Martinko, J.M. 2006. *Brock Biology of Microorganisms*. Upper Saddle River, NJ: Prentice Hall.

Mojarrad Moghanloo, G.M., Fatehifar, E., Saedy, S., Aghaeifar, Z., Abbasnezhad, H. 2010. Biological oxidation of hydrogen sulfide in mineral media using a biofilm airlift suspension reactor. *Bioresource Technology* 101:8330–8335.

Moosa, S., Nemati, M., Harrison, S.T.L. 2002. A kinetic study on anaerobic reduction of sulphate. Part I: Effect of sulphate concentration. *Chemical Engineering Science* 57:2773–2780.

Muyzer, G., Stams, A.J.M. 2008. The ecology and biotechnology of sulphate-reducing bacteria. *Nature Reviews Microbiology* 6:441–454.

Naicker, K., Cukrowska, E., McCarthy, T.S. 2003. Acid mine drainage arising from gold mining activity in Johannesburg, South Africa and environs. *Environmental Pollution* 122:29–40.

Okabe, S., Ito, T., Sugita, K., Satoh, H. 2005. Succession of internal sulfur cycles and sulfur-oxidizing bacterial communities in microaerophilic wastewater biofilms. *Applied and Environmental Microbiology* 71:2520–2529.

Oyekola, O.O. 2008. An investigation into the relationship between process kinetics and microbial community dynamics in a lactate-fed sulphidogenic CSTR as a function of residence time and sulphate loading. PhD thesis, University of Cape Town, South Africa.

Pagella, C., De Faveri, D.M. 2000. H_2S gas treatment by iron bioprocess. *Chemical Engineering Science* 55:2185–2194.

Postgate, J.R. 1984. *The Sulphate-Reducing Bacteria*. Cambridge, UK: Cambridge University Press.

Pulles, W., Howie, D., Otto, D., Easton, J. 1995. A manual on mine water treatment and management practices in South Africa. Water Research Commission, report no. TT 80/96.

Rabus, R., Hansen, T.A., Widdel, F. 2006. Dissimilatory sulfate- and sulfur-reducing. In *The Prokaryotes*, edited by M. Dworkin, S. Falkow, E. Rosenberg, K.H. Schleifer, E. Stackebrandt, 659–768. New York: Springer.

Robertson, L.A., Kuenen, J.G. 2006. The genus *Thiobacillus*. In *The Prokaryotes*, edited by M. Dworkin, S. Falkow, E. Rosenberg, K.H. Schleifer, E. Stackebrandt, 812–827. New York: Springer.

Sanchez-Andrea, I., Sanz, J.L., Bijmans, M.F.M., Stams, A.J.M. 2014. Sulfate reduction at low pH to remediate acid mine drainage. *Journal of Hazardous Materials* 269:98–109.

Shen, Y., Buick, R. 2004. The antiquity of microbial sulfate reduction. *Earth Science Reviews* 64:243–272.

Sheng, Y., Cao, H., Li, Y., Zhang, Y. 2010. Effects of various pretreatments on biological sulfate reduction with waste activated sludge as electron donor and waste activated sludge diminution under biosulfidogenic condition. *Journal of Hazardous Materials* 179:918–925.

Tang, K., Baskaran, V., Nemati, M. 2009. Bacteria of the sulphur cycle: An overview of microbiology, biokinetics and their role in petroleum and mining industries. *Biochemical Engineering Journal* 44:73–94.

United Nations Environment Programme (UNEP). 1990. GEMS/Water data summary (1985–1987). Burlington, Ontario, Canada Centre for Inland Waters; United Nations Environment Programme, Global Environment Monitoring System, GEMS/Water Programme Office.

United States Department of Health, Education and Welfare (US DHEW). 1962. Drinking water standards—1962. Washington, DC, US Department of Health, Education and Welfare, Public Health Service; US Government Printing Office (Publication No. 956).

United States Environmental Protection Agency (US EPA). 1985. National primary drinking water regulations; synthetic organic chemicals, inorganic chemicals and microorganisms; proposed rule. US Environmental Protection Agency. *Federal Register* 50, no. 219:46936.

Visser, J.M., Stefess, G.C., Kuenen, J.G. 1997. Thiobacillus sp.w5, the dominant autotroph oxidizing sulfide to sulfur in a reactor for aerobic treatment of sulfidic wastes. *Antonie van Leeuwenhoek* 72, no. 2:127–134.

Wakao, N., Takahashi, T., Sakurai, Y., Shiota, H. 1979. A treatment of acid mine water using sulfate-reducing bacteria. *Journal of Fermentation Technology* 57:445–452.

Wang, X., Zhang, Y., Zhang, T., Zhou, J., Chen, M. 2016. Waste activated sludge fermentation liquid as carbon source for biological treatment of sulfide and nitrate in microaerobic conditions, *Chemical Engineering Journal* 283:167–174.

Wani, A.H., Lau, A.K., Branion, R.M.R. 1999. Biofiltration control of pulping odors–hydrogen sulfide: Performance, macrokinetics and coexistence effects of organo-sulfur species. *Journal of Chemical Technology and Biotechnology* 74:9–16.

White, C., Gadd, G.M. 1996. Mixed sulphate-reducing bacterial cultures for bio-precipitation of toxic metals: Factorial and response surface analysis of the effects of dilution rate, sulphate and substrate concentration. *Microbiology* 142:2197–2205.

Widdel, F. 1988. Microbiology and ecology of sulphate and sulphur reducing bacteria. In *Biology of Anaerobic Microorganisms,* edited by A.J.B. Zehnder, 469–586. New York: Wiley Interscience.

Widdel, F., Hansen, T.A. 1992. The dissimilatory sulphate and sulfur reducing bacteria. In *The Prokaryotes* (2nd ed.), edited by P. Mortumer, 583–624. New York: Springer.

Xu, X.-J., Chen, C., Wang, A.-J., Fang, N., Yuan, Y., Ren, N.-Q., Lee, D.-J. 2012. Enhanced elementary sulfur recovery in integrated sulfate reducing, sulfur-producing rector under micro-aerobic condition. *Bioresource Technology* 116:517–521.

Zhang, M., Wang, H. 2014. Organic wastes as carbon sources to promote sulfate reducing bacterial activity for biological remediation of acid mine drainage. *Minerals Engineering* 69:81–90.

CHAPTER 8

Enhancement of Anaerobic Digestion of Pharmaceutical Effluent

J. Kanimozhi, Arunachalam Bose Sathya,

Anant Achary, and Velmurugan Sivasubramanian

CONTENTS

Abstract

Pharmaceutical industries generate large quantities of liquid and solid wastes. Considering their impacts on the environment, wastes should be properly treated. Among the many available methods of pharmaceutical effluent treatment, the most recognized and approved method is the anaerobic digestion process. Treatment of high-strength wastewater is feasible through an anaerobic route and it provides a cost-effective alternative to the aerobic process by saving energy, nutrient addition, and reactor volume and higher volumetric loading rates. The present investigation is focused on enhancing anaerobic digestion with immobilized mixed anaerobic bacteria and optimization of physicochemical parameters, such as immobilized cell concentration, temperature, and pH, to increase the fermentation efficiency. Anaerobic bacteria from a pharmaceutical sludge were isolated under anaerobic mesophilic conditions and immobilized using calcium alginate. The performance of anaerobic digestion was studied through the reduction of chemical oxygen demand. The maximum reduction of chemical oxygen demand (COD) is achieved after 76 hours at 0.8% (w/v) immobilized cell, 38°C temperature, and 6.8 pH. The performance of immobilized cells were stable after storing at 4°C for 25 days and the immobilized cells can be reused eight times. The rate of COD reduction of immobilized cells were also maintained for up to four cycles.

8.1 INTRODUCTION

The pharmaceutical manufacturing industries produce a great amount of liquid and solid wastes. These wastes are generally complex due to the presence of toxic and inhibitory substances (Chelliapan et al. 2011), and their composition varies depending on the product manufactured, the materials used in the process, and other process details. These wastewaters are hazardous to human populations and the environment, and must be handled prior to disposal into streams, lakes, oceans, and land surfaces (de-Bashan and Bashan 2010). Moreover, pharmaceutical wastewaters can contain high levels of chemical oxygen demand (COD) as high as 80,000 mg/L (Nandy and Kaul 2001). Figure 8.1 illustrates the development of anaerobic digestion of pharmaceutical wastewater with an immobilized biosystem. Anaerobic digestion of biomass wastes could have an immense impact on renewable energy demands. It is best suited to convert organic wastes from agriculture, livestock, industries, municipalities, and other human activities into energy and fertilizer, and it has become popular in developing countries (Manyi-Loh et al. 2013). The treatment systems developed by industry are frequently regarded as a regulatory obligation, increasing capital and running costs and yielding negative economic returns. While complying with environmental legislations, the treatment system should not inevitably contribute to the creation of additional costs but can alternatively provide a secondary source of income (Department for Environment, Food and Rural Affairs 2014).

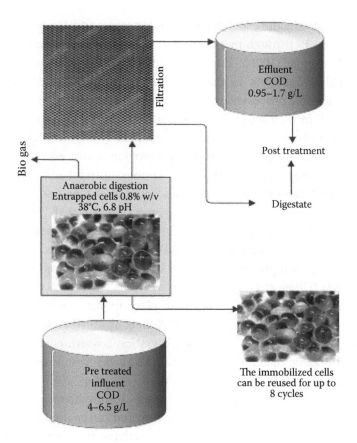

FIGURE 8.1   Schematic representation of anaerobic digestion by immobilized cells.

Biological treatments can be conducted to reduce the COD of industrial wastewater either aerobically or anaerobically. Nevertheless, due to high COD levels, it is unfeasible to treat some pharmaceutical wastewater through an aerobic process (Paul 2008). Another possible route to treat wastewater is the anaerobic digestion process (Samir and Huang 2003; Chen et al. 2008). Before being digested, the feedstocks have to undergo pretreatment. There are various types of pretreatment depending on the feedstocks availability. The aim of such pretreatment is to mix different feedstock and remove unwanted materials, such as large items and inert materials, to allow a better digestate quality (Oz et al. 2004).

Anaerobic treatment processes have been considered as one of the most effective and economical methods in treating effluent, owing to the capacity to treat slowly degradable substrates at high concentration, very low sludge production, low energy requirements, and the possibility of energy recovery through methane generation (Prakasham and Satyavani 2001; Chen et al. 2008). Therefore, anaerobic processes have become increasingly in demand in the treatment of complex industrial wastewater. Anaerobic processes, which can tolerate a wide variety of toxins, are effective in treating waste containing synthetic organics, which are refractory under aerobic conditions. Treatment of high-strength waste is feasible through an anaerobic route and provides a cost-effective alternative to the aerobic process with energy savings, nutrient addition, reactor volume, rapid response to substrate additions after long periods without feeding, and higher volumetric loading rates (Zeeman et al. 2000). Hence, wastes containing significant quantities of organics are a valuable source of renewable energy (Zeb et al. 2013). Despite these benefits, poor operational stability still prevents anaerobic digestion from being widely commercialized (Dupla et al. 2004).

Anaerobic digestion processes take place in many situations where organic material is available (Holm-Nielsen et al. 2009) and redox potential is down (zero oxygen). Anaerobic digestion is a fermentation process in which organic material is degraded and biogas (composed of mainly methane and carbon dioxide) is produced (Zeb et al. 2013). It seems to be a promising technology to gain revenue from methane gas (biogas), which is generated from anaerobic digestion and can be used as renewable energy (Saghafi et al. 2010). The yield of methane from wastes is receiving renewed attention as it can potentially bring down CO_2 emissions via production of renewable energy and limit the emission of the greenhouse gas (Manyi-Loh et al. 2013).

8.2 BIOLOGY OF ANAEROBIC DIGESTION PROCESS

An anaerobic digester contains a community of microorganisms to carry out the process of fermenting organic matter into methane (biogas), and the process is mediated by methanogens, bacteria, fungi, and protozoa. The digestion process consists of a series of biochemical events that convert organic compounds to methane, carbon dioxide, and new bacterial cells (Manyi-Loh 2013). The Anaerobic Digestion and Bioresource Association (ADBA) (Department for Environment, Food and Rural Affairs 2014) has defined the process as a controlled breakdown of organic matter to produce a combustible gas. This process can be divided into four phases: hydrolysis, fermentation (acidogenesis), acetogenesis, and methanogenesis, in

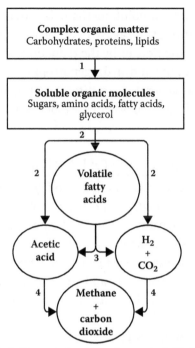

1. Hydrolysis 2. Fermentation 3. Acetogenesis 4. Methanogenesis

FIGURE 8.2   Biology of the anaerobic digestion process.

which hydrolytic, fermentative bacteria, acetogens, and methanogens play distinct roles, respectively, as shown in Figure 8.2.

In the first step, hydrolysis takes place, in which the large, complex soluble and insoluble polymeric matter is hydrolyzed to monomers. For example, hydrolytic enzymes (lipases, proteases, cellulases, amylases, etc.) secreted by microbes convert cellulose to sugars or alcohols, and proteins to peptides or amino acids, and make them available for other bacteria (Madigan et al. 2008). The digestion systems are defined by two major steps depending on the nature of the substrates. Hydrolysis is often limited if the substrate is of complex organic solids while in the digestion of soluble organic matter (Björnsson et al. 2008). Later, fermentative or acidogenic bacteria convert hydrolyzed products into organic acids (primarily acetic acid), hydrogen, and carbon dioxide. Acetate, carbon dioxide, formate, methylamines, methyl sulfide, acetone, and methanol produced in this phase can be directly utilized by methanogens for methanogenesis. At a final point, methane is produced by methanogens in two ways: either through cleavage of acetic acid molecules to produce methane and carbon dioxide, or by the reduction of carbon dioxide with hydrogen by acetotrophic and hydrogenotrophic methanogens (Manyi-Loh 2013). The biogas generated constitutes mainly methane (50%–75%), CO_2 (25%–45%), and traces of other gases such as CO, H_2S, NH_3, O_2, and water vapor.

8.3 REACTOR DESIGN AND OPERATIONAL CONDITIONS

There are several types of reactors in use today, and the design is related to the material to be digested. The principal biological processes used for anaerobic wastewater treatment can be divided into two main groups: suspended growth and attached growth (or biofilm) processes. The operation of these processes depends on the performance of the microorganisms, the specific reactions and their reaction kinetics, the need for nutrients, and other environmental factors affecting their behavior (Tchobanoglous et al. 1993).

In suspended growth systems, the bacteria are suspended in the digester through some kind of mixing. Attached growth anaerobic treatment reactors can be divided into two groups: upflow and downflow of treated water. Upflow attached growth anaerobic treatment reactors differ in the type of packing and the degree of bed expansion (Ward et al. 2008). Downflow attached growth reactors differ only in the packing material used, and these can be random or tubular plastic. There are three main types of reactor designs available for anaerobic digestion:

1. Batch reactors

2. One-stage continuously fed systems

3. Two-stage or multistage continuously fed systems

Batch reactors are the simplest. They are filled with the pretreated wastewater that is allowed to go through all the degradation sequentially, and left for a period that is called the hydraulic retention time, after which the reactors are emptied (Cosenza 2014). The second type is the one-stage continuously fed system (Angenenta 2002), where all the biochemical reactions take place in one reactor. In the two-stage (or even multistage) continuously fed system the hydrolysis/acidification and acetogenesis/

methanogenesis processes are separated (Chua and Fung 1996). The attractive features and drawbacks of anaerobic digester are detailed next.

Attractive features

- Able to degrade recalcitrant natural compounds (e.g., lignin) and xenobiotic compounds.

- Control of some filamentous organisms through recycling of sludge and supernatant.

- Improved dewater ability of sludge.

- Production of methane (biogas).

- Use of bio solids as a soil additive or conditioner.

- Suitable for treatment of high-strength industrial wastewater.

- Reduction in malodors.

- Reduction in number of pathogens.

- Reduction in sludge handling and disposal costs.

- Reduction in volatile content of sludge.

Limitations

- The high sensitivity of methanogenic bacteria to a large number of chemical compounds.

- The first start-up of an installation without the presence of proper seed sludge can be time consuming due to the low growth yield of anaerobic bacteria.

- When treating waste(water) containing sulfurous compounds, the anaerobic treatment can be accompanied by odor due to the formation of sulfide.

- The main one is that it requires a more stringent process control and only reduces the organic pollution by 85% to 90%, which means a second step is usually needed to guarantee high effluent quality. This is usually an aerobic stage for polishing before discharge.

The complexity of the bioconversion processes affecting the performance of an anaerobic digester has been investigated. These can be divided into three main classes: (1) feedstock characteristics, (2) reactor design, and (3) operational conditions (Cioabla et al. 2012). Among the operational conditions, temperature, organic loading rate (OLR), and pH are the most important parameters. High-rate anaerobic digestion can be achieved by meeting three conditions: a high concentration of anaerobic bacterial sludge must be retained under high organic load, high hydraulic loading conditions (Zeeman et al. 2000), and maximum contact must occur between the incoming feedstock and the bacterial mass (Appels 2008). However, these reactors often face difficulties, such as a low methane value in the biogas (Amon et al. 2007), self-heating of the reactors (Lindorfer 2007), unstable processes, long start-up times, slow kinetics at low temperatures, and the odors from fermentation (Speece 1996).

The organic loading rate must be increased and the start-up time needs to be abbreviated, thus the reactor can be operated by raising the influent flow rate or by increasing the influent concentration. However, each of these methods presents a challenge. An increase of the influent flow rate can cause sludge washout, which will reduce the number of bacteria in the reactor, slowing the treatment process, as well as decrease the pH, which can also reduce the number of bacteria present (Imai et al. 2007).

To enable an anaerobic reactor system to suit a high organic loading rate for treating pharmaceutical wastewater, the reactor must retain a high amount of viable sludge under the operational conditions, which can be accomplished by immobilizing the biomass (Callander and Barford 1983).

Moreover, the doubling time of hydrolytic and acetogeneic bacteria is about 24 to 36 hours, while methanogens need about 1 to 15 days (Miron et al. 2000). This shows that methanogens require a long time to convert acetic acid; meanwhile, these bacterial populations of the anaerobic digestion process can be reduced and easily taken away from the process by sludge removal. Furthermore, the bacteria are very sensitive to operating conditions, and the process requires a comparatively long start-up period to make it a stable operation (Lindorfer 2007). The digestion process can be improved by retaining the high rate bacterial population within the batch digester by immobilizing bacteria. Immobilization of bacteria reduces the amount of microorganisms washed out of the digester (Lan et al. 2009), which will protect the biomass from deteriorating and toxic chemicals (Zhou et al. 2008; Liu et al. 2012). The immobilization practice has long been in use in the treatment of wastewater, especially wastewater from industry where the large throughput volumes would otherwise take a very large reactor (Ward et al. 2008).

The present investigation is focused on enhancing anaerobic digestion in batch reactors with immobilized mixed anaerobic bacteria and optimization of operational parameters such as immobilized cell concentration, temperature, and pH to increase the digestion efficiency. Immobilizing biomass protects the cells to form dense granules, which settle in the digester, for example, in upflow anaerobic sludge blanket (UASB) reactors. The pharmaceutical effluent was collected from various pharmaceutical-based industries before being fed into the UASB reactor and the effluent was used as feed. The biological oxygen demand/chemical oxygen demand (BOD/COD) ratio gives an indication of the fraction of pollutants in the effluent that is biologically degradable. Table 8.1 describes the characteristics of pharmaceutical effluent collected from various pharmaceutical-based industries. The effluent is complex in nature as it had a low BOD/COD ratio (0.38), low COD/SO4^{-2} ratio (3.75), and high Total Dissolved Inorganic Solids (TDIS) (12 g/L) that must be treated biologically.

The COD level indicates the amount of dissolved organic pollutants that can be taken out in chemical oxidation by adding strong acids. The oxygen equivalent of organic matter that can be oxidized is measured using a strong chemical oxidizing agent in an acidic medium and expressed in milligrams per liter. The sample is refluxed with an excess of potassium dichromate for two hours (APHA 1995). After digestion, the excess dichromate is titrated against standard ferrous ammonium sulfate. The COD level represents the concentration of contaminants in industrial wastewater. The amount of COD removal efficiency is used to quantify the efficiency of anaerobic digestion. Moreover, COD is a good index of

TABLE 8.1   Characteristics of Pharmaceutical Effluent

Particulars	Range
Chemical oxygen demand (COD)	6500–4000 mg/L
Biological oxygen demand (BOD)	2500–1700 mg/L
pH	6–9
Temperature	28°C–29°C
Total dissolved solids	9000–8000 mg/L
Total suspended solids	600–700 mg/L
Sulfates	1200–900 mg/L
Chlorides	4600–3700 mg/L
Total alkalinity	900–700 mg/L

the degree of completeness of the degradation process, as any undigested material will require oxygen (in an aerobic environment) to complete degradation.

8.4 CELL IMMOBILIZATION

Immobilization of microbial cells has received increasing interest in the area of waste treatment (Martins et al. 2013), and immobilization of cells has been explored as promising for wastewater treatment in the past few decades for its function as biocatalysts, using several types of reactors such as feed lot and semicontinuous packed bed reactors (Zhou et al. 2008). Immobilization is the imprisonment of all types of biocatalysts including enzymes, cellular organelles, and animal and plant cells in a distinct phase that allows exchange with, but is separated from, the bulk phase or the external environment. Whole cells or biomass can be immobilized either in a viable or nonviable form (Souza 2002).

Immobilized cells have some advantages over free cells, such as a higher fermentation rate, higher cell density per volume of the reactor, easier separation from the reaction medium, continuous operation without the cells being carried away downstream, reduction of the adaptation phase (lag phase), a higher substrate conversion, less inhibition by products, reduced reaction time, and control of cell replication (Souza 2002). Immobilization of biomass is the one of the important and effective techniques that has been employed for protecting the microbial population against inhibition; harsh conditions such as pH, temperature, organic solvents, and toxic components present in wastewater (Pishgar 2011); and stability against shear force when used in reactors (Liu et al. 2012; Martins et al. 2013). The environmental applications of immobilized microbial cells and biomass are listed in Table 8.2.

Encapsulation refers to a physicochemical or mechanical process to entrap a substance in a material, thus the produced particles have diameters of a few nanometers to a few millimeters (Park and Chang 2000). Thus here, the active agents are entrapped within the polymer matrix or the capsules. In such cases, cells are restricted by the membrane walls that are in a capsule but free-floating within the core space. This technique creates a protective barrier around the entrapped microbes, which ensures their prolonged viability during processing and storage in polymers. Among all the methods, entrapment is the most extensively studied method due to its simplicity and operationally convenient work-up (de-Bashan and Bashan 2010), and the entrapped cells maintain a higher sludge retention time

TABLE 8.2   Environmental Applications of Immobilized Microbial Cells and Biomass

Biosystem	Immobilization	Application	Reference
Pseudomonas sp., *Fusarium flocciferum*	Polyacrylamide hydrazide (PAAH) and alginate	Phenol degradation	Bettmann and Rehm 1984; Anselmo et al. 1985
Pseudomonas sp.	Sodium alginate	Dehalogenation of 3-chlorobenzoate	Sahasrabudhe et al. 1988
Mixed microbial cells	Mono-carrier (cellulose triacetate) and bi-carrier (combined calcium alginate and cellulose triacetate)	Synthetic organic wastewater (glucose and phenol)	Yang et al. 1988
Flavobacterium	Polyurethane	Degradation of pentachlorophenol	Oreilly and Crawford 1989
Pimelobacter sp.	Polyvinyl alcohol	Biodegradation of pyridine	Rhee et al. 1994
Aspergillus niger	Polyurethane sponge cubes	Olive mill wastewater treatment	Vassilev et al. 1997
Activated sludge	Alginate	Degradation of phenol	Joshi and D'Souza 1999
Gordona sp., *Nocardia* sp.	Celite	Desulfurization of light gas oil	Chang et al. 2000
Pseudomonas sp.	Polyurethane foam	Degradation of naphthalene	Manohar et al. 2001
Bacillus sp.	Polyurethane foam or alginate	Degradation of dimethylphthalate	Niazi and Karegoudar 2001
Activated sludge	Polyacrylamide	Wastewater treatment	Horan and Huo 2004
Mycobacterium goodii	Calcium alginate, carrageenan, agar, polyvinyl alcohol, polyacrylamide, and gelatin-glutaraldehyde	Desulfurization of gasoline	Li et al. 2005
Arthrobacter citreus	Alginate and agar	Degradation of phenol	Karigar et al. 2006
Sphingomonas sp.	Magnetically immobilized cells with agar, alginate, κ-carrageenan, and gellan gum	Degradation of carbazole	X. Wang et al. 2007
Klebsiella oxytoca	Alginate and Cellulose triacetate	Treatment of cyanide wastewater	Chen et al. 2007

(Continued)

TABLE 8.2 (CONTINUED)   Environmental Applications of Immobilized Microbial Cells and Biomass

Biosystem	Immobilization	Application	Reference
Acinetobacter sp.	Polyvinyl alcohol	Biodegradation of phenol	Y. Wang et al. 2007
Pseudomonas sp.	Calcium alginate	Biodegradation of rice mill effluent	Manogari et al. 2008
Anammox bacteria	Polyvinyl alcohol	Swine wastewater treatment	Vanotti et al. 2010
Phanerochaete chrysosporium	Alginate	Decolorization of paper mill effluent	Gomathi et al. 2012
Activated sludge	Alginate, chitosan and carboxymethyl cellulose	Enhanced biogas production	Youngsukkasam et al. 2012
Rhizopus sp.	Polyvinyl alcohol and sodium alginate	Repeated reduction of Cr(VI)	Liu et al. 2012

(SRT), which is indispensable for the functioning of a stable and efficient biological treatment process (Yang 1988). Commonly used polymers for entrapment of cells are calcium alginate, polyvinyl alcohol, chitosan, and zeolites. Alginate is the most suitable biomaterial for the entrapment technique due to its abundance, excellent biocompatibility, and biodegradability properties (Nussinovitch 2010). It is better to immobilize cells rather than enzymes because the immobilized microbial cell eliminates the often tedious, time-consuming, and expensive steps involved in isolating and purifying the enzymes. Use of encapsulated cells in anaerobic digestion improves the operation of the continuous process by maintaining high cell densities and cuts process costs by reusing immobilized cells for the next cyclic operations.

Generally, digested sludge from a running biogas plant or a municipal digester, material from well-rotted manure pit, or cow dung slurry is used as starter cultures. The addition of inoculum tends to improve both the biogas production and methane content in biogas (Yadvika et al. 2004) with high COD removal, when high sludge retention time at relatively short hydraulic retention time is maintained (Wellinger 1999). Anaerobic bacteria (mixed culture) present in pharmaceutical second-stage sludge was enriched and grown anaerobically. The cells were immobilized (entrapped) within alginate beads, and the digestion efficiency of the pharmaceutical effluent of immobilized cells at different temperature and pH was studied and assessed as percentage of COD removal. The COD removal efficiency of free calls and immobilized cells is depicted in Figure 8.3.

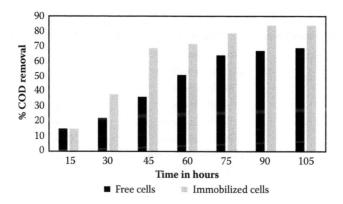

FIGURE 8.3   COD removal efficiency by free cells and immobilized cells.

8.5 EFFECT OF TEMPERATURE

Generally, temperature has a significant effect on the intracellular and extracellular environment of bacteria, and it also works as an accelerator of the conversion procedures. The temperature within the anaerobic digester has a major effect on the digestion and biogas production process (Yadvika et al. 2004). Anaerobic digestion can be performed under two ideal ranges for the performance of anaerobic bacteria: one at 30°C to 40°C of mesophilic microorganisms (optimum temperature 37°C) and one at 45°C to 60°C for thermophilic microorganisms (optimum temperature 55°C) (Mata-Alvarez et al. 1993). Performance of a digester in the mesophilic range is more stable, as these microbial communities can tolerate greater changes in environmental parameters and consume less energy. Nevertheless, the mesophilic microorganisms are more tedious and hence a longer retention time in the digester is needed to maximize biogas yield and digestion process (Juanga 2005).

Thermophilic digesters are more effective in terms of retention time, loading rate and higher digestion, pathogen destruction, and gas production, but less feasible in a developing country. Additional heat energy inputs needed to maintain thermophilic conditions necessitates high energy and certainly high operating costs (Gomec 2010). Other disadvantages are the low stability of the process and a greater sensitivity to operational and environmental variables (Veeken and Hamelers 1999). Thus, the treatment systems are designed to operate in the mesophilic range. The optimum temperature of digestion process may vary depending on the influent composition (COD level), mesophilic bacteria (optimum growth temperature of some mesophilic bacteria) listed in Table 8.3 and

TABLE 8.3   Optimal Growth Temperatures of Methanogenic Bacteria

Temperature Range	Methanogens	Optimum Temperature (°C)
Mesophilic	Methanobacterium	37–45
	Methanobrevibacter	37–40
	Methanosphaera	35–40
	Methanolobus	35–40
	Methanococcus	35–40
	Methanosarcina	30–40
	Methanocorpusculum	30–40
	Methanoculleus	35–40
	Methanogenium	20–40
	Methanoplanus	30–40
	Methanospirillum	35–40
	Methanococcoides	30–35
	Methanolobus	35–40
	Methanohalophilus	35–45
Thermophilic	Methanohalobium	50–55
	Methanosarcina	50–55

Source: Adapted from Gerardi, M.H., 2003, *The Microbiology of Anaerobic Digesters*, Hoboken, NJ: John Wiley & Sons.

type of digester, but in most digestion processes it should be studied in the laboratory and production scale, and maintained relatively constant to sustain the digestion rate and biogas production rate. The small changes in temperature, 35°C to 30°C and from 30°C to 32°C, have proven to reduce the biogas production rate (Chae et al. 2008). It can be seen from Figure 8.4 that an increasing temperature of the mesophilic system had a negative

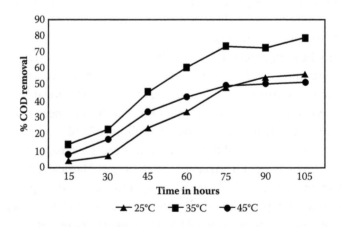

FIGURE 8.4   Effect of temperature on anaerobic digestion by immobilized cells.

effect on anaerobic digestion. The increased temperature may affect the biochemical reaction of encapsulated mesophilic anaerobic bacteria. The efficiency of COD removal of anaerobic digestion of pharmaceutical effluent increased from 36% to 76% when the temperature increased from 25°C–35°C has positively influencing COD removal and further increase to 45°C.

The consequence of temperature on anaerobic digestion of encapsulated anaerobic bacteria of pharmaceutical effluent was investigated at different temperatures. The COD removal efficiency was relatively low at 45°C at 50% and at 25°C was 55%. The maximum COD reduction was observed at 35°C. Based on these results a second slot of analysis with a temperature range of 34°C to 40°C was carried out for a period of 105 hours. The maximum COD removal efficiency is achieved at temperature 38°C.

8.6 EFFECT OF pH

pH is an expression of the intensity of the basic or acid condition of a liquid, a quantity of the acidity of a solution. The beginning steps of anaerobic digestion can take place in a broad range of pH values, while methanogenesis only proceeds when the pH is neutral (Lettinga and Haandel 1993). The optimum pH for a generally stable anaerobic digestion process and high biogas yield lies in the range of 6.5 to 7.5 (Mata-Alvarez 2002; Khalid et al. 2011). During digestion, the processes of hydrolysis and acidogenesis occur at acidic pH levels (pH 5.5–6.5) as compared to the methanogenic phase (pH 6.5–8.2). An alkalinity level of approximately 3000 mg/L should be available at all times to maintain sufficient buffering capacity. The immobilized cells are less sensitive than that of free cells against pH variation.

During acetogenesis (fermentation), production of acetic acid, lactic acid, and propionic acid leads to fall of pH. Under normal conditions, the bicarbonate that is produced by methanogens buffers this pH reduction. Under adverse environmental conditions, the buffering capacity of the system can be upset, eventually stopping the production of methane. Acidity is inhibitory to growth of methanogens than of acidogenic bacteria. An increase in volatile acid levels, thus serves as an early indicator of system upset. Monitoring the ratio of total volatile acids (as acetic acid) to total alkalinity (as calcium carbonate) has been suggested to ensure that it remains below 0.1 (Hema Krishna 2013).

A sufficient amount of hydrogen carbonate (frequently denoted as bicarbonate alkalinity) in the solution is important to maintain the optimal pH

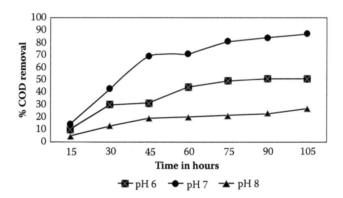

FIGURE 8.5   Effect of pH on anaerobic digestion by immobilized cells.

range required. Alternatively, sodium bicarbonate can also be used for pH adjustment. Lime is usually much cheaper and there might be free sources of spent lime solutions from local industry. Lime, however, often leads to precipitations and clogging of pipes when used in larger amounts. Sodium bicarbonate and sodium hydroxide are fully soluble and commonly do not lead to precipitations, but on the other hand, they may contribute to higher cost. For immediate action, the addition of sodium salts is recommended. Nevertheless, lime might be the best choice for pH adjustment (Vögeli et al. 2014).

Considering the effect of pH on anaerobic digestion, it should be maintained at 6.4 to 7.2. The study was performed at three different pH levels (6, 7, and 8) to maximize the COD removal efficiency of immobilized cells. Figure 8.5 demonstrates the effect of pH on COD reduction in anaerobic digestion by immobilized cells. The COD removal efficiency was low at pH 6 and pH 8 relative to pH 7. At pH 7 the COD removal efficiency is around 90% and at pH 8 it was around 23%. Hence the maximum COD reduction was observed at pH 7. Based on these results a second slot of analysis with a pH range of 6.4 to 7.4 was carried out for a period of 90 hours. The maximum COD removal efficiency is achieved at pH 6.8.

8.7 STORAGE AND REUSABILITY

The stability and reusability during long-term storage and operation is an essential factor for practical application of an immobilized cell system. In order to investigate the storage stability of immobilized cells of pharmaceutical effluent digestion, the immobilized cells were stored in 150 ml biological saline at 4°C for 25 days and their reusability efficiency was also investigated.

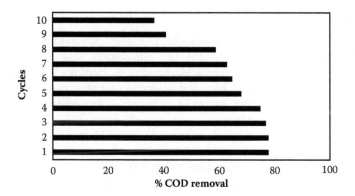

FIGURE 8.6   Stability of immobilized cells.

Thereafter, digestion was performed with immobilized cells under the optimum conditions. The COD removal rates of pharmaceutical effluent by immobilized and free cells maintained at 76% and 51% respectively after being stored for 25 days at 4°C, suggesting that storage stability of immobilized cells are better than free cells. In addition, the digestion efficiency of immobilised cells were above 60% up to 7 cycles of application after reuse for 12 cycles of immobilized cells, as shown in Figure 8.6. The immobilized cells maintained physiological stability up to the fourth cycle, and the Ca-alginate beads has high mechanical strength and operational stability.

8.8 CONCLUSION

An anaerobic treatment system with immobilized cells is a promising alternative for pharmaceutical industrial wastewater treatment, and it could be cost effective since the cells can be used several times without significant loss of activity. Anaerobic digestion of pharmaceutical effluent was investigated using immobilized mixed cultures. Immobilized cells have been a successful approach in batch mesophilic anaerobic digestion of pharmaceutical wastewater in an effluent concertation of 6500 mg/L. When the process operated at 38°C utilizing a 76 hour hydraulic retention time, constant COD removal efficiencies of 70% to 80% was achieved. The performance of immobilized cells was stable after storing at 4°C for 25 days and could be reused up to 10 cycles though significant loss of COD removal efficiency was recorded after 8 cycles. The rate of COD reduction of immobilized cell is also maintained up to the fourth cycle. Anaerobic digestion with immobilized cells is a proven technique and at present applied to a variety of wastewater treatment process.

REFERENCES

American Public Health Association (APHA). 1995. *Standard Method for the Examination of Water and Wastewater*. 19th edition. Washington, D.C.

Amon, T., Amon, B., Kryvoruchko, V., Machmuller, A., Hopfner-Sixt, K., Bodiroza, V., Hrbek, R. et al. 2007. Methane production through anaerobic digestion of various energy crops grown in sustainable crop rotations. *Bioresource Technology* 98:3204–3212.

Angenenta, L.T., Sungb, S., and Raskina, L. 2002. Methanogenic population dynamics during start-up of a full-scale anaerobic sequencing batch reactor treating swine waste. *Water Research* 36:4648–4654.

Anselmo, A.M., Mateus, M., Cabral, J.M.S., and Novais, J.N. 1985. Degradation of phenol by immobilized cells of *Fusarium flocciferum*. *Biotechnology Letters* 7:889–894.

Appels, L., Baeyens, J., Degrève, J., and Dewil, R. 2008. Principles and potential of the anaerobic digestion of waste-activated sludge. *Progress in Energy and Combustion Science* 34:755–781.

Bettmann, H., and Rehm, H.J. 1984. Degradation of phenol by polymer entrapped microorganisms. *Applied Microbiology and Biotechnology* 20:285–290.

Björnsson, L., Kivaisi, A.K., and Mshandete, A.M. 2008. Performance of biofilm carriers in anaerobic digestion of sisal leaf waste leachate. *Electronic Journal of Biotechnology* 11, no. 1:1–8.

Callander, I.J., and Barford, J.P. 1983. Recent advances in anaerobic digestion technology. *Process Biochemistry* 18, no. 4:24–30.

Chae, K.J., Jang, A., Yim, S.K., and Kim, S.I. 2008. The effects of digestion temperature and temperature shock on the biogas yields from the mesophilic anaerobic digestion of swine manure. *Bioresource Technology* 99:1–6.

Chang, J.H., Chang, Y.K., Ryu, H.W., and Chang, H.N. 2000. Desulfurization of light gas oil in immobilized-cell systems of *Gordona* sp. CYKS1 and *Nocardia* sp. CYKS2. *FEMS Microbiology Letters* 182, no. 2:309–312.

Chelliapan, S., Yuzir, A., Md Din, M.F., and Sallis, P.J. 2011. Anaerobic pretreatment of pharmaceutical wastewater using packed bed reactor. *International Journal of Chemical Engineering and Applications* 2, no. 1: 32–37.

Chen, Y., Cheng, J.J., and Creamer, K.S. 2008. Inhibition of anaerobic digestion process: A review. *Bioresource Technology* 99:4044–4064.

Chua, H., and Fung, J.P.C. 1996. Hydrodynamics in the packed bed of anaerobic fixed film reactor. *Water Science and Technology* 33, no. 8:1–6.

Ciobla, A.E., Ionel, I., Dumitrel, G.-A., and Popescu, F. 2012. Comparative study on factors affecting anaerobic digestion of agricultural vegetal residues. *Biotechnol Biofuels* 5:39.

Cosenza, A. 2014. Enzyme enhanced microbial anaerobic digester. http://www.c2biotechnologies.com/index.php/products/5-enzyme-enhanced-microbial-anaerobic-digester-eemad.

de-Bashan, L.E., and Bashan, Y. 2010. Immobilized microalgae for removing pollutants: Review of practical aspects. *Bioresource Technology* 101: 1611–1627.

Department for Environment, Food and Rural Affairs. 2014. Anaerobic Digestion Strategy and Action Plan Annual Report 2014. United Kingdom.

Dupla, M., Conte, T., Bouvier, J., Bernet, N., and Steyer, J. 2004. Dynamic evaluation of a fixed bed anaerobic digestion process in response to organic overloads and toxicant shock loads. *Water Science Technology* 49:61–68.

Gerardi, M.H. 2003. *The Microbiology of Anaerobic Digesters.* Hoboken, NJ: John Wiley & Sons.

Gomathi, V., Cibichakravarthy, B., Ramanathan, A., Sivaramaiah, N., Ramanjaneya, V., Mula, R., and Jayasimha Rayalu, D. 2012. Decolourization of paper mill effluent by immobilized cells of Phanerochaete chrysosporium. *International Journal of Plant, Animal, and Environmental Sciences* 2, no. 1:141–146.

Gomec, C.Y. 2010. High-rate anaerobic treatment of domestic wastewater at ambient operating temperatures: A review on benefits and drawbacks. *Journal of Environmental Science and Health A* 45, no. 10:1169–1184.

Hema Krishna, R. 2013. Role of factors influencing on anaerobic process for production of bio hydrogen: Future fuel. *International Journal of Advanced Chemistry* 1, no. 2:31–38.

Holm-Nielsen, J.B., Al Seadi, T., and Oleskowicz-Popiel, P. 2009. The future of anaerobic digestion and biogas utilization. *Bioresource Technology* 100: 5478–5484.

Horan, N.J., and Huo, C.X. 2004. Production and characterization of immobilized-biomass carriers, using polyacrylamide. *Environmental Technology* 25, no. 6:667–672.

Imai, T., Li, F., Ukita, M., Yuasa, A., and Zhou, W. 2007. Effect of loading rate on the granulation process and granular activity in a bench scale UASB reactor. *Bioresource Technology* 98, no. 7:1386–1392.

Joshi, N.J., and D'Souza, S.F. 1999. Immobilization of activated sludge for the degradation of phenol. *Journal of Environmental Science And Health Part A* 34, no. 8:1689–1700.

Juanga, J.P. 2005. Optimizing dry anaerobic digestion of organic fraction of municipal solid waste. (Master thesis No. EV-05-11, Asian Institute of Technology, 2005). Bangkok: Asian Institute of Technology.

Karigar, C., Mahesh, A., Nagenahalli, M., and Yun, D.J. 2006. Phenol degradation by immobilized cells of *Arthrobacter citreus. Biodegradation* 17, no. 1: 47–55.

Khalid, A., Arshad, M., Anjum, M., Mahmood, T., and Dawson, L. 2011. The anaerobic digestion of solid organic waste: Review. *Waste Management* 31, no. 8:1737–1744.

Lan, W.U., Gang, G.E., and Jinba, W.A.N. 2009. Biodegradation of oil wastewater by free and immobilized *Yarrowia lipolytica* W29. *Journal of Environmental Sciences* 21:237–242.

Lettinga, G., and Haandel, A.C. 1993. Anaerobic digestion for energy production and environmental protection. In *Renewable Energy: Sources for Fuels and Electricity*, edited by T.B. Johansson, H. Kelly, A.K.N. Reddy, and R.H. Williams, 817–839. Washington, D.C.: Island Press.

Li, F.L., Xu, P., Feng, J.H., Meng, L., Zheng, Y., Luo, L.L., and Ma, C.Q. 2005. Microbial desulfurization of gasoline in a *Mycobacterium goodii* X7B immobilized-cell system. *Applied and Environmental Microbiology* 71:276–281.

Lindorfer, H. 2007. Optimised Digestion of Energy Crops and Agricultural Residues in Rural Biogas Plants. Thesis, University of Natural Resources and Applied Life Sciences, Vienna.

Liu, H., Guo, L., Liao, S., and Wang, G. 2012. Reutilization of immobilized fungus *Rhizopus* sp. LG04 to reduce toxic chromate. *Journal of Applied Microbiology* 12:651–659.

Madigan, M.T., Martinko, J.M., Stahl, D., and Clark, D.P. 2008. *Brock Biology of Microorganisms*. Reading, MA: Benjamin-Cummings.

Manogari, R., Daniel, D., and Krastanov, A. 2008. Biodegradation of rice mill effluent by immobilised *Pseudomonas* sp. cells. *Ecological Engineering and Environment Protection* 1:30–35.

Manohar, S., Kim, C.K., and Karegoudar, T.B. 2001. Enhanced degradation of naphthalene by immobilization of *Pseudomonas* sp. strain NGK1 in polyurethane foam. *Applied Microbiology and Biotechnology* 55, no. 3:311–336.

Manyi-Loh, C.E., Mamphweli, S.N., Meyer, E.L., Okoh, A.I., Makaka, G., and Simon, M. 2013. Microbial anaerobic digestion (bio digesters) as an approach to the decontamination of animal wastes in pollution control and the generation of renewable energy. *International Journal of Environmental Research Public Health* 10, no. 9: 4390–4417.

Martins, S.C.S., Martins, C.M., Fiúza, L.M.C.G., and Santaella, S.T. 2013. Immobilization of microbial cells: A promising tool for treatment of toxic pollutants in industrial wastewater. *African Journal of Biotechnology* 12, no. 28:4412–4418.

Mata-Alvarez, J. 2002. Fundamentals of the anaerobic digestion process. In *Biomethanization of the Organic Fraction of Municipal Solid Wastes*, edited by J. Mata-Alvarez, 1–20. London: IWA Publishing.

Mata-Alvarez, J., Cecchi, F., Pavan, P., and Bassetti, A. 1993. Semi-dry thermophilic anaerobic digestion of fresh and pre-composted organic fraction of municipal solid waste (MSW): Digester performance. *Water Science & Technology* 27, no. 2:87–96.

Miron, Y., Zeeman, G., van Lier, J.B., and Lettinga, G. 2000. The role of sludge retention time in the hydrolysis and acidification of lipids, carbohydrates and proteins during digestion of primary sludge in CSTR systems. *Water Resources* 34, no. 5:1705–1713.

Nandy, T., and Kaul, S.N. 2001. Anaerobic pre-treatment of herbal–based wastewater using fixed–film reactor with recourse to energy recovery. *Water Research* 35:351–362.

Niazi, J.H., and Karegoudar, T.B. 2001. Degradation of dimethylphthalate by cells of *Bacillus* sp. immobilized in calcium alginate and polyurethane foam. *Journal of Environmental Science and Health—Part A, Toxic/hazardous Substances and Environmental Engineering* 36, no. 6:1135–1144.

Nussinovitch, A. 2010. Bead formation, strengthening and modification. In *Polymer Macro- and Micro-Gel Beads: Fundamentals and Applications,* 27–52. New York: Springer.

Oreilly, K.T., and Crawford, R.L. 1989. Degradation of pentachlorophenol by polyurethane immobilized Flavobacterium cells. *Applied and Environmental Microbiology* 55:2113–2118.

Oz, N.A., Ince, O., and Ince, B.K. 2004. Effect of wastewater composition on methanogenic activity in an anaerobic reactor. *Journal of Environmental Science and Health—Part A Toxic/Hazardous Substances and Environmental Engineering* 39:2941–2953.

Park, J.K., and Chang, H.N. 2000. Microencapsulation of cells. *Biotechnology Advances* 18, no. 4:303–319.

Paul, J. 2008. Anaerobic Digestion—A Feel Good Strategy or a Sustainable Manure Management Strategy? http://www.transformcompostsystems.com/articles/Is%20Anaerobic%20digestion%20a%20good%20idea%20Dec%2008.pdf, (accessed 5.30.12).

Pishgar, R., Najafpour, G., Neya, B.N.N., Mousavi, N., and Bakhshi, Z. 2011. Anaerobic biodegradation of phenol: Comparative study of free and immobilized growth. *Iranica Journal of Energy & Environment* 2, no. 4:348–355.

Prakasham, S., and Satyavani, R.S. 2001. Bio treatability studies of pharmaceutical waste waters using an anaerobic digestion. *Waters Science Technology* 43:271–276.

Rhee, S.K., Lee, S.T., and Lee, G.M. 1994. Biodegradation of pyridine by freely suspended and immobilized *Pimelobacter* sp. *Applied Microbiology and Biotechnology* 41:652–657.

Saghafi, S., Bakhshi, Z., Najafpour, G.D., Kariminezhad, E., and Rad, H.A. 2010. Biodegradation of toluene and xylene in an UAPB bioreactor with fixed film of *Pseudomonas putida*. *American-Eurasian Journal of Agricultural & Environmental Sciences* 1:1–7.

Sahasrabudhe, S.R., Modi, A.J., and Modi, V.V. 1988. Dehalogenation of 3- chlorobenzoate by immobilized *Pseudomonas* sp. B13 cells. *Biotechnology and Bioengineering* 31:889–893.

Samir, K., and Huang, J.C. 2003. ORP-based oxygenation for sulfide control in anaerobic treatment of high-sulfate wastewater. *Water Research* 37, no. 10:2053–2062.

Souza, S.F.D. 2002. Trends in immobilised enzyme and cell technology. *Indian Journal of Biotechnology* 1:321–328.

Speece, R.F. 1996. Anaerobic biotechnology for industrial wastewater. *App Enviro Microbiol* 58:365–370.

Tchobanoglous, G., Theisen, H., and Vigil, S. 1993. Integrated solid waste manage-ment. Engineering principles and management issues. McGraw-Hill inter-national editions. ISBN: 0-07-063237-5.

Vanotti, M.B., Szogi, A.A., and Rothrock, M.J. 2010. Novel anammox (anaerobic ammonium oxidation) bacterium. US Patent Application SN 61/298,952, filled 1/28/2010. US Patent and Trademark Office, Washington, D.C.

Vassilev, N., Fenice, M., Federici, F., and Azcon, R. 1997. Olive mill waste water treatment by immobilized cells of Aspergillus niger and its enrichment with soluble phosphate. *Process Biochemistry* 32, no. 7:617–620.

Veeken, A., and Hamelers, B. 1999. Effect of temperature on the hydrolysis rate of selected biowaste components. *Bioresource Technology* 69, no. 3:249–255.

Vögeli, Y., Lohri, C.R., Gallardo, A., Diener, S., and Zurbrügg, C. 2014. *Anaerobic Digestion of Biowaste in Developing Countries: Practical Information and Case Studies.* Dübendorf, Switzerland: Swiss Federal Institute of Aquatic Science and Technology (Eawag).

Wang, X., Gai, Z., Yu, B., Feng, J., Xu, C., Yuan, Y., Lin, Z., and Xu, P. 2007. Degradation of carbazole by microbial cells immobilized in magnetic gellan gum gel beads. *Applied and Environmental Microbiology* 73, no. 20:6421–6428.

Wang, Y., Tian, Y., Han, B., Zhao, H.-B., Bi, J.-N., and Cai, B.-L. 2007. Biodegradation of phenol by free and immobilized *Acinetobacter* sp. strain PD12. *Journal of Environmental Sciences* 19:222–225.

Ward, A.J., Hobbs, P.J., Holliman, P.J., and Jones, D.L. 2008. Optimization of the anaerobic digestion of agricultural resources. *Bioresource Technology* 99:7928–7940.

Wellinger, A. 1999. *Anaerobic digestion: A waste treatment technology. Critical Reports on Applied Chemistry* 3.

Yadvika, Santosh, Sreekrishnan, T.R., Kohli, S., and Rana, V. 2004. Enhancement of biogas production from solid substrates using different techniques—A review. *Bioresource Technology* 95:1–10.

Yang, P.Y., Cai, T., and Wang, M.L. 1988. Immobilized mixed microbial cells for wastewater treatment. *Biological Wastes* 23, no. 4:295–312.

Youngsukkasam, S., Sudip, K.R., and Mohammed, J.T. 2012. Biogas production by encapsulated methane-producing bacteria. *Bacterial Capsules-Bio Resources* 7, no. 1:56–65.

Zeb, B.S., Mahmood, Q., and Pervez, A. 2013. Characteristics and performance of anaerobic wastewater treatment (a review). *Journal of Chemical Society of Pakistan* 35:217–232.

Zeeman, G., Sanders, W.T.M., and G. Lettinga. 2000. Feasibility of the on-site treat-ment of sewage and swill in large buildings. *Water Science and Technology* 41, no. 1:9–16.

Zhou, L., Guiying, L., Taicheng, A., Jiamo, F., and Guoying, S. 2008. Recent patents on immobilized microorganism technology and its engineering application in wastewater treatment. *Recent Patents on Engineering* 2:28–35.

CHAPTER 9

Treatment of Textile Dyes Using Biosorption and Bioaccumulation Techniques

Sahadevan Renganathan,

Saravanan Panneerselvam, Baskar Gurunathan,

Vijay Mani, and Shyamasundari Murugan

CONTENTS

Abstract

Increased population, industrialization, and urbanization are responsible for environmental pollution. Textile effluents containing dyes are usually discharged into aquatic systems with or without treatment, leading to environmental pollution. Soluble pollutants present in the aqueous environment interact with biological materials either by binding to cellular surfaces via biosorption or by accumulating inside the cells via bioaccumulation. Biosorption using dead biomass are exceptionally efficient for the removal of synthetic dyes from industrial effluents. Bioaccumulation can be used for the removal of different kinds of textile dyes if growing cells. This chapter discusses biosorption and bioaccumulation processes. In the biosorption method, various kinds of biosorbent, and parameters like sorbent dosage, pH, and dye concentration are discussed. In the bioaccumulation method, growth of cells with dye, initial cells, pH, and dye concentration are discussed. Characterization of sorbent and color removal are analyzed with scanning electron microscopy (SEM), Fourier transform infrared spectroscopy (FTIR), and ultraviolet-visible spectroscopy.

9.1 INTRODUCTION

Biochemical engineering is a multidisciplinary technique for applying microorganisms to solve many environmental problems. Colored effluents produced by textile, paper, cosmetic, food, pharmaceutical, leather, and plastic industries cause significant damage to the environment (Groff 1993). Textile azo dyes are let into the water as effluent and they are highly visible, undesirable, reduce light penetration, and inhibit photosynthesis. These dyes are often recalcitrant to microbial degradation because they contain substitutions such as azo and sulfo groups.

Biotechnological approaches are used to eliminate environmental problems such as industrial pollution in the form of liquid, gaseous, and solid wastes.

Increased population, industrialization, and urbanization are responsible for environmental pollution. Large amounts of toxic waste have been dispersed in thousands of contaminated sites spread across our nation. Thus every one of us is being exposed to contamination from past and present industrial practice. This pollutant fits into two major classes: inorganic and organic. Dyes are organic compounds used for coloring final products from different industries. World dyestuff and dye transitional production is estimated to be around 7×10^8 kg per annum (Vijayakumar et al. 2010). Development of industrial activity may lead to increase in the pollution level to the environment. Textile effluents containing dyes are usually discharged into aquatic systems with and without treatment leading to environmental pollution. Dyes used in the textile industries are exposed to light, water, oxidizing agents, and microbial attack (Wesenberg et al. 2002).

Dyes are highly toxic for most living organisms and disturb the ecosystem (Ozkahraman et al. 2011). Pollution with dyes leads to reduction of sunlight penetration, which decreases both photosynthetic activity and dissolved oxygen concentration affecting the aerobic organisms (Abdallah and Taha 2012).

Dyes can cause allergic dermatitis, cancer, skin irritation, and mutation (Khaled et al. 2009). Furthermore, they can have a severe effect on humans by damaging the liver, kidneys, brain, central nervous system, and reproductive system. The impact of textile-dye-containing wastewater in terms of genotoxicity, carcinogenicity, and toxicity has made the removal of color from the wastewater inevitable when compared with the removal of other soluble organic substances (Ademola and Ogunjobi 2012). Physical/chemical methods are expensive processes and produce large amounts of sludge. More studies are now focused on their biodegradation. Compared with biological processes, chemical/physical methods have received more interest because of their cost effectiveness, lesser sludge production, and environmental friendliness (Liu et al. 2006).

Most of the textile manufacturers are located on the riverbanks and coastal areas because of the low cost transportation and available source of water. The release of effluents from these industrial units to various water bodies (canals, lakes, rivers, etc.) leads to water pollution. As a result, many governments have traditional environmental restrictions with regard to the quality of colored effluents, and have forced dye houses to decolorize their effluents before discharging (Shen et al. 2009). Industrial countries have toxic wastewater mainly due to industrial proliferation and current agricultural technologies, which are mostly addressed through refining wastewater treatment techniques.

Adsorption techniques using adsorbents and biosorption using live or dead biomass procedures are able to remove synthetic dyes from industrial effluents (Corso and Maganha de Almeida 2009). The major advantages of an adsorption system for water pollution control are good elimination performance, simplicity of design, initial cost, flexibility, easy operation, and insensitivity to toxic pollutants as compared with conventional biological process (alive microbial) treatments (Ansari and Mosayebzadeh 2010).

The most popular and widely used adsorbent material is activated carbon. It has a high surface area, adsorption capacity, and degree of surface reactivity in addition to a microporous structure, but high operating costs and problems with regeneration hamper its large-scale application (Akar et al. 2006). Dead cells are apparently preferable for dye removal since they are not affected by chemicals and toxic wastes, and do not pollute the environment by releasing toxins (Prigione et al. 2008).

The soluble pollutants present in the aqueous environment interact with biological materials either by binding to cellular surfaces via biosorption or by accumulating inside the cells via bioaccumulation. Under controlled conditions, such processes can find application in industrial practices for biological wastewater treatment and in bioremediation technologies. One such handling practice to utilize these processes is the removal of dyes from industrial wastewater (Chojnacka 2010).

Microbial decolorization and degradation is an ecofriendly and cost-competitive alternative to chemical decomposition processes (Banat et al. 1996). Bioremediation is a research area within the environmental sciences. In such approaches, bacteria acclimatize themselves to the toxic wastes and convert various toxic chemicals into less harmful forms (Saratale et al. 2011).

9.2 TEXTILE DYES

Azo dyes contain the azo chromophore ($N = N - R$, where R can be aryl hetero aryl or alkyl derivatives). Azo dyes with one azo group are called monoazo dyes and azo dyes with two azo groups are called diazo dyes, followed by triazo and polyazo dyes. An excellent review has been made on dye house wastewater treatment (Cooper 1993). So the removal of these dyes from the effluent is of great importance (Davis et al. 1977). These dyes are both anionic and cationic in nature. The chromophores in anionic and cationic dyes are azo groups and responsible for the formation of toxic amines in the effluent. Azo dyes used in textile industries are mutagenic or carcinogenic (Heiss et al. 1992).

Dyes that are synthetic in origin have complex molecular structures that are classified as anionic-direct, acid and reactive dyes, non-ionic-disperse dyes, and cationic-basic dyes. Usually dyes are azo-based chromophores combined with different types of reactive groups like chlorotriazine, vinyl sulfone, diflurochloropyrimidine, and trichloropyrimidine. Water-soluble dyes are considered to be problematic because those dyes enter into wastewater systems (Sadettin and Donmez 2006). Reactive dyes contain risky compounds among other dyes in textile wastewater. Reactive dyes are extensively used in the textile industry because of their wide variety of high wet-fastness properties, color shades, ease of application, and minimal energy use. Unfortunately, azo dyes present in the wastewater are normally unaffected by conventional treatment processes. Their diligence is mainly due to sulfo and azo groups, which do not occur naturally (Kulla et al. 1983).

Azo dyes account for up to 70% of dyestuffs applied in textile processing, due to the ease and cost-effectiveness in their synthesis, stability, and availability of variety of colors compared to natural dyes. Azo dyes discharged during manufacturing processes can cause severe water pollution in nearby areas. Generally, the azo dyes exhibit great structural variety, and they are not uniformly susceptible to microbial attack (Meyer 1981).

9.3 TREATMENT METHODS

Dyes contained in wastewater can be treated and detoxified using one or a combination of the following physical or chemical treatment processes: flocculation and precipitation, membrane filtration, zonation, chemical oxidation, adsorption using resin and lysozyme crystals, electrochemical destruction, chemical degradation, ultrasound degradation, and biodegradation. The removal of color from the textile effluent using the aforementioned methods is not economically feasible because of high chemical consumption (flocculation, precipitation, chemical oxidation, and chemical degradation), power (electrochemical destruction), adsorbent consumption (adsorption), and initial investment cost (membrane filtration, biodegradation ozonation, ultrasound degradation). They are also not adaptable to a wide range of dyes, and the initial and operational costs for treatment are too high (Banat et al. 1996).

9.3.1 Physical and Chemical Methods

Among the physical and chemical methods, adsorption by activated carbon has been found to be an efficient process, but this adsorbent is too expensive and it is not useful in the case of large-scale microbial treatment of textile dyes (El-Geundi 1991). So activated carbon has been replaced by low-cost adsorbent such as china clay, wood shavings, agricultural byproducts, eucalyptus bark, charred plants, banana pith, cellulose-based wastes, biogas waste slurry, bottom ash, sugar industry mud, red mud, and activated sludge. But these low-cost adsorbents have low adsorption capacities, which means large quantities of the adsorbents are required. So, new economical, and easily available and highly effective biological adsorbents are required.

Many physical and chemical processes for color removal have been applied including membrane, coagulation, oxidizing agents, photodecomposition, and electrochemical. The merits and demerits of each technique have been extensively reviewed (Robinson et al. 2001).

Physical and chemical methods have been used for the treatment of industrial textile wastewater including coagulation/flocculation, membrane, adsorption, chemical reduction, flotation, advanced oxidation (chlorination) and ion exchange. However, many of these technologies are cost prohibitive, particularly when applied for treating grand waste streams. Filtration methods such as ultrafiltration, nanofiltration, and reverse osmosis have been used for water reuse and chemical recovery. In the textile industry, the utilization of membranes provides interesting possibilities for the separation of hydrolyzed dyestuffs and dyeing auxiliaries that simultaneously decrease the color. Chemical oxidation methods enable the obliteration or decomposition of dye molecules, and such approaches use a variety of oxidizing agents, such as ozone (O_3), hydrogen peroxide (H_2O_2), and permanganate (MnO) (Anjaneyulu et al. 2005).

9.3.2 Biological Methods

Biological methods are used to solve this trouble with low-cost biosorbents such as fungal, bacterial, yeast and algae, biomass which has been used for the removal of dyes from the waste water. Among these fungal biomass can be produced cheaply and abundantly by using low cost growth medium. Previous research has been reported with the use of fungus for biodegradation of dyes, bioaccumulation, and biosorption. The reported studies using *Phanerochaete chrysosporium*, *Irpex lacteus*, and *Geotrichum candidum* for decolorization processes using enzymatic activity are available (Lee et al. 2000).

Bioremediation is a very useful technique for wider application in the area of ecological protection. Bioremediation can be used for the treatment of contaminants present in soil, groundwater, surface water, and sediments. In biological wastewater treatment, biodegradation, bioaccumulation, and biosorption methods are included. The biosorption development involves the liquid phase (solvent, normally water) and solid phase containing a dissolved material to be sorbed (adsorbate, metal/dyes). The quantity of adsorbent affinity intended for the adsorbate determines allocation among the solid and liquid phases.

Dead cells as adsorbents were found to be more advantageous when compared to living cells. Dead cells might be stored or used for complete periods at room temperature without putrefaction. Their action is easy and their regeneration is effortless. Microbial biosorbents are essentially small particles with low density, little rigidity, and poor mechanical strength. Their positive qualities include high biosorption capacity, rapid equilibrium attainment, low production cost, and high-quality particle mass transfer (Vijayaraghavan et al. 2008). Moreover, dead cells have been shown to collect pollutants equal to or greater than the amounts of growing or latent cells (Aksu and Tezer 2005).

Bioaccumulation is defined as the accumulation of pollutants by actively growing cells (Sadettin and Donmez 2006). Bioaccumulation can be used for the removal of different kinds of textile dyes if the growing cells find a sufficient amount of carbon and nitrogen sources.

The application of microorganisms for the biodegradation of synthetic dyes is considered to be a good and effortless method. Unfortunately, the majority of dyes are chemically stable and resistant to microbiological attack. Anaerobic reduction and decolorization most often generate aryl amines, which are generally more poisonous than the parent compounds. The separation of new strains or the variation of presented ones to the breakdown of dyes will probably increase the efficiency of microbiological degradation of dyes.

9.3.2.1 Biological Methods Using Viable Microbial Biomass

The main advantage of viable culture in the decolorization process is the avoidance of separate biomass production processes, for instance, activation, harvesting, drying, processing, and storage when compared to using nonviable microbial biomass. The bioaccumulation mechanism plays an imperative role in the decolorization of dye through living fungi. Color removal by *Trametes versicolor* mycelium (Benito et al. 1997) and *Aspergillus niger* (Miranda et al. 1996) have been considered. The adsorption accounted for 5%–10% and 10%–25% of the total color removal, respectively.

Among the various fungal strains, *Schizophyllum commune* is a white rot fungus, which was used for the decolorization of lignin (Belsare and Prasad 1988). *S. commune* fungal biomass was used for the removal of textile dye based on the physical adsorption technique (Renganathan et al. 2006).

9.3.2.2 Biological Methods Using Dead Microbial Biomass

Fungal biomass can be produced easily by using simple fermentation techniques. Fungal biomass, which is treated as wastes by many industries, also can be used for the removal of dyes from wastewater (Zhou and Banks 1993). The biosorption capacity of dyes can be increased by treating the living cells using physical treatment methods, which mainly include autoclaving. Gallagher et al. (1997) reported that dye adsorption was found to be increased by the autoclaving process, which results in the disruption of fungal surface and exposure of the active latent sites.

9.3.2.3 Bioaccumulation of Dyes Using Yeast

Yeast is a raw biosorbent material used for the removal of rash dye or heavy metal ions due to their unicellular scenery and high growth rate. Yeast cells are able to be easily cultivated into low-cost growth media and are a readily existing source of biomass that has potential for bioremediation of wastes at worse pH values. The live yeast *Saccharomyces cerviceae*, *Candida albicans*, *Candida utilis*, *Candida oleophila*, *Candida tropicalis*, *Pichia stipites*, *Trichosporan porosum*, *Pichia fermentans*, *Rhodotorula mucilaginosa*, and *Williopsis californica* have been used for the removal of contaminants from the wastewater. In recent years, a number of studies have focused on some yeast that is able to bioaccumulate azo dyes in wastewater (Aksu and Donmez 2003; Donmez 2002).

A wide selection of microorganisms as well as fungi, algae, and bacteria are able to decolorize the dye via aerobic, anaerobic, and sequential anaerobic–aerobic treatment processes. In recent years, the utilization of white-rot fungi in the decolorization of textile dyes has been investigated; however, the requirement for the intrusion of other carbon sources is considered as the major drawback (Swamy and Ramsay 1999). Therefore, there is a still demand to develop more effective and more economical alternative means of dye decolorization. Although bacteria play an important role in the treatment of dyes, recent studies indicate that in addition to providing oxygen for an aerobic bacterial biodegradation or bioaccumulation of dyes, microalgae can also be used directly. In this respect, there are some recent reports available for the treatment of textile dyes by algae.

9.4 BIOACCUMULATION

9.4.1 Bioaccumulation of Dyes Using Microalgae

Microalgae are considered as eukaryotic, photosynthetic, unicellular, and fresh or marine habitant. Microalgae are used in various ways such as single-cell proteins, biofertilizers, biofuels, and bioactive compounds. Algae have been shown to be capable of removing color from various dyes through mechanisms such as biosorption, bioconversion, and bioagulation.

Biological decolorization of triphenylmethane dye (malachite green, MG) using three freshwater microalgae (*Chlorella*, *Cosmarium*, and *Euglena* species) was investigated by Khataee et al. (2010). Development parameters such as algal concentration, reaction time, pH, and temperature, in addition to initial dye concentration, were optimized. The maximum color removal was observed as 92%, 91%, and 87% for *Chlorella*, *Cosmarium*, and *Euglena* species, respectively, at an optimum conditions of 45°C, contact time of 180 min, pH value of 9.0, inoculum concentration of 9×10^6 cells/mL, and initial dye concentration of 10 mg/L. Removal of monoazo dye (Tectilon Yellow 2G) by *Chlorella vulgaris* was investigated (Acuner and Dilek 2004). Removal efficiencies were determined as 69%, 66%, and 63% for the initial Tectilon Yellow 2G concentration of 50, 200, and 400 mg/L, respectively, while acclimation of *C. vulgaris* caused them toward increase to 88%, 87%, and 88%, respectively.

The removal of malachite green using a viable freshwater alga *Cosmarium species* was investigated (Daneshvar et al. 2007). The effects of prepared parameters (pH, algal concentration, dye concentration, and temperature) on decolorization were examined. Optimum initial pH value was determined as 9. The maximum decolorization was obtained as 92.4% at optimum conditions.

The ability of *Nostoc lincki*, *Volvox aureus*, *Chlorella vulgaris*, *Oscillatoria rubescens*, *Elkatothrix viridis*, and *Lyngbya lagerlerimi* to decolorize and eliminate Orange II, methyl red, basic cationic, G-Red (FN-3G), and basic fuchsin was investigated by El-Sheekh et al. (2009). These algae showed dissimilar efficiency for color removal. Basic cationic and basic fuchsin dyes were the most suitable dyes for decolorization. *C. vulgaris* displayed the removal efficiency as 43.7% and 59.12% for Orange II and G-Red dyes, whereas *V. aureus* showed the removal efficiency as 5.02% and 3.25% for orange II and G-Red dyes, respectively.

9.4.2 Bioaccumulation of Dyes Using Fungi

Fungi are eukaryotic; they are classified as molds and yeast based on their structural morphology. Molds are filamentous in nature, multicellular, and spore-forming fungi. Yeast are unicellular, and circular or oval in shape. Fungi are used in many fields including enzyme production, antibiotic production, and pigment production. They have also been also in color removal studies.

Aspergillus flavus was tested for the elimination of Reactive Red 198 (Esmaeili and Kalantari 2012). Effect of contact time, pH of solution, inoculum volume, and initial dye concentration was optimized for the maximum color removal. The maximum color removal was reported as 84.96% with 50 mg/L of initial dye concentration, pH valve of 4, inoculum volume of 60 mL/L, and 24 h of contact time.

The fungi *Aspergillus oryzae* was used for the bioaccumulation and biosorption of Procion Red HE7B and Procion Violet H3R dye at different pH values (2.50, 4.50, and 6.50) (Corso and Maganha de Almeida 2009). The best pH for biosorption was determined as 2.50. The autoclaved fungi demonstrated a higher biosorption capacity when compared to nonautoclaved fungi at the same initial pH value of 2.50. After the dye solution was treated with fungi, the liquid was subjected to dye toxicity for the survival of *Daphnia similis* plankton.

Erum and Ahmed (2011) reported that the physicochemical cultural conditions were optimized for azo dye removal using Acid Red 151 as a mold dye. Three fungal strains (*Aspergillus flavus* SA2, *Aspergillus niger* SA1, and *Aspergillus terreus* SA3) were used for the dye removal abilities. *Aspergillus nidulans* in minimal media proved to be most effectual for utmost decolorization of AR 151 dye. However, the uppermost percentage of decolorization was shown by *A. flavus* SA2 (92.56%) at a lower concentration of dye (50 ppm) and a larger concentration of dye showed a negative result on decolorization fraction of all the experienced fungal strains. Process optimization for Acid Blue 120 dye removal using *Aspergillus lentulus* was investigated by Kaushik and Malik (2011). The interaction between three variables—urea, glucose, and initial dye concentration—was studied and modeled for two responses: dye removal and biomass production. An elevated dye removal efficiency (99.97%) and high uptake capacity (97.54 mg/g) was obtained in 24 h by means of optimum variables.

Selvam et al. (2003) reported that a white-rot fungus, *Fomes lividus*, was tested for decolorization of azo dyes such as Orange G, Congo red, and Amido Black 10B. The results showed that the fungus removed only 30.8% of orange G in the synthetic solution, while Congo red and Amido Black 10B removed 74.0% and 98.9%, respectively.

Decolorization of Orange II dye using *Phanerochaete chrysosporium* was investigated by Sharma et al. (2009). The maximum decolorization was obtained as 85% at optimized conditions of 28°C–30°C and pH value of 5 in liquid cultures under shaking aerobic incubation.

The ability of the fungus *Trichophyton rubrum* for decolorization of textile azo dyes was evaluated by Yesiladal et al. (2006). After two days of dye addition, the fungus was found to decolorize 83% of Remazol Tiefschwarz, 86% of Remazol Blue RR, and 80% of Supranol Turquoise GGL in liquid cultures. The dyes Remazol Tiefschwarz and Remazol Blue could be removed by fungal biodegradation. The fungus *T. rubrum* was found to efficiently degrade and adsorb textile dyes.

The effect of Acid Red 18, Acid Orange 7, and Reactive Black 5 on the expansion and color removal properties of *Schizophyllum commune* was studied by Renganathan et al. (2006) with respect to the original pH varying between 1 to 6 and initial dye concentration (10–100 mg/L). The best pH value was established as 2 for both growth and color elimination of the above mentioned azo dyes. Increasing the concentration of azo dyes reduced the growth of *S. commune*. It was observed that *S. commune* was found to remove Acid Red 18, Acid Orange 7, and Reactive Black 5 with a maximum specific uptake ability of 44.23, 127.53, and 180.17 mg/g in the initial dye concentration of 100 mg/L. A higher proportion of color removal was observed at lower concentrations for all the dyes. Finally it was established that the percentage of color removal was found to be more in the case of Reactive Black 5 dye when compared to the other two dyes (Acid Orange 7 and Acid Red 18).

The growth of Acid Red 88 dyes and Acid Orange 7 by *Schizophyllum commune* and *Trametes versicolor* in various inorganic acid media was studied by Renganathan et al. (2008) with the influence of initial pH (1–6) and initial dye concentration (10–100 mg/L). The most favorable pH for the accumulation of the dye was determined as 2. The maximum color removal and dye uptake capacity was observed to be more in phosphoric acid when compared to other acid medium such as sulfuric, nitric, and hydrochloric acid. The maximum percentage color removal was determined as 34.3%

(*S. commune*) and 31.9% (*T. versicolor*) for Acid Orange 7; and 72.1% (*T. versicolor*) and 72.7% (*S. commune*) for Acid Red 88.

Eichlerova et al. (2006) studied the white decay fungus *Ischnoderma resinosum* for the removal of synthetic dyes. The strain efficiently decolorized Orange G, Remazol Brilliant Blue R, Amaranth, and Poly R-478 on agar plates and Cu-phthalocyanin in liquid culture. Malachite green and crystal violet was decolorized at the low concentration of 0.1 g/L. Decolorization power of *I. resinosum* was found to be more when compared to *Pleurotus ostreatus* or *Trametes versicolor* and *Phanerochaete chrysosporium*. From the literature it was observed that *I. resinosum* was found to be able to decolorize a wide spectrum of synthetic dyes.

The use of *Fusarium oxysporum*, *Penicillium chrysogenum*, *Penicillium lanosum*, *Aspergillus fumigates*, and *Ganoderma resinaceum* on the biodegradation of five rash textile dyes (namely, Reactive Orange 16, Reactive Yellow 160, Reactive Blue 21, Reactive Blue 16, and Black 5) was evaluated by Abd El-Zaher (2010). Abedin (2008) studied the potential of *Fusarirum solani* for the decolorization of crystal violet (CV) and malachite green (MG) dyes. Maximum decolorization of CV (98%) and MG (96%) was found to be achieved through bioaccumulation after two days of incubation under shaking conditions, a nutrient medium containing 2.5 mg dye/L, supplemented with 0.5% starch for CV decolorization and 0.5% glucose for MG decolorization as an additional carbon source, 0.03% $NaNO_2$ as the nitrogen source at a pH value between 8 and 9 and 30°C. Maximum decolorization of CV (93%) and MG (94%) could be achieved through biosorption after two days of incubation at 30°C and pH value of 2.

Decolorization, degradation, and detoxification of four textile dyes (Market Red, Pagoda Red, Market Blue, and Madonna Blue) by four *Aspergillus* species were established (Ademola and Ogunjobi 2012). The decolorization/degradation capacity of the isolate was analyzed on the fifth day by a UV/visible spectrophotometer and a Fourier transform infrared (FTIR) spectrophotometer, while detoxification of the dyes was analyzed by a phytotoxicity test. At the initial concentration of 200 mg/L of the dyes, the amount of decolorization was observed to be from 71.38% to 84.76% for Market Red, 80.89% to 86.26% for Madonna Blue, 70.46% to 79.46% for Market Blue, and 60.68% to 74.82% for Pagoda Red. A decrease in the decolorization of the dyes was observed as the concentration of dye reduced slowly from 100 to 500 mg/L. Decolorization of Pagoda Red reduced from 66.47% to 19.71%, 60.68% to 10.31%, and 74.82% to 26.19% with *A. fumigatus* (3E), *A. ustus* (3D), and *A. fumigatus* (8F), respectively.

Kaushik and Malik (2012) used fungal species for the removal of dyes present in sewage with dual mode biosorption and bioaccumulation. The outcome of dye toxicity on the growth of the fungal biomass was studied through phase difference and scanning electron microscopy. Dye biosorption was studied by first- and second-order kinetic models. Effects of factors influencing adsorption and isotherm were also studied. In addition to this, fungal biomass showed toxicity response toward Methylene Blue by means of producing better aggregate of fungal pellet. To overcome the limitations of bioaccumulation, dye removal in biosorption mode was studied. Within this mode, significant removal was observed for cationic (35.4%–90.9%) and anionic (96.7%–94.3%) dyes within 24 h.

9.4.3 Bioaccumulation of Dye Using Yeast

Vitor and Corso (2008) isolated *Candida albicans* from industrial effluents for the decolorization of textile dye Direct Violet 51. In their study, the pH value was maintained as 2.5, 4.5, and 6.5. The equilibrium time was obtained for pH value of 2.5, 4.5, and 6.5 as 72, 240, and 280 h, respectively. The maximum decolorization was found to be 73.2% at pH value of 2.5 in live condition and 87.26% in autoclaved conditions. Degradation of Direct Violet 51 was confirmed by UV-spectrophotometer and FTIR analysis.

Saccharomyces cerevisiae was used for the bioaccumulation of three diazo reactive dyes such as, Remazol Black B, Remazol Blue, and Remazol Red RB by Aksu (2003). The maximum bioaccumulation capacity was found to be 88.5 mg/g for Remazol Black B, 84.6 mg/g for Remazol Blue, and 48.8 mg/g for Remazol Red RB at pH value of 3. Remazol Black B bioaccumulation through yeast on all initial dye concentrations studied was significantly higher when compared to the other two dyes studied (Remazol Blue and Remazol Red RB).

Donmez (2002) has studied bioaccumulation of reactive textile dyes by *Candida tropicalis* growing in molasses medium. Dyes used in that study were Remazol Blue, Reactive Black, and Reactive Red. The operational parameter such as pH and initial dye concentrations were optimized. The maximum specific bioaccumulation capacity of *C. tropicalis* was found to be 111.9 mg/g for Remazol Blue, 101.9 mg/g for Reactive Black, and 79.3 mg/g for Reactive Red at approximately 700 mg/L initial dye concentration. With the combination of molasses sucrose and textile dye on the growth and dye bioaccumulation property of *Candida tropicalis* was investigated by Donmez (2002).

Immobilized cells of *Candida tropicalis* for the removal of Basic Violet 3 was investigated by Charumathi and Das (2010). Yeast cells were immobilized with various matrixes, including CMC, sodium alginate, agar, agarose, and polyvinyl alcohol. The maximum color removal was found to be higher with the use of sodium alginate as a matrix. The sodium alginate matrix concentrations varied from 1% to 4%; the maximum color removal was found at 3% matrix concentration. Bead size and cell concentrations were found to be optimum at 2 mm and 3 mg/L, respectively. Arora et al. (2005) reported that *Candida tropicalis* cell decolorized 93% of monoazo dispersed dye 1 at aerobic and optimized conditions.

Bioaccumulation properties of *Candida utilis* for the removal of Remazol Turquoise Blue-G (RTBG) reactive dye was studied by Gonen and Aksu (2009) with the effect of initial sucrose concentration and initial dye concentration. The initial sucrose concentrations in growth media varied from 5 to 15 g/L. The initial dye concentrations varied from 50 to 500 mg/L. The optimization studies were performed with the use of response surface methodology (RSM). RSM confirmed that *C. utilis* was capable of bioaccumulating RTBG with the maximum uptake yield of 82.0% in 15 g/L of sucrose and 50 mg/L of dye containing growth medium.

Das et al. (2010) studied the mutual effects of sugarcane bagasse extract and synthetic dyes on the growth and bioaccumulation properties of *Pichia fermentans*. Direct Red 28, Acid Blue 93, and Basic Violet 3 dyes were utilized for studying the bioaccumulation capacity of *P. fermentans*. The influence of pH on bioaccumulation of dyes was investigated by optimizing the pH (3, 5, 7, and 9) value of the aqueous extract. The combined effect of sugarcane bagasse extract (8 to 24 mg/L) and dye concentration (10 to 30 mg/L) was studied. The maximum bioaccumulation was noted at a pH value of 5 for all the dyes. The optimum combination predicted through RSM confirmed that *P. fermentans* was competent for bioaccumulation of Basic Violet 3 dye with 69.8% at 10 mg/L of dye plus 24 g/L sugar extract from sugarcane bagasse.

Thirty-nine yeast isolates from Las Yungas rainforest (Tucumán, Argentina) soil were obtained using an isolation scheme for both dye tolerance and assimilation. Eight isolates were chosen according to their absorption of dyes using a sole carbon/nitrogen source. Selected yeasts fit into the basidiomycetous genus. Enzymatic production of manganese peroxidase, tyrosinase, and laccase was studied in liquid media for decolorization.

9.4.4 Bioaccumulation of Dyes Using Bacteria

Bacteria are prokaryotic and unicellular, and have been investigated for their color removal abilities. The decolorization of azo-metal complex dyes using *Halobacillus* sp. was investigated by Demirci et al. (2011). Lanaset Brown B and Lanaset Navy R complex dyes were used. A decolorization experiment was carried out at 120 mg/L concentration of both dyes at room temperature with an acidic pH of 4.5. Lanaset Brown B was decolorized at a higher percentage of 96.12% with 78 h duration. However, Lanaset Navy R was decolorized with the percentage removal of 60.66% after 3 h duration.

Bacterial decolorization of azo dyes using *Rhodopseudomonas palustris* was investigated by Liu et al. (2006). The property of environmental factors, such as pH (4–10), temperature (30°C–45°C), additional carbon sources (peptone, yeast extract, sucrose, etc.), and dye concentrations (0–1250 mg/L), on bacterial decolorization were investigated. The specific decolorization uptake of Acid Red B, Reactive Blue GL, Acid Red G, and RBR X-3B were obtained as 14.9, 11.8, 7.6, and 3.4 mg/g cell/h, respectively.

Çelik et al. (2012) reported that the sulfonated Reactive Red 195 dye (RR195) was decolorized by the extensively spread, ecological, photohetherothrophic strain 51ATA that belong to *Rhodopseudomonas palustris*. This bacterium mineralized the dye under anaerobic conditions with 100% efficiency.

A strain of *Pseudomonas putida* was utilized by Chen et al. (2007) for the decolorization of crystal violet (CV) that effectively degraded CV up to 80% of 60 μM CV, while the sole carbon source was degraded in liquid media for 1 week. They proposed that CV degradation occurs through a stepwise demethylation process to produce mono-, di-, tri-, tetra-, penta-, and hexa-demethylated CV species.

Telke et al. (2008) used *Rhizobium radiobacter* bacterial culture for the removal of Reactive Red 141. This strain decolorized 90% of Reactive Red 141 with 50 mg/L initial concentration and 0.807 mg of dye removed/g of dry cells/h of specific decolonization speed in stationary anoxic state at the most favorable conditions of a pH of 7.0 and temperature of 30°C.

Sani and Banerjee (1999) screened several organisms for their ability to decolorize triphenylmethane collection of dyes. A *Kurthia* sp. was selected on the basis of fast dye decolorization. Under aerobic conditions, 98% color removal was achieved by this study. A number of triphenylmethane dyes, including crystal violet, brilliant green, pararosaniline malachite green, magenta, and ethyl violet, were used. The percentage of decolonization of ethyl magenta as 92%, violet as 8%, malachite green as 96%, pararosaniline as 100%, brilliant green as 100%, and crystal violet as 96% was observed.

Thermophilic microbial strains in northwestern Spain displayed outstanding decolonization ability, and this was established by Deive et al. (2010). The research work was carried out for the degradation of Reactive Black dye as 70% at a neutral pH.

Wang et al. (2012) isolated a bacterium from a soil sample obtained from a textile dye factory. Phenotypic and phylogenetic analysis of the 16S rRNA indicated *Bacillus* sp. It shows the ability to decolorize reactive textile dyes. Static circumstances by means of pH 7.0 at 40°C were observed to be the best parameters for decolorizing RB-5.

Mabrouk and Yusef (2008) isolated *Bacillus subtilis* bacterium from a dye-contaminated sample. The bacterium exhibited color removal capability over a range of dye concentration (12.5–125 mg/L), pH (5–9), and temperatures (25°C–40°C). The maximum decolorization was observed at almost 99% after 6 hours of incubation.

Decolorization of azo dyes using *Rhodobacter sphaeroides* was investigated by Song et al. (2003). *Rhodobacter sphaeroides* AS1.1737 was found to decolorize more than 90% of azo dyes at 200 mg/L within 24 h. The most favorable culture conditions were observed as anaerobic lighting (1990 1×), peptone as the carbon source, a temperature between 35°C and 40°C, and pH value of 7–8.

Ozdemir et al. (2008) studied the decolorization of Acid Black 210 through a bioluminescent bacterium, *Vibrio harveyi* TEMS1, isolated from coastal seawater from Izmir Bay, Turkey. The maximum speed of decolorization of Acid Black 210 was observed when Luria Bertani medium was used. Decolorization of Acid Black 210 was observed as 38.9% and 93.9% at 24 h under shaking and static situations, respectively. The most favorable dye-decolorizing activity of the bacterium was obtained at 100 ppm initial dye concentration at the temperature of 20°C.

9.5 BIOSORPTION

9.5.1 Biosorption of Dyes Using Microalgae

Biosorption of three reactive dyes—namely, Remazol Black B, Remazol Red RR, and Remazol Golden Yellow RNL—onto dried *Chlorella vulgaris* was investigated within a batch system (Aksu and Tezer 2005). The algal biomass exhibit the maximum dye uptake ability at the preliminary pH value of 2.0 for all dyes. The outcome of temperature on maximum equilibrium sorption capacity was obtained at 35°C for Remazol Black B and at 25°C for both Remazol Red RR and Remazol Golden Yellow RNL. Biosorption capacity of alga was found to be increased with an increase in initial dye concentration up to 800 mg/L for both Remazol Black B and Remazol Red RR dyes, and up to 200 mg/L for Remazol Golden Yellow RNL dye. Among the three dyes, Remazol Black B was found to be adsorbed with uptake capacity of 419.5 mg/g.

The green microalgae *Scenedesmus quadricauda* was immobilized in alginate gel beads. The immobilized active (live) (IASq) and heat inactivated (dead) *S. quadricauda* (IHISq) was used for the elimination of Reactive Blue 19 (Remazol Brilliant Blue R or RBBR) from aqueous solution in the concentration range of 25 to 200 mg/L (Ergene et al. 2009). At 150 mg/L initial dye concentration, the IASq and IHISq exhibited the maximum dye uptake capacity at 30°C and at the initial pH value of 2.0. At the equivalent initial dye in the batch system, the adsorption capacity was found for IASq as 45.7 mg/g with 300 min. The adsorption capacity for IHISq was observed as 48.3 mg/g in 300 min. After 300 min, the adsorption capacity was not further increased for 24 h contact time. The Freundlich, Langmuir, Temkin, Dubinin–Radushkevich, and Flory–Huggins isotherm models were used to fit the equilibrium biosorption data. The Langmuir, Freundlich, and Dubinin–Radushkevich equations showed better coefficient of determination than the Temkin and Flory–Huggins equations. The monolayer biosorption capacity of the biomass was established to be 68 and 95.2 mg/g for IASq and IHISq, respectively. The experimental information was also tested in conditions of kinetic individuality and it was observed that the biosorption data was well fitted with pseudo-second-order kinetics.

Tsai and Chen (2010) studied *Chlorella*-based biomass as a low-cost biosorbent. The dye malachite green (MG) was removed in batch experiments with the influence of a variety of parameters, including initial dye concentration, agitation speed, biosorbent loading, initial pH, and temperature. The experimental data revealed that the rapid removal of cationic solute by means of the dead microalgae was significant depending on the initial MG concentration and algal loading in addition the biosorption kinetics obeyed the pseudo-second-order rate equation. According to the biosorption behaviors of MG from aqueous solution with *Chlorella*-based biomass, the uptake capacity of the biomass was established to be more when compared to the uptake capacity of commercial-activated carbon. This work showed that the *Chlorella* biomass was able to be efficiently used as an inexpensive biosorbent for the removal of MG from aqueous solutions.

9.5.2 Biosorption of Dyes Using Fungi

Biosorption of methyl violet, basic fuchsin, and them combined using *Aspergillus niger* dead fungal biomass was studied. The parameters such as initial pH, dye concentration and biosorbent concentration were optimized. The maximum uptake capacities of both dyes were reported as 25 mg/g at a preliminary dye concentration of 30 mg/L and at pH value of 5.

The possible use of dried out biomasses of *Rhizopus stolonifer*, *Geotrichum* sp., *Fusarium* sp., and *Aspergillus fumigatus* as biosorbents for the removal of bromophenol blue (BPB) dye from aqueous solution was conducted (Zeroual et al. 2006). The effect of the solution pH and initial dye concentration was optimized. The fungal biomasses exhibited the maximum dye biosorption on a pH value of 2. The kinetic data was well fitted in the pseudo-second-order model. The equilibrium data was found to be well fitted with the Langmuir adsorption isotherm model when compared to the Freundlich model.

Erden et al. (2011) studied the lyophilized *Trametes versicolor* biomass. This biomass was used as a sorbent for biosorption of textile dye (Sirius Blue K-CFN) from aqueous solution. The batch sorption was studied with respect to pH, adsorbent dose, dye concentration, and equilibrium time. The maximum dye uptake capacity was reported as 62.62 mg/g at an optimal initial pH of 3.0, equilibrium time of 2 h, initial dye concentration of 100 mg/L, biomass concentration of 1.2 mg/L, and a 26°C temperature.

Fungal stain *Cladosporium* sp. used for the removal of Azure Blue dye from an aqueous solution was studied (Fan et al. 2012) with kinetic and equilibrium studies. The biosorption studies were established with the influence of pH, initial dye concentration, and biosorbent dosage. The adsorbent amount of 0.1 g in 20 mL was found to be optimum for maximum dye uptake. The equilibrium data fit well with the Langmuir isotherm model when compared to the Freundlich model. The utmost monolayer biosorption competence of the biosorbent was observed as 51.4 mg/g. The pseudo-second-order kinetic model adequately describes the kinetic data.

Biosorption of methylene blue (MB) via dead fungal biomass, *Aspergillus fumigates*, was investigated by Abdallah and Taha (2012). For a preliminary concentration of MB at 5 mg/L, the maximum uptake was obtained at pH 7 and 30°C. The equilibrium was reached after 90 min. Different chemical and physical pretreatments were employed to improve the adsorption capacity. The maximum biosorption capacity was observed as 125 mg/g. A desorption study was conducted altering the pH value from 3 to 10. The highest desorption percentage of MB was found to be 80% at pH 3. The modeling of the experimental information at equilibrium was performed with Freundlich, Langmuir, Dubinin–Radushkevich (D–R), Langmuir–Freundlich, and Temkin isotherms. The consequence has revealed that the biosorption was positive and agreed with the Langmuir and Freundlich models, with correlation coefficient as 1.00 in both cases.

Biosorption of Acid Red 57 (AR57) on *Neurospora crassa* was studied by Akar et al. (2006) with variation of contact time, pH, biosorbent concentration initial dye concentration, and temperature to describe stability and kinetic models. The equilibrium time was attained as 40 min for the biosorption of AR57. Langmuir and Dubinin–Radushkevich (D–R) isotherm models were used to analyze the equilibrium data for the biosorption of AR57 at different temperatures. The kinetic data for the biosorption of AR57 was analyzed and rate constants were determined. The overall biosorption procedure was well described with a pseudo-second-order kinetic model. The change in Gibbs free energy and enthalpy in addition to entropy of biosorption was also evaluated for the biosorption of AR57 on *N. crassa*. The results indicated that the biosorption was observed to be impulsive and exothermic in the environment.

Trichoderma sp. and *Aspergillus niger* biomasses were used as biosorbent for biosorption of an azo dye (Sivasamy and Sundarabal 2011). Batch biosorption studies were performed for the removal of Orange G from aqueous solution with the effect of various parameters like initial dye concentration, initial aqueous phase, pH, and biomass dosage. From the results, it was observed that the highest biosorption was observed at pH 2. Equilibrium data was analyzed by model equations such as Freundlich and Langmuir isotherms, and it was found that both isotherm models fitted the adsorption data well. The monolayer adsorption capacity was obtained as 0.48 mg/g for *Aspergillus niger* and 0.45 mg/g for *Trichoderma* sp. Biosorption kinetic data was analyzed by means of pseudo-first-order and pseudo-second-order rate equations, and it was observed that the pseudo-second-order model fitted the kinetic data well for both biomasses.

Yang et al. (2011) investigated the biosorption of nonviable *Penicillium* YW 01 biomass for elimination of acerbic Black 172 metal-complex dye (AB) as well as Congo Red (CR) in aqueous solutions. Utmost biosorption capacity of 225.38 mg/g for AB and 411.53 mg/g for CR under innovative dye absorption at 800 mg/L, pH 3.0, and 40°C conditions were observed. Biosorption data were effectively described by Langmuir isotherm as well as the pseudo-second-order kinetic model. The Weber-Morris model investigation indicated that intraparticle dispersal was the controlling step intended for biosorption of AB and CR against biosorbent. Analysis based on the artificial neural complex and genetic algorithms hybrid model indicate that preliminary dye concentration along with temperature appear to exist the most influential parameter for biosorption procedure of AB as well as CR against biosorbent, respectively.

Caner et al. (2011) reported that the biosorption of dried *Penicillum restrictum* for Reactive Yellow 145 was studied with respect to equilibrium time, pH, and temperature to decide equilibrium and kinetic models. The most suitable pH as well as equilibrium time was observed as 1.0 and 75 min, respectively, next to a biomass amount of 0.4 mg/L with 20°C. Data obtained from batch studies was analyzed with the Dubinin–Radushkevich (D–R), Freundlich, and Langmuir isotherm models. Maximum uptake capacity (q_m) for the dye was found to be 109.7, 115.2, and 116.5 mg/g for different temperatures of 20°C, 30°C, and 40°C, respectively. The kinetic data was found to be fitted well with the pseudo-second-order kinetic model.

9.5.3 Biosorption of Dyes Using Yeast

Aksu and Donmez (2003) studied the biosorption capacity of yeast cultures *Saccharomyces cerevisiae, Candida* sp., *Candida tropicalis, Kluyveromyces marxianus, Candida lipolytica, Candida utilis, Schizosaccharomyces pombe, Candida quilliermendii,* and *Candida membranaefaciens* for the removal of Remazol Blue dye from aqueous solution. Biosorption studies were carried out for the effect of initial solution pH and initial dye concentration for the removal of Remazol Blue dye. When compared with the nine yeast species, *C. lipolytica* exhibited the highest dye uptake capacity of 250 mg/g.

Rhodotorula glutinis has been described as a promising producer of commercially valuable products such as carotenoids and exopolysaccharides. The commercial production and extraction of these products from *R. glutinis* would generate large quantities of waste biomass, which may be employed as inexpensive biosorbent (Cho et al. 2011).

9.5.4 Biosorption of Dyes Using Bacteria

A biosorption study for the removal of Acid Blue 225 and Acid Blue 062 dyes from aqueous solutions by means of *Bacillus amyloliquefaciens* was investigated by Yenikaya et al. (2010). A batch biosorption study was conducted with the effect of parameters such as pH, initial dye concentration, contact time, and adsorbent dosage. The maximum adsorption capacity of biomass for Acid Blue 225 and Acid Blue 062 were obtained as 111.15 and 112.19 mg/g, correspondingly.

Atar et al. (2008) researched thermodynamic, kinetic, and equilibrium studies for the biosorption of Basic Blue 41 (BB 41) from aqueous solution using *Bacillus macerans*. Environmental parameters such as temperature, pH, contact time, and biomass dosage in addition to initial dye concentration were optimized. The maximum adsorption capacity was established to be 89.2 mg/g under optimal conditions of pH (10.0) as well as temperature (25°C).

The ability of *Pseudomonas* sp. for the adsorption of Acid Black 172 was studied (Du et al. 2012) to decide the kinetics as well as mechanisms involved in biosorption of the dye. Live and heat-treated cultures were used. Scanning electron microscopy (SEM) and atomic force microscopy (AFM) were used to study the surface characteristics of strain DY1 (*Pseudomonas* sp.). The structural characteristics of live and heat-treated biomass of strain DY1 were analyzed by transmission electron microscopy (TEM). Kinetic data for biosorption was found to fit well with the pseudo-second-order model. The maximum uptake capacity was reported as 2.98 mmol/g of heat-treated biomass of the *Pseudomonas* sp.

Vijayaraghavan et al. (2008) investigated the amino acid fermentation of industrial waste using *Corynebacterium glutamicum*. It was established that *C. glutamicum* had outstanding biosorption capacity toward the removal methylene blue (MB) dye. Owing to practical difficulties in solid–liquid separation and biomass regeneration, *C. glutamicum* was immobilized in a polysulfone matrix. The optimization of pH on biosorption was exposed and the neutral or alkaline pH favored MB biosorption. Isotherm experiments indicated that *C. glutamicum* exhibited somewhat lesser dye uptake for the immobilized state when compared to free biomass. The maximum dye uptake capacity was reported as 339.2 mg/g at pH value of 8.

Further study using *Corynebacterium glutamicum* was employed as a biosorbent for the elimination of Reactive Black 5 (RB5) from aqueous solution (Vijayaraghavan and Yun 2007). In the study, pretreatment of biomass on the biosorption capacity of *C. glutamicum* toward RB5, using more than few chemical agents, such as HNO_3, HCl, NaOH, Na_2CO_3, H_2SO_4, $CaCl_2$, and NaCl was established. From this study, 0.1 M HNO_3

gave the highest uptake of the RB5 as 195 mg/g at pH 1 at an initial RB5 concentration of 500 mg/L. The solution pH and temperature were established to affect the biosorption capacity. The biosorption isotherms were studied at different pH and temperatures. The optimum pH was observed as 1 at the temperature of 35°C. The equilibrium data was found to fit very well with three-parameter models (Redlich–Peterson and Sips models) when compared to two-parameter models (Freundlich and Langmuir models). Maximum RB5 uptake of 419 mg/g was obtained at pH 1 and temperature of 35°C, according to the Langmuir model. The kinetic data was analyzed using pseudo-first-order and pseudo-second-order models and kinetic data was found to be fitted very well with the pseudo-second-order kinetic model. *C. glutamicum* biomass was used as a biosorbent for the removal of Reactive Yellow 2, Reactive Orange 16, Reactive Red 4, and Reactive Blue 4.

9.5.5 Advantages of Using Yeast for Color Removal Studies

Yeast is one of the major biomasses that could be used for the bioremediation of textile dyes at lower pH. Yeasts are low cost, and readily and abundantly available. Additionally, yeast cells could be used for a broad range of textile dyes under a wide range of environmental parameters. Yeast has been utilized for rapid accumulation of metal ions from solution, but little research work has been completed for the accumulation of textile dyes in wastewater (Aksu 2003). Compared to algae, filamentous fungi, and bacteria, yeasts exhibit potential individuality (Gonen and Aksu 2009). Yeast cells are easily inoculated with growth media in a laboratory scale and are readily available for bioaccumulation of contaminants from wastewater at lower pH values (Ertugrul et al. 2009). Yeast needs light source for their growth. The biomass production per unit time was found to be more in yeast culture when compared to fungi and bacteria.

9.6 CONCLUSION

An enormous amount of textile dyes have been dispersed across our nation and worldwide and pollution control is a prime concern. Textile dyes are usually dispersed in aquatic streams, leading to environmental pollution. These dyes can cause allergic, skin irritation, cancer, mutation, liver damage, and damage to the reproductive system. These dyes have been treated using various treatment physical and chemical methods. But these methods are not economically feasible. Treatment of textile dyes using activated carbon was found to be an efficient method, but the production of activated carbon was not found to be economically viable. Low-cost adsorbents were used instead of activated carbon. But low-cost adsorbents have low adsorption capacity. So new economical and easily available adsorbents and methods are required.

Biological methods have been utilized using various viable microbial biomass and dead microbial biomass. Bioaccumulation of dyes using yeast, microalgae, fungi, and bacteria have been studied. But the viable microbial species requires suitable process conditions for the survival. The higher concentration of dye leads to the inhibition for the survival of viable microbial species. Thus, the biosorption method was used for the treatment of textile dyes. In biosorption, microalgae, fungi, yeast, and bacteria were used. In the biosorption method process, conditions could be maintained based on the removal efficiency. Compared with living microbial species, dead species biomass could be stored for extended periods and their operation is easy and their regeneration is simple. Therefore, dead cells can be considered efficient biosorbents.

REFERENCES

Abd El-Zaher, E.H.F. 2010. Biodegradation of reactive dyes using soil fungal isolates and *Ganoderma resinaceum*. *Ann. Microbiol.* 60:269–278.

Abdallah, R., and Taha, S. 2012. Biosorption of fumigates blue from aqueous solution by nonviable *Aspergillus fumigates*. *Chem. Eng. J.* 195–196:69–76.

Abedin, R.M.A. 2008. Decolorization and biodegradation of crystal violet and malachite green by *Fusarium solani* (Martius) Saccardo. A comparative study on biosorption of dyes by the dead fungal biomass. *Am.-Euras. J. Bot.* 1, no. 2:17–31.

Acuner, E., and Dilek, F.B. 2004. Treatment of tectilon yellow 2G by *Chlorella vulgaris*. *Process Biochem.* 39:623–631.

Ademola, E.A., and Ogunjobi, A.A. 2012. Decolourization, degradation and detoxification of textile dyes by *Aspergillus* species. *Environmentalist* 32:19–27.

Akar, T., Demir, T.A., Kiran, I., Ozcan, A., Ozcan, A.S., and Tunali, S. 2006. Biosorption potential of *Neurospora crassa* cells for decolorization of acid red 57 (AR57) dye. *J. Chem. Technol. Biotechnol.* 81:1100–1106.

Aksu, Z., 2003. Reactive dye bioaccumulation by *Saccharomyces cerevisiae*. *Process Biochem.* 38:1437–1444.

Aksu, Z., and Donmez, G. 2003. A comparative study on the biosorption characteristics of some yeasts for Remazol Blue reactive dye. *Chemosphere* 50:1075–1083.

Aksu, Z., and Tezer, S. 2005. Biosorption of reactive dyes on the green alga *Chlorella vulgaris*. *Process Biochem.* 40:1347–1361.

Anjaneyulu, Y., Sreedhara Chary, N., and Raj, S.S.D. 2005. Decolourization of industrial effluents—Available methods and emerging technologies—A review. *Rev. Environ. Sci. Biotechnol.* 4:245–274.

Ansari, R., and Mosayebzadeh, Z. 2010. Removal of basic dye methylene blue from aqueous solutions using sawdust and sawdust coated with polypyrrole. *J. Iranian Chem. Soc.* 7:339–350.

Arora, S., Saini, H.S., and Singh, K. 2005. Decolorisation of a monoazo disperse dye with *Candida tropicalis. Color. Technol.* 121:298–303.

Atar, N., Olgun, A., and Çolak, F. 2008. Thermodynamic, equilibrium and kinetic study of the biosorption of Basic Blue 41 using *Bacillus macerans. Eng. Life Sci.* 8, no. 5:499–506.

Banat, I.M., Nigam, P., Singh, D., and Marchant, R. 1996. Microbial decolorization of textile-dye containing effluents—A review. *Bioresour. Technol.* 58: 217–227.

Belsare, D.K., and Prasad, D.Y. 1988. Decolorization of effluent from the bagasse based pulp mills by white rot fungus *Schizophyllum commune. Appl. Microbiol. Biotechnol.* 28:301–304.

Benito, G.G., Miranda, M.P., and de los Santos, D.R. 1997. Color removal by *Trametes versicolor* mycelium. *Bioresource Technol.* 61, no. 1:33–37.

Caner, N., Kiran, I., Ilhan, S., Pinarbasi, A., and Iscen, C.F. 2011. Biosorption of Reactive Yellow 145 dye by dried *Penicillum restrictum*: Isotherm, kinetic, and thermodynamic studies. *Sep. Sci. Technol.* 46:2283–2290.

Çelik, L., Öztürk, A., and Abdullah, M.I. 2012. Biodegradation of Reactive Red 195 azo dye by the bacterium *Rhodopseudomonas palustris* 51ATA. *Afr. J. Microbiol. Res.* 6, no. 1:120–126.

Charumathi, D., and Das, N. 2010. Removal of synthetic dye basic violet 3 by immobilised *Candida tropicalis* grown on sugarcane bagasse extract medium. *Int. J. Eng. Sci. Technol.* 2, no. 9:4325–4335.

Chen, C., Liao, H., Cheng, C., Yen, C., and Chung, Y. 2007. Biodegradation of crystal violet by *Pseudomonas putida. Springer* 29, no. 3:391–396.

Cho, D.H., Chu, K.H., and Kim, E.Y. 2011. Loss of cell components during rehydration of dried *Rhodotorula glutinis* and its implications for lead uptake. *Eng. Life Sci.* 11, no. 3:283–290.

Chojnacka, K. 2010. Biosorption and bioaccumulation—The prospects for practical applications. *Environ. Int.* 36:299–307.

Cooper, P. 1993. Removing color from dye house wastewaters a critical review of technology available. *J. Soc. Dyers Col.* 109:97–100.

Corso, C.R., and Maganha de Almeida, A.C. 2009. Bioremediation of dyes in textile effluents by *Aspergillus oryzae. Microb. Ecol.* 57:384–390.

Daneshvar, N., Ayazloo, M., Khataee, A.R., and Pourhassan, M. 2007. Biological decolorization of dye solution containing malachite green by microalgae *Cosmarium sp. Bioresour. Technol.* 98:1176–1182.

Das, D., Charumathi, D., and Das, N. 2010. Combined effects of sugarcane bagasse extract and synthetic dyes on the growth and bioaccumulation properties of *Pichia fermentans* MTCC 189. *J. Hazard. Mater.* 183:497–505.

Davis, G.M., Koon, J.H., and Adams, J.C.E. 1977. Treatment of two textile dye house wastewaters. *Proceedings of the 32nd Industrial Waste Conference*, Purdue University, Lafayette, Indiana, 981–997.

Deive, F.J., Domínguez, A., Barrio, T., Moscoso, F., Morán, P., Longo, M.A., and Sanromán, M.A. 2010. Decolorization of dye Reactive Black 5 by newly isolated thermophilic microorganisms from geothermal sites in Galicia (Spain). *J. Hazard. Mater.* 182:735–742.

Demirci, A., Mutlu, M.B., Guven, A., Korcan, E., and Guven, K. 2011. Decolorization of textile azo-metal complex dyes by a halophilic bacterium isolated from Camalti Saltern in Turkey. *Clean* 39, no. 2:177–184.

Donmez, G. 2002. Bioaccumulation of the reactive textile dyes by *Candida tropicalis* growing in molasses medium. *Enzyme Microb. Technol.* 20:363–366.

Du, L.N., Wang, B., Li, G., Wang, S., Crowley, D.E., and Zhao, Y.H. 2012. Biosorption of the metal-complex dye acid black 172 by live and heat-treated biomass of *Pseudomonas* sp. strain DY1: Kinetics and sorption mechanisms. *J. Hazard. Mater.* 205–206:47–54.

Eichlerova, I., Homolka, L., and Nerud, F. 2006. Evaluation of synthetic dye decolorization capacity in *Ischnoderma resinosum. J. Ind. Microbiol. Biotechnol.* 33:759–766.

El-Geundi, M.S. 1991. Colour removal from textile effluents by adsorption techniques. *Wat. Res.* 25:271–273.

El-Sheekh, M.M., Gharieb, M.M., and Abou-El-Souod, G.W. 2009. Biodegradation of dyes by some green algae and cyanobacteria. *Int. Biodeterior. Biodegrad.* 63:699–704.

Erden, E., Kaymaz, Y., and Pazarlioglu, N.K., 2011. Biosorption kinetics of a direct azo dye Sirius blue K-CFN by *Trametes versicolor. Electron. J. Biotechnol.* doi: 10.2225/vol14-issue2-fulltext-8.

Ergene, A., Ada, K., Tan, S., and Katircioğlu, H. 2009. Removal of Remazol brilliant blue R dye from aqueous solutions by adsorption onto immobilized *scenedesmus quadricauda*: Equilibrium and kinetic modeling studies. *Desalination* 249:1308–1314.

Ertugrul, S., San, N.O., and Donmez, G. 2009. Treatment of dye (Remazol Blue) and heavy metals using yeast cells with the purpose of managing polluted textile wastewaters. *Ecol. Eng.* 35:128–134.

Erum, S., and Ahmed, S. 2011. Comparison of dye decolorization efficiencies of indigenous fungal isolates. *Afr. J. Biotechnol.* 10, no. 17:3399–3411.

Esmaeili, A., and Kalantari, M. 2012. Bioremoval of an azo textile dye, Reactive Red 198, by *Aspergillus flavus. World J. Microbiol. Biotechnol.* 28:1125–1131.

Fan, H., Yang, J.S., Gao, T.G., and Yuan, H.L. 2012. Removal of a low-molecular basic dye (azure blue) from aqueous solutions by a native biomass of a newly isolated *Cladosporium* sp.: Kinetics, equilibrium and biosorption simulation. *J. Taiwan Inst. Chem. Eng.* 43:386–392.

Gallagher, K.A., Healy, M.G., and Allen, S.J. 1997. Biosorption of synthetic dye and metal ions from aqueous effluents using fungal biomass. In *Global Environmental Biotechnology*, edited by D.L. Wise, 27–50. Elsevier, UK.

Gonen, F., and Aksu, Z. 2009. Predictive expressions of growth and Remazol turquoise blue-G reactive dye bioaccumulation properties of *Candida utilis*. *Enzyme Microb. Technol.* 45:15–21.

Groff, K.A. 1993. Textile waste–textile industry waste-water waste-disposal review. *Wat. Environ. Res.* 65:421–423.

Heiss, G.S., Gowan, B., and Dabbs, E.R. 1992. Cloning of DNA from a *Rhodococcus* strain conferring the ability to decolorize sulfonated azo dyes. *FEMS Microbiol. Lett.* 99:221–226.

Kaushik, P., and Malik, A. 2011. Process optimization for efficient dye removal by *Aspergillus lentulus* FJ172995. *J. Hazard. Mater.* 185:837–843.

Kaushik, P., and Malik, A. 2012. Comparative performance evaluation of *Aspergillus lentulus* for dye removal through bioaccumulation and biosorption. *Environ. Sci. Pollut. Res.* doi:10.1007/s11356-012-1190-8.

Khaled, A., El Nemr, A., El-Sikaily, A., and Abdelwahap, O. 2009. Removal of direct n-blue-106 from artificial textile dye effluent using activated carbon from orange peel: Adsorption isotherm and kinetic studies. *J. Hazard. Mater.* 165:100–110.

Khataee, A.R., Zarei, M., and Pourhassan, M. 2010. Bioremediation of malachite green from contaminated water by three microalgae: Neural network modeling. *Clean* 38, no. 1:96–103.

Kulla, H.G., Klausener, F., Meyer, U., Lüdeke, B., and Leisinger, T. 1983. Interference of aromatic sulfo groups in the microbial degradation of the azo dyes Orange I and Orange II. *Arch. Microbiol.* 135:1–7.

Lee, T.H., Aoki, H., Sugano, Y., and Shoda, M. 2000. Effect of molasses on the production and activity of dye decolorizing peroxidase from *Geotrichum candidum*. *J. Biosci. Bioeng.* 89:545–549.

Liu, G.F., Zhou, J.T., Wang, J., Song, Z.Y., and Qv, Y.Y. 2006. Bacterial decolorization of azo dyes by *Rhodopseudomonas palustris*. *World J. Microbiol. Biotechnol.* 22:1069–1074.

Mabrouk, M.E.M., and Yusef, H.H. 2008. Decolorization of fast red by *Bacillus subtilis* HM. *J. Appl. Sci. Res.* 4, no. 3:262–269.

Meyer, U. 1981. Biodegradation of synthetic organic colorants. *FEMS Symp.* 12:371–385.

Miranda, M.P., Benito, G.G., San Cristobal, N., and Nieto, C.H. 1996. Color elimination from molasses wastewater by *Aspergillus niger*. *Bioresour. Technol.* 57:229–235.

Ozdemir, G., Pazarbasi, B., Kocyigit, A., Omeroglu, E.E., Yasa, I., and Karaboz, I. 2008. Decolorization of acid black 210 by *Vibrio harveyi* TEMS1, a newly isolated bioluminescent bacterium from Izmir Bay, Turkey. *World J. Microbiol. Biotechnol.* 24:1375–1381.

Ozkahraman, B., Bal, A., Acar, I., and Guclu, G. 2011. Adsorption of brilliant green from aqueous solutions onto crosslinked chitosan graft copolymers. *Clean* 39, no. 11:1001–2011.

Prigione, V., Varese, G.C., Casieri, L., and Marchisio, V.F. 2008. Biosorption of simulated dyed effluents by inactivated fungal biomasses. *Bioresour. Technol.* 99:3559–3567.

Renganathan, S., Gautam, P., Karthik, V., Miranda, L.R., and Velan, M. 2008. A comparative study on the accumulation of azo dyes using *Schizophyllum commune* and *Trametes versicolor* in different inorganic acid containing growth medium. *Asia-Pac. J. Chem. Eng.* 3:400–407.

Renganathan, S., Thilagaraj, W.R., Miranda, L.R., Gautam, P., and Velan, M. 2006. Accumulation of Acid Orange 7, Acid Red 18 and Reactive Black 5 by growing *Schizophyllum commune*. *Bioresour. Technol.* 97:2189–2193.

Robinson, T., McMullan, G., Marchant, R., and Nigam, P. 2001. Remediation of dyes in textile effluent: A critical review on current treatment technologies with a proposed alternative. *Bioresour. Technol.* 77, no. 3:247–255.

Sadettin, S., and Donmez, G. 2006. Bioaccumulation of reactive dyes by thermophilic cyanobacteria. *Process Biochem.* 41:836–841.

Sani, R.K., and Banerjee, U.C. 1999. Decolorization of triphenylmethane dyes and textile and dye-stuff effluent by *Kurthia* sp. *Enzyme Microb. Technol.* 24, no. 7:433–437.

Saratale, R.G., Saratale, G.D., Chang, J.S., and Govindwar, S.P. 2011. Bacterial decolorization and degradation of azo dyes: A review. *J. Taiwan Inst. Chem. Eng.* 42:138–157.

Selvam, K., Swaminathan, K., and Chae, K.S. 2003. Microbial decolorization of azo dyes and dye industry effluent by *Fomes lividus*. *World J. Microbiol. Biotechnol.* 19:591–593.

Sharma, P., Singh, L., and Dilbaghi, N. 2009. Biodegradation of Orange II dye by *Phanerochaete chrysosporium* in simulated wastewater. *J. Sci. Ind. Res.* 68:157–161.

Shen, D.Z., Fan, J.X., Zhou, W.Z., Gao, B.Y., Yue, Q.Y., and Kang, Q. 2009. Adsorption kinetics and isotherm of anionic dyes onto organo-bentonite from single and multisolute systems. *J. Hazard. Mater.* 172:99–107.

Sivasamy, A., and Sundarabal, N. 2011. Biosorption of an azo dye by *Aspergillus niger* and *Trichoderma* sp. fungal biomasses. *Curr. Microbiol.* 62:351–357.

Song, Z.Y., Zhou, J.T., Wang, J., Yan, B., and Du, C.H. 2003. Decolorization of azo dyes by *Rhodobacter sphaeroides*. *Biotechnol. Lett.* 25:1815–1818.

Swamy, J., and Ramsay, J.A. 1999. Effects of glucose and NH_4 concentration on sequential dye decolorization by *Trametes versicolor*. *Enzyme Microb. Technol.* 25:278–284.

Telke, A., Kalyani, D., Jadhav, J., and Govindwar, S. 2008. Kinetics and mechanism of Reactive Red 141 degradation by a bacterial isolate *Rhizobium radiobacter* MTCC 8161. *Acta Chim. Slov.* 55:320–329.

Tsai, W.T., and Chen, H.R. 2010. Removal of malachite green from aqueous solution using low-cost chlorella-based biomass. *J. Hazard. Mater.* 175:844–849.

Vijayaraghavan, K., and Yun, Y. 2007. Chemical modification and immobilization of *Corynebacterium glutamicum* for biosorption of reactive black 5 from aqueous solution. *Ind. Eng. Chem. Res.* 46, no. 2:608–617.

Vijayakumar, G., Yoo, C.K., Elango, K.G.P., and Dharmendirakumar, M. 2010. Adsorption characteristics of rhodamine B from aqueous solution onto barite. *Clean* 38, no. 2:202–209.

Vijayaraghavan, K., Mao, J., and Yun, Y.S. 2008. Biosorption of methylene blue from aqueous solution using free and polysulfone-immobilized *Corynebacterium glutamicum*: Batch and column studies. *Bioresour. Technol.* 99:2864–2871.

Vitor, V., and Corso, C.R. 2008. Decolorization of textile dye by *Candida albicans* isolated from industrial effluents. *J. Ind. Microbiol. Biotechnol.* 35:1353–1357.

Wang, Z.W., Liang, J.S., and Liang, Y. 2012. Decolorization of reactive black 5 by a newly isolated bacterium *Bacillus* sp. YZU1. *Int. Biodeterior. Biodegrad.*, http://dx.doi.org/10.1016/j.ibiod.2012.06.023.

Wesenberg, D., Buchon, F., and Agathos, S.N. 2002. Degradation of dye-containing textile effluent by the agaric white-rot fungus *Clitocybula dusenii*. *Biotechnol. Lett.* 24:989–993.

Yang, Y., Wang, G., Wang, B., Li, Z., Jia, X., Zhou, Q., and Zhao, Y. 2011. Biosorption of acid black 172 and congo red from aqueous solution by non-viable *Penicillium* YW 01: Kinetic study, equilibrium isotherm and artificial neural network modeling. *Bioresour. Technol.* 102:828–834.

Yenikaya, B., Atar, E., Olgun, A., Atar, N., Ilhan, S., and Colak, F. 2010. Biosorption study of anionic dyes from aqueous solutions using *Bacillus amyloliquefaciens*. *Eng. Life Sci.* 10, no. 3:233–241.

Yesiladal, S.K., Pekin, G., Bermek, H., Arslan-Alaton, I., Orhon, D., and Tamerler, C. 2006. Bioremediation of textile azo dyes by *Trichophyton rubrum* LSK-27. *World J. Microbiol. Biotechnol.* 22:1027–1031.

Zeroual, Y., Kim, B.S., Blaghen, M., Kim, C.S., and Lee, K.M. 2006. A comparative study on biosorption characteristics of certain fungi for bromophenol blue dye. *Appl. Biochem. Biotechnol.* 134:51–60.

Zhou, J.L., and Banks, C.J. 1993. Mechanism of humic acid color removal from natural: Waters by fungal biomass biosorption. *Chemosphere* 27:607–620.

CHAPTER 10

Production of Biosurfactants and Its Application in Bioremediation

C. Vigneshwaran, K. Vasantharaj, M. Jerold,
and Velmurugan Sivasubramanian

CONTENTS

Abstract

Biosurfactants are amphiphilic compounds produced by micro-organisms containing both hydrophilic and hydrophobic groups. These molecules reduce surface tension between aqueous solutions and hydrocarbon mixtures. The types of biosurfactants are glyco-lipids, phospholipids, lipoproteins or lipopeptides, and polymeric. *Pseudomonas* sp. produce a biosurfactant called rhamnolipids, based on the disaccharide (Rhamnose). *Candida* sp., which is one of the few yeasts to produce biosurfactants, produces a high yield of sophorolip-ids from vegetable oil and carbohydrates, whereas *Bacillus* sp. produces a lipopeptide usually referred to as surfactin. Cheap substrates can be used for production, including carbohydrates, vegetable oils, food wastes, cooked oils, fruit juices, and used engine oil. Biosurfactants have a wide range of industrial applications, including oil recovery, oil spill cleanup, textiles, pharmaceuticals, and cosmetics. This chapter illustrates biosurfactant production methods, and the application of biosurfactants in the remediation of hydrocarbon-contaminated soil and recovery of oil spills.

10.1 INTRODUCTION

Biosurfactants are amphiphilic compounds that are produced on the surface of microorganisms or they may also be secreted extracellularly, and they contain both hydrophilic and hydrophobic ends, which reduce the surface and interfacial tension of the surface and interface, respectively. Since biosurfactants and bioemulsifiers both exhibit emulsification properties, bioemulsifiers are frequently considered with biosurfactants even though emulsifiers may not lower surface tension. Biosurfactants are classified as glycoloipids, mycolic acids, polysaccharide–lipid composites, lipoprotein/lipopeptides, phospholipids, or the microbial cell surface itself (Lin et al. 1994). Some research has indicated that biosurfactants produce a surface-active molecule leading to their potential application in the food industry, oil industry, and pharmacology (Ramana and Karanth 1989). The type and amount of biosurfactant produced depends mainly on the producer organism, factors such as carbon and nitrogen source, aeration, temperature, and trace elements.

The carbon source serves as an important organic compound for the growth and metabolism of the microorganism. Suppose the carbon source is in insoluble form, microbes produce biosurfactants that diffuses inside breaking the hydrocarbon (insoluble form) creating food for the microorganism. Some of the bacteria and yeast excrete ionic surfactants that break the hydrocarbon substance in the production medium. Some of the examples of this group of biosurfactant are rhamnolipids that are produced by different *Pseudomonas* sp. (Guerra-Santos et al. 1986) or sophorolipids (Cooper and Paddlock 1983). The structure of the cell wall can be changed by certain microorganisms, producing nonionic or lipopolysaccharides in the cell wall. A few examples of this group are *Candida* sp. (Fukui and Tanaka 1981) and *Mycobacterium* sp. Biosurfactants like lipoprotein are otherwise called surfactins and are produced by *Bacillus substilis* (Cooper et al. 1981).

The potential application of biosurfactants is based on functional properties, which include emulsification, demulsification, solubilization, foaming, and the ability to reduce the viscosity in heavy crude oil (Fiechter et al. 1992). Biosufactants have many applications such as bioremediation, enhanced oil recovery, dispersion of oil spills, and degradation of crude oil. Other applications of biosurfactants are in the food, health care, cosmetics, and pharmaceutical industries, and the removal of toxic chemicals. This chapter will describe biosurfactant production methods and their application in remediation of hydrocarbon-contaminated soil and recovery of oil spills.

10.2 TYPES, CLASSIFICATION, AND MICROBIAL ORIGIN OF BIOSURFACTANTS

Biosurfactants are classified into two groups: one is the high molecular weight substance called a bioemulsifier and it is used for the emulsification of hydrocarbon; and other one is the low molecular weight substance called a biosurfactant. Biosurfactants are further divided into six groups known as glycolipids, lipopolysaccharides, lipoproteins/lipopeptides, phospholipids, hydroxylated fatty acids, and cross-linked fatty acids. Biosurfactants produced by variety of microorganisms, mainly bacteria, fungi, and yeast, are shown in Table 10.1. The chemical composition, nature, and quantity of biosurfactants depend upon the type of microorganism produced by a particular biosurfactant.

10.3 STUDIES ON PHYSICOCHEMICAL CHARACTERIZATION OF BIOSURFACTANTS

The surface free energy per unit area influences surfactant effectiveness, which is the measure of work required to bring down the molecules

TABLE 10.1   Biosurfactant and Microorganism Involved

Biosurfactant	Microorganism
Glycolipids	*Pseudomonas aeruginosa*
Rhamnolipids	*Rhodococcus erithropolis*
Trehalose lipids	*Arthobacter* sp.
Sophorolipids	*Candida bombicola, C. apicola*
Mannosylerythritol lipids	*C. antartica*
Lipopeptides	*Bacillus subtilis*
Surfactin	*P. fluorescens*
Viscosin	*B. licheniformis*
Lichenysin	*Serratia marcescens*
Polymeric surfactants	*Acinetobacter calcoaceticus*
Emulsan	*A. radioresistens*
Liposan	*C. lipolytica*
Lipomanan	*C. tropicalis*
Alasan	

Source: Adapted from Lourith, N., Kanlayavattanakul, M., *Int. J. Cosmet.* 29:225–261, 2009.

from bulk phase to the surface (for example, surface tension of water from 72 to 35 mN/m) (Rosen 1978). Surface tension relates the concentration profile of biosurfactants until the critical micelle concentration (CMC) is reached (Figure 10.1). Thus an efficient surfactant will have a lower CMC.

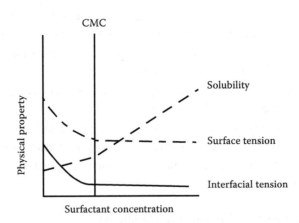

FIGURE 10.1   Surface tension, interfacial tension, and solubility as a function of surfactant concentration. CMC, critical micelle concentration. (Adapted from Mulligan, C.N., Gibbs, B.F., *Proc. Indian Nat. Sci. Acad. B*, 70, 1:31–35, 2004.)

Properties including critical micelle concentration, hydrophilic and lipophilic balance, chemical structure, charge, and properties from origin source are used for characterization of surfactants (Van Hamme et al. 2006). At critical micelle concentration, the surfactant molecules form aggregates, whereas at low concentration surfactants exist as single molecules or monomers. The CMC depends on surfactant structure, pH, temperature, ionic strength, and the organic additives in the solution (Fuguet et al. 2005). A physicochemical property, such as surface tension, interfacial tension, and solubility, tends to vary at the CMC of surfactant concentration (Hanna et al. 2005).

10.4 SUBSTRATE FOR PRODUCTION OF BIOSURFACTANTS

Biosurfactants have properties such as biodegradability, low toxicity, flexibility in operation, and ecofriendliness. The cost of production of chemical surfactants is cheap when compared to biological methods. Several studies have been carried out to reduce the production cost of biological methods by using agro waste or molasses, or using optimizing parameters of the bioprocess such as optimum temperature and pH. Release of hazardous material has become common in industrialized cities, but these wastes can be properly utilized for the production of various reliable products. These wastes can be obtained from agriculture such as peels, hull, sweet potatoes, residues from cooking oil industries, coffee residues, crude oil residues from petroleum industries, and carbohydrates (Table 10.2).

TABLE 10.2   Substrate for Biosurfactants

Biosurfactants	Producing Organism	Substrate
Cellobioselipids	*Ustilago* sp.	Vegetable oil
Corynomycolates	*Arthrobacter* sp.	Different sugars
Mannosylerythriol lipids	*Candida* sp.	Glucose, soybean oil
Rhamnolipids	*Pseudomonas* sp.	n-Alkanes, glycerol
Sophorolipids	*Torulopsis* sp.	Glucose, vegetable oil
Trehalose corynomycolates	*Rhodococcus erythropolis*	n-Alkanes, carbohydrates
Mono		
Di		
Tri		
Lipopeptides	*Bacillus licheniformis*	Glucose
Surfactin	*Bacillus subtilis*	Glucose

Source: Adapted from Sineriz, F., Hommel, R.K., Kleber, H.-P., 2001, Production of biosurfactants. In *Encyclopedia of Life Support Systems*, Vol. 5, Oxford: Eolss Publishers.

10.5 PRODUCTION OF BIOSURFACTANTS

Rhamnolipids, lipopeptides, and sophorolipids are the better biosurfactants compared to others because of their environmental application such as biodegradability of hydrocarbons and emulsifying properties (Bezza and Chirwa 2015; Heyd et al. 2008; Mulligan 2005).

10.5.1 Medium Optimization

The culture conditions depend upon type, quality, and quantity of biosurfactant production. The conditions are pH, temperature, agitation, aeration, and the nature of the carbon and nitrogen source. There are many studies regarding biosurfactant production relating the optimization of their physiological properties (Sarubbo et al. 2001). The *Pseudomonas* species plays a vital role in the production of rhamnolipids, which can be altered by feeding a variety of carbon sources such as vegetable oils (e.g., olive, corn, and soy), glucose, glycerol and n-alkanes (Santos et al. 2002). In fact carbon and nitrogen are important growth factors for the production of biosurfactants (Henkel et al. 2012). Usually these are produced in the stationary growth phase because they are secondary metabolites (Deziel et al. 1999; Santa Anna et al. 2001; Venkata-Ramana and Karanth 1989). Moreover, use of inexpensive substrates, such as crude or waste materials, affects production cost of biosurfactants (Henkel et al. 2012). Various waste substrates, such as fatty acid from soybean oil refinery wastes, glycerin from biodiesel production waste, and sunflower oil refinery waste, can be used for rhamnolipid production (Abalos et al. 2001; Benincasa et al. 2002; De Sousa et al. 2011). The waste material have tendency to influence the field of microbial production. Because it is cheaper, it increases the efficiency of utility during production process and makes it ecofriendly (Henkel et al. 2012).

Sources like nitrogen, carbon–nitrogen ratio, and mineral salts are important factors to increase rhamnolipid production. Recent research by Muller et al. (2012) proposed that nitrogen is the main source in the form of nitrate; it is used for the higher production of rhamnolipids. Increasing the carbon–nitrogen ratio and the carbon–phosphorous (inorganic form) ratio in the production media helps to increase rhamnolipid production. Also a limited amount of trace element and multivalent ions such as K, Na, and Mg are used in the media to help increase rhamnolipid production (Guerra-Santos et al. 1986).

10.5.2 Bioreactors for Production

Various operating conditions for the production of biosurfactants studied by researches include batch, fed-batch, continuous cultivation, and biomolecular (Muller et al. 2012). The production of rhamnolipids and other biosurfactants is possible using shake flask, batch, fed-batch, continuous, and integrated microbial–enzymatic processes (Muller et al. 2012). In comparison of all the methods, the optimization of culture conditions in shake flasks is practiced most due to its simplicity. Culture cultivation in bioreactors offers high control over vital parameters such as oxygen transfer, temperature, and pH, and is also suitable for further scale up investigation (Vegilo et al. 1998). Continuous or fed-batch fermentation in biosurfactant production is advantageous over batch process, because of the suppression of several steps regarding inoculum preparation, sterilization, and finalization of the production process by carrying out the production for long periods (Kronemberger et al. 2010). The performance of reactors such as continuous stirred tank reactors (CSTRs) and sequencing batch reactors (SBRs) was compared for biosurfactant production in slurry-contaminated soil. The production of biosurfactants was obtained in SBRs but not in CSTRs. CMC values were over 70. To control foam formation, the cycling time in SBRs was reduced from 4 days to 1 day (Cassidy et al. 2000). Biosurfactant production is mainly carried out at small scale using the shake flask method; to reduce production, the scale-up process was investigated using bioreactors. Rhamnolipid produced by *Pseudomonas aeruginosa* was cultivated in fed-batch reactors. The concentration of rhamnolipids obtained was up to 16.9 g/L (Kronemberger et al. 2010). Biosurfactant production by *Pseudomonas aeruginosa* SP4 was carried out in SBRs. The oil-to-glucose ratio was 40:1 and the cycling time was 2 days/cycle. The CMC of biosurfactant was found to be 150 mg/L (Pansiripat et al. 2010). In a previous study it was reported that biosurfactant produced by *Pseudomonas aeruginosa* SP4 using the batch fermentation process showed a CMC of about 200 mg/L (Pornsunthorntawee et al. 2008). This present study indicated that glucose slightly affected the biosurfactant production but strongly enhanced the growth of microbes in the SBR unit.

10.5.3 Environmental Factors Affecting Biosurfactant Production

Environmental factors are very important for a higher production yield and characterization of biosurfactants. To obtain large quantities of biosurfactants, the process conditions, such as pH, temperature, aeration, and agitation speed, should be optimized.

10.5.3.1 pH

pH plays an important role in biosurfactant production (Gobbert et al. 1984). The effect of pH in biosurfactant production by *Pseudomonas aeruginosa* PBSC1 has been investigated with pH values varying from 5 to 8.5. The highest surface tension reduction was 29.19 mN/m for pH 6.5 and emulsification activity is 75.12% for pH 7, and the highest biosurfactant production at pH 7 was observed (Joice and Parthasarathi 2014). The best biosurfactant production occurred at pH 8.0, which is the natural pH of seawater (Zinjarde and Pant 2002). The highest activity of biosurfactant production was observed at pH 7, though low activity was obtained at pH 6, 8, and 9, according to Jagtap et al. (2010). Auhmin and Mohamed (2013) showed that pH 7 was optimum for higher biosurfactant production from *Azotobacter chrococcum*. Mahdy et al. (2012) studied that optimum pH 7 is best for production of biosurfactant by *Candida* strains. The effect of pH in biosurfactant production by *V. salarius* has been investigated by varying pH values from 4 to 12. The biosurfactant production is found to increase when the pH of the medium is increased from 4 to 11 and decreases when the pH is increased to 12. The maximum production of biosurfactant occurred at pH 9 (Elazzazy et al. 2015). Hamzah et al. (2013) agreed with the previous result that maximum production of biosurfactant was obtained at pH 9.0.

10.5.3.2 Temperature

Most biosurfactants showed higher production in the temperature range of 25°C to 30°C. Biosurfactants were obtained in the culture medium of *Azotobacter chrococcum* at the temperature of 30°C (Auhmin and Mohamed 2013). Joice and Parthasarathi (2014) observed that highest biosurfactant production by *Pseudomonas aeruginosa* PBSC1 at 30°C, and for Elazzazy et al. (2015) a higher yield of biosurfactant production was performed by *V. salarius* (KSA-T) at 40°C. The optimum temperature for bioemulsifier production was 37°C, which can depend on the character of organism (Jagtap et al. 2010). The optimum temperature of 30°C is best for production of biosurfactant from *Candida tropicalis* (Sudha et al. 2010). The optimum temperature for biosurfactant production for all *Candida* strains was usually 20°C (Mahdy et al. 2012).

10.5.3.3 Aeration and Agitation

Aeration and agitation play a very important role in the production of biosurfactants, as they both facilitate the oxygen transfer from the gas phase to the aqueous phase. The higher yield of biosurfactant (45.5 g l^{-1}) was obtained when the airflow rate was 1 vvm and the dissolved oxygen concentration was obtained at 50% saturation (Adamczak and Bednarski 2000). By increasing the airflow rate, foam formation is high and it affects biosurfactant production (Guilmanov et al. 2002). Konishi et al. (2011) evaluated cell growth amid mannosylerythritol lipids by varying agitation speed from 100 rpm to 250 rpm. From this experiment they found that there is a nominal change in cell growth with respect to agitation. Mannosylerythritol lipid production increased with agitation speed up to 200 rpm and then it decreased. The overall results show that agitation speed of about 150 to 200 rpm produces more mannosylerythritol lipids. The 250 rpm agitation speed produced slightly fewer mannosylerythritol lipids than 200 rpm, but higher than 100 rpm. By increasing the dissolved oxygen concentration by agitation it helps to increase the biosurfactant production (Yeh et al. 2005). Increasing the agitation speed to more than 350 rpm reduces biosurfactant production by the formation of foam (Shaligram and Singhal 2010). Also aeration affects biosurfactant production by the accumulation of foam (Shaligram and Singhal 2010). Sen (1997) noticed through response surface methodology that the airflow rate at 0.75 vvm is best for biosurfactant production.

10.6 SEPARATION AND RECOVERY OF BIOSURFACTANT

The selection of raw materials and cultivation of organisms are important factors for reducing the downstream cost. The biomass, crude biosurfactant, and their applications are important factors to decide the purification cost. Acid and ammonium sulfate precipitation, crystallization, centrifugation, and solvent extractions are the common methods that have been followed in industries for the separation of biosurfactants. Some novel methods employed for biosurfactant recovery include ion exchange chromatography, ultrafiltration, and foam fractionation (Table 10.3).

In recent years, cheap and less toxic solvents such as methyl tertiarybutyl ether have been used to recover biosurfactant. San Keskin et al. (2015) tested that purification of biosurfactant carried out using a preparative HPLC System, and a Zorbax Eclipse column XDB C18 was used for

TABLE 10.3   Methods and Mechanism for the Recovery of Biosurfactants

Method	Mechanism
Adsorption on wood	Adsorption
Adsorption on polysterene	Adsorption
Ion exchange chromatography	Charge separation
Thin layer chromatography	Difference in relative flow against solvent
Solvent extraction	Dissolve in organic solvents
Organic solvent extraction	Solubility in organic solvents
Centrifugation	Centrifugal force
Acid precipitation	Insoluble at low pH
Membrane ultrafiltration	Micelles formation
Selective crystallization	Redissolution in organic solvents
Ammonium sulfate precipitation	Salting out of protein
Foam fractionation	Surface activity
Dialysis	Difference in solute concentration
Lyophilisation	Cryodesiccation
Isoelectric focusing	Electric charge difference

Source: Adapted from Saharan, B.S. et al., *Gen. Eng. Biotechnol. J.* 2011:10–14, 2011.

separation. These types of low cost, less toxic, and highly available solvents are used for recovery and help to minimize the environmental hazards (Kuyukina et al. 2001).

10.7 APPLICATION OF MICROBIAL SURFACTANTS

The microbial surfactants are characterized based on chemical structure and functional properties such as emulsification, de-emulsification, wetting, foaming, solubilization, and viscosity reduction of crude oil (Fietcher 1992). Biosurfactants are one of the most important substances for many fields of industry including pharmacy, food industry, design of washing agents, petroleum industry, agriculture, environmental protection, and remediation. Nowadays the petroleum industries, chemical industries, and paper industries are generating various organics as by-products that are released to the environment or spilled accidently and then get accumulated into soil and water resulting in environmental pollution. By-product mixtures containing aromatic and chlorinated derivatives are generated by the petroleum industries (e.g., polycyclic aromatic hydrocarbons), chemical industries (e.g., phenols), and pulp industries (e.g., dioxins). The various classes of biosurfactant and their applications are shown in Table 10.4 (Matvyeyeva et al. 2014). This chapter illustrates applications

TABLE 10.4   Biosurfactants and Their Applications in Bioremediation

Class of Biosurfactant	Microorganism	Application
Rhamnolipids	*Pseudomonas aeruginosa, Pseudomonas* sp.	Enhancement of the degradation and dispersion of different classes of hydrocarbons; emulsification of hydrocarbons and vegetable oils
Trehalolipids	*Mycobacterium tuberculosis, Rhodococcus erythropolis, Arthrobacter* sp., *Nocardia* sp., *Corynebacterium* sp.	Enhancement of the bioavailability of hydrocarbons
Sophorolipids	*Torulopsis bombicola, Torulopsis petrophilum, Torulopsis apicola*	Recovery of hydrocarbons from dregs and muds; enhancement of oil recovery
Corynomycolic acid	*Corynebacterium lepus*	Enhancement of bitumen recovery
Surfactin	*Bacillus subtilis*	Enhancement of the biodegradation of hydrocarbons
Lichenysin	*Bacillus licheniformis*	Enhancement of oil recovery
Emulsan	*Acinetobacter calcoaceticus* RAG-1	Stabilization of the hydrocarbon-in-water emulsions
Alasan	*Acinetobacter radioresistens* KA-53	
Liposan	*Candida lipolytica*	Stabilization of hydrocarbon-in-water emulsions
Mannoprotein	*Saccharomyces cerevisiae*	

Source: Adapted from Matvyeyeva, O.L. et al., *Int. J. Environ. Bioremed. Biodegrad.* 2, no. 2:69–74, 2014.

in remediation of hydrocarbon-contaminated soil and recovery of oil spills.

10.7.1 Role of Biosurfactants in Hydrocarbon Remediation

Biosurfactants are better potential agents for hydrocarbon remediation. There are two mechanisms to enhance hydrocarbon remediation. The first mode of action is direct interfacial accession, which involves contact of cells with hydrocarbon droplets; and the other one is biosurfactant mediated transfer, in which cell contact takes place with hydrocarbons enhancing degradation by emulsification or solubilization and mobilization. The mobilization process takes place at concentrations below the biosurfactant critical micelle concentration. At these concentrations biosurfactant reduces the surface and interfacial tension between phases of differing polarity. The solubilization process takes place at concentrations above the biosurfactant critical micelle concentration. At such concentrations biosurfactant molecules connect to form micelles and solubility of oil increases as shown in Figure 10.2 (Pacwa-Plociniczak et al. 2011).

Emulsification is the process of breaking down large fat molecules into smaller ones. High molecular weight biosurfactants are potential emulsifiers used as an additive for hydrocarbon remediation. Biosurfactant produced by *Pseudomonas aeruginosa* UG2 is mixed in soil contaminated with a hydrocarbon mixture of hexadecane, tetradecane, 2-methylnapthalene, and pristine. After incubation it was observed that biosurfactant enhanced the degradation of hydrocarbon mixtures except 2-methylnapthalene (Jain et al. 1992). In another experiment, *Pseudomonas* ML2 and *Acinetobacter haemoliticus* were inoculated in hydrocarbon-contaminated soil and the reduction of hydrocarbon was compared with the same soil. After two months of incubation, 39% to 71% reduction of hydrocarbon was achieved by *Acinetobacter haemoliticus*, whereas the *Pseudomonas* ML2 showed 11% to 71% reduction. This result suggested that using cell-free biosurfactant stimulated reduction by indigenous microorganisms (Banat 1995). The purified form of rhamnolipids biosurfactant produced by *Pseudomonas aeruginosa* was applied to remove oil from contaminated sandy soil. The researchers optimized the biosurfactant and the oil concentration in the removal of oil by using a statistical experimental design tool. By mixing a reduced amount of biosurfactant concentration (6.3–7.9 g/L) to the oil-contaminated sandy soil removes 91% and 78% of aromatic and paraffinic hydrocarbon, respectively (Santa Anna et al. 2007). From oil-contaminated seawater, *Pseudomonas aeruginosa* was isolated. After 28 days of incubation it is capable of degrading hexadecane, heptadecane,

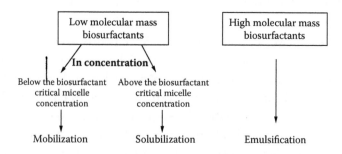

FIGURE 10.2 Hydrocarbon removal by biosurfactants. (Adapted from Pacwa-Plociniczak, M. et al., *Int. J. Mol. Sci.* 12:633–654, 2011.)

TABLE 10.5   Reduction of Hydrocarbon by Biosurfactant-Producing Organism

Biosurfactant-Producing Organism	Reduction of Hydrocarbon from Oil-Contaminated Soil (%)	Reference
Pseudomonas ML2	11–71	Banat 1995
Acinetobacter haemoliticus	39–71	Banat 1995
C. bombicola ATCC 22214	85–97	Kang et al. 2010
Pseudomonas aeruginosa	25–70 (silt-loam soil), 40–80 (sandy-loam soil)	Van Dyke et al. 1993
Pseudomonas aeruginosa	91 (aromatic hydrocarbon), 78 (paraffinic hydrocarbon)	Santa Anna et al. 2007

octadecane, and nonadecane present in seawater. Presence of biosurfactants in culture was shown by tensiometric measurement (Shafeeq et al. 1989). The rate of aliphatic and aromatic hydrocarbon degradation was increased in the range of 85% to 97% using sophorolipids under lab-scale conditions (Kang et al. 2010). Van Dyke et al. (1993) noticed that rhamnolipids produced by *Pseudomonas aeruginosa* were capable of removing hydrocarbon from silt-loam soil and sandy-loam soil. The rate of hydrocarbon reduction reported by different authors is shown in Table 10.5.

10.7.2 Role of Biosurfactant in Oil Recovery

It was found that biosurfactant produced from *Bacillus* sp. had the ability to release oil from oily sand at a concentration of 0.04 mg/mL (Eliseev et al. 1991). Urum et al. (2006) evaluated that the efficiency of removing crude oil from contaminated soil using different surfactant solutions. They obtained the results that crude oil removal is high by synthetic surfactant-sodium doedecyl sulfate (46%) and in rhamnolipid (44%) compared to saponin-natural surfactants (27%). Within 24 hours, recovery of 85% motor oil from contaminated sand by *Bacillus subtilis* CN2 producing biosurfactant. After 18 days of incubation, biosurfactant produced by *Bacillus subtilis* CN2 enhanced up to 82% degradation of used motor oil polycyclic aromatic hydrocarbon components (Bezza and Chirwa 2015).

The recovery of oil ranged from 30.22% to 34.19% of the water flood residual oil saturation using crude biosurfactant produced from *Bacillus* strains by sand packed column (Joshi and Desai 2013). Bezza and Chirwa (2015) showed the results that recovery of used motor oil from the contaminated sands by *Bacillus subtilis* CN2 is up to 84.6% ± 7.1%, whereas the control (distilled water) recovered 15% ± 3%. It was observed that there was a 17% to 26% oil recovery using crude biosurfactant produced by *Bacillus subtilis* B30 (Al-Wahaibi et al. 2014). The present study shows that

TABLE 10.6   Recovery of Oil by Biosurfactant-Producing Organism

Biosurfactant-Producing Organism	Recovery of Oil from Oil Contaminated Soil (%)	Reference
Bacillus subtilis CN2	84.6 ± 7.1	Bezza and Chirwa 2015
Bacillus subtilis BS-37	96	Liu et al. 2015
Bacillus strain	30.22–34.19	Joshi and Desai 2013
Bacillus subtilis B 30	17–26	Al-Wahaibi et al. 2014
Candida sphaerica	75 (clay soil), 92 (silty soil)	Sobrinho et al. 2008
Candida Tropicalis	78–97	Batista et al. 2010
Candida glabrata UCP 1002	92.6	Gusmao et al. 2010
Candida sphaerica UCP 0995	95	Luna et al. 2011

biosurfactant produced by *Bacillus subtilis* CN2 has potential application in oil recovery (Bezza and Chirwa 2015). The rate of oil recovery reported by different authors is shown in Table 10.6.

10.8 CONCLUSION

This chapter provides basic methodical information about the production of biosurfactants and their application. Nowadays the interest in biosurfactant production has been increasing and it can be used for bioremediation purposes. Production of biosurfactants at laboratory scale was well established, whereas up-scaling the process to pilot levels was affected by numerous factors such as nutrients (micro and macro) and environmental factors. However, the high cost of production processes and recovery of biosurfactants limits industrial use. The economics of biosurfactant production will depend on inexpensive carbon substrate. In this chapter we have presented detailed investigations about cheaper sources of substrates used for production to reduce process costs. Novel microorganisms used for biosurfactant production, parameter optimization studies, and various types of bioreactors used for production were discussed. From this chapter we conclude that when compared to other reactors, the sequencing batch reactor is suitable for a higher yield of biosurfactant production, which helps in hydrocarbon removal and recovery of oil from the contaminated soil. Oil is used in many fields such as industries, automobiles, and households and it is also one of the major pollutants. Oil pollution causes serious problems in the environment. Hydrocarbons penetrate from topsoil to the subsoil slowly, leading to groundwater contamination. To overcome these problems, ecofriendly technologies must be used to clean the environment. Bioremediation has been accepted as an important method for the treatment of hydrocarbon contaminants by biosurfactants.

REFERENCES

Abalos, A., Pinazo, A., Infante, M.R., Casals, M., Garcia, F., Manresa, A. 2001. Physiochemical and antimicrobial properties of new rhamnolipids produced by *Pseudomonas aeruginosa* AT10 from soyabean oil refinery waste. *Langmuir* 17:1367–1371.

Adamczak, M., Bednarski, W. 2000. Influence of medium composition and aeration on the synthesis of biosurfactants produced by *Candida antartica*. *Biotechnol. Lett.* 22:313–316.

Al-Wahaibi, Y., Joshi, S., Al-Bahry, S., Elshafie, A., Al-Bemani, A., Shibulal, B. 2014. Biosurfactant production by *Bacillus subtilis* B30 and its application in enhancing oil recovery. *Colloids Surf. B Biointerfaces* 114:324–333.

Auhmin, H.S., Mohamed, A.I. 2013. Effect of different environmental and nutritional factors on biosurfactant production from *Azotobacterchrococcum*. *Int. J. Adv. Pharm. Biol. Chem.* 2, no. 3:477–481.

Banat, I.M. 1995. Biosurfactants production and possible uses in microbial enhanced oil recovery and oil pollution remediation: A review. *Bioresour. Technol.* 51:1–12.

Batista, R.M., Rufino, R.D., Luna, J.M., de Souza, J.E.G., Sarubbo, L.A. 2010. Effect of medium components on the production of a biosurfactant from *Candida tropicalis* applied to the removal of hydrophobic contaminants in soil. *Water Environ. Res.* 82:418–425.

Benincasa, M., Contiero, J., Manresa, M.A., Moraes, I.O. 2002. Rhamnolipid production by *Pseudomonas aeruginosa* LBI growing on soapstock as the sole carbon source. *J. Food Eng.* 54, no. 4:283–288.

Bezza, F.A., Chirwa, E.M.N. 2015. Production and application of lipopeptide biosurfactant for bioremediation and oil recovery by *Bacillussubtilis* CN2. *Biochem. Eng. J.* 101:168–178.

Cassidy, D.P., Ependiev, S., White, D.M. 2000. A comparison of CSTR and SBR bioslurry reactor performance. *Water Res.* 34:4333–4342.

Cooper, D.G., Paddlock. 1983. *Torulopsis petrophilum* and surface activity. *Appl. Environ. Microbiol.* 46:1426–1429.

Cooper, D.G., Zajic, J.E., Denis, C. 1981. Surface active properties of a biosurfactant from *Corynebacterium lepus. J. Am. Oil Chem. Soc.* 58:77–80.

De Sousa, J.R., Da Costa Correia, J.A., De Almeida, J.G.L., Rodrigues, S., Pessoa, O.D.L., Melo, V.M.M., Gonçalves, L.R.B. 2011. Evaluation of a co-product of biodiesel production as carbon source in the production of biosurfactant by *Pseudomonas aeruginosa* MSIC02. *Process Biochem.* 46, no. 9:1831–1839.

Deziel, E., Lepine, F., Dennie, D., Boismenu, D., Mamer, D.A., Villemur, R. 1999. Liquid chromatography/mass spectrometry analysis of mixtures of rhamnolipids produced by *Pseudomonas aeruginosa* strain 57 RP grown on manitol or naphthalene. *Biochem. Biophys. Acta* 1440:244–252.

Elazzazy, A.M., Adelmoneim, T.S., Almaghrabi, O.A. 2015. Isolation and characterization of biosurfactant production under extreme environmental conditions by alkali-halo-thermophilic bacteria from Saudi Arabia. *Saudi J. Biol. Sci.* 22:466–475.

Eliseev, S.A., Vildanova-Martsishin, R., Shluga, A., Shabo, A., Turovsky, A. 1991. Oil washing bioemulsifier produced by *Bacillus* sp. *Microbiol. J.* 53:61–66.

Fiechter, A. 1992. Biosurfactant moving towards industrial application. *Tibtech* 10: 208–217.

Fuguet, E., Rafols, C., Roses, M., Bosch, E. 2005. Critical micelle concentration of surfactant in aqueous buffered and unbuffered systems. *Anal. Chim. Acta* 548:95–100.

Fukui, S., Tanaka, A. 1981. Metabolism of alkanes by yeasts. *Adv. Biochem. Eng.* 19:217–237.

Gobbert, U., Lang, S., Wagner, F. 1984. Sophorose lipid formation by resting cells of *Torulopsisbombicola*. *Biotechnol. Lett.* 6:225–230.

Guerra-Santos, L.H., Kappeli, O., Fletcher, A. 1986. Dependence of *Pseudomonas aeruginosa* continuous culture biosurfactant production on nutritional and environmental factors. *Appl. Microbiol. Biotechnol.* 24:443–448.

Guilmanov, V., Ballistreri, A., Impallomeni, G. 2002. Oxygen transfer rate and sophorose lipid production by *Candida bombicola*. *Biotechnol. Bioeng.* 77:489–495.

Gusmao, C.A.B., Rufino, R.D., Sarubbo, L.A. 2010. Laboratory production and characterization of a new biosurfactant from *Candida glabrata* UCP 1002 cultivated in vegetable fat waste applied to the removal of hydrophobic contaminant. *World J. Microbiol. Biotechnol.* 26:1683–1692.

Hamzah, A., Sabturani, N., Radiman, S. 2013. Screening and optimization of biosurfactant by the hydrocarbon degrading bacteria. *Sain Malaysiana* 42, no. 5:615–623.

Hanna, K., Denoyel, R., Beurroies, I., Dubes, J.P. 2005. Solubilization of pentachlorophenol in micelles and confined surfactant phases. *Coll. Surf. A: Physicochem. Eng. Aspects* 254:231–239.

Henkel, M., Muller, M.M., Kugler, J.H., Lovagilo, R.B., Contiero, J., Syldatk, C., Hausmann, R. 2012. Rhamnolipids as biosurfactants from renewable resources: Concepts for next generation rhamnolipid production. *Process Biochem.* 47, no. 8:1207–1219.

Heyd, M., Kohnert, A., Tan, T.H., Nusser, M., Kirschhöfer, F., Brenner-Weiss, G., Franzreb, M., Berensmeier, S. 2008. Development and trends of biosurfactant analysis and purification using rhamnolipids as an example. *Anal. Bioanal. Chem.* 391, no. 5:1579–1590.

Jagtap, S., Yavankar, S., Pardesi, K., Chopade, P. 2010. Production of bioemulsifier by *Acinetobacter* species isolated from healthy human skin. *Indian J. Exp. Biol.* 48:70–76.

Jain, D.K., Lee, H., Trevors, J.T. 1992. Effect of addition of *Pseudomonas aeruginosa* UG2 inocula or biosurfactants on biodegradation of selected hydrocarbons in soil. *J. Ind. Microbiol.* 10:87–93.

Joice, P.A., Parthasarathi, R. 2014. Optimisation of biosurfactant production from *Pseudomonas aeruginosa* PBSC1. *Int. J. Curr. Microbiol. App. Sci.* 3, no. 9:140–151.

Joshi, S.J., Desai, A.J. 2013. Bench-scale production of biosurfactants and their potential in ex-situ MEOR application. *Soil Sediment Contam.* 22, no. 6:701–715.

Kang, S.W., Kim, Y.B., Shin, J.D., Kim, E.K. 2010. Enhanced biodegradation of hydrocarbon in soil by microbial biosurfactant, Sophorolipids. *Appl. Biochem. Biotechnol.* 160, no. 3:780–790.

Konishi, M., Nagahama, T., Fukuoka, T., Morita, T., Imura, T., Kitamoto, D., Hatada, Y. 2011. Yeast extract stimulates production of glycolipids biosurfactants, mannosylerythritol lipids by *Pseudozymahubeiensis* SY62. *J. Biosci. Bioeng.* 111, no. 6:702–705.

Kronemberger, F.D.A., Borges, C.P., Freire, D.M.G. 2010. Fed-batch biosurfactant production in a bioreactor. *Int. Rev. Chem. Eng.* 2, no. 4:513–518.

Kuyukina, M.S., Ivshina, I.B., Philp, J.C., Christofi, N., Dunbar, S.A., Ritchkova, M.I. 2001. Recovery of Rhodococcus biosurfactants using methyl tertiary butyl ether extraction. *J. Microbiol. Methods* 46:149–156.

Lin, S.C., Minton, M.A., Sharma, M.M., Georgiou, G. 1994. Structural and immunological characterization of a biosurfactant produced *Bacillus licheniformis* JF-2. *Appl. Environ. Microbiol.* 60:31–38.

Liu, Q., Lin, J., Wang, W., Huang, H., Li, S. 2015. Production of surfactin isoforms by *Bacillus subtilis* BS-37 and its applicability to enhanced oil recovery under laboratory conditions. *Biochem. Eng. J.* 93:31–37.

Lourith, N., Kanlayavattanakul, M. 2009. Natural surfactants used in cosmetics: Glycolipids. *Int. J. Cosmet.* 29:225–261.

Luna, J.M., Rufino, R.D., Sarubbo, L.A., Rodrigues, L.R.M., Teixeira, J.A.C., Campos-Takaki, G.M. 2011. Evalution antimicrobial and antiadhesive properties of biosurfactant lunasan produced by *Candida sphaerica* UCP 0995. *Curr. Microbiol.* 62:1527–1534.

Mahdy, H.M., Fareid, M.A., and Hamdan, M.N. 2012. Production of biosurfactants from certain *Candida* strains under special conditions. *Researcher* 4, no. 7:39–55.

Matvyeyeva, O.L., Vasylchenko, O.A., Aliivea, O.R. 2014. Microbial biosurfactants role in oil products degradation. *Int. J. Environ. Bioremed. Biodegrad.* 2, no. 2:69–74.

Muller, M.M., Kugler, J.H., Henkel, M., Gerlitzki, M., Hormann, B., Pohnlein, M., Syldatk, C., Hausmann, R. 2012. Rhamnolipids—Next generation surfactants? *J. Biotechnol.* 162, no. 4:366–380.

Mulligan, C.N. 2005. Environmental applications for biosurfactants. *Environ. Pollut.* 133:183–198.

Mulligan, C.N., Gibbs, B.F. 2004. Types, production and applications of biosurfactants. *Proc. Indian Nat. Sci. Acad. B*70 1:31–35.

Pacwa-Plociniczak, M., Plaza, G.A., Piotrowska-Seget, Z., Cameotra, S.S. 2011. Environmental applications of biosurfactants: Recent advances—A review. *Int. J. Mol. Sci.* 12:633–654.

Pansiripat, S., Pornsunthorntawee, O., Rujiravanit, R., Kitiyanan, B., Somboonthanate, P., Chavadej, S. 2010. Biosurfactant production by *Pseudomonas aeruginosa* SP4 using sequencing batch reactors: Effect of oil-to-glucose ratio. *Biochem. Eng. J.* 49:185–191.

Pornsunthorntawee, O., Wongpanit, P., Chavadej, S., Abe, M., Rujiravanit, R. 2008. Structural and physicochemical characterization of crude biosurfactant produced by *Pseudomonas aeruginosa* SP4 isolated from petroleum contaminated soil. *Bioresour. Technol.* 99:1589–1595.

Ramana, K.V., Karanth, N.G. 1989. Production of biosurfactants by the resting cells of *Pseudomonas aeruginosa* CFTR-6. *Biotechnol. Lett.* 11:437–442.

Rosen, M.J. 1978. *Surfactants and Interfacial Phenomena.* New York: John Wiley & Sons.

Saharan, B.S., Sahu, R., Sharma, D. 2011. A review on biosurfactants: Fermentation, current developments and perspectives. *Gen. Eng. Biotechnol. J.* 2011:10–14.

San Keskin, N.O., Han, D., Ozkan, A.D., Angun, P., Umu, O.C.O., Tekinay, T. 2015. Production and structural characterization of biosurfactant produced by newly isolated *staphylococcus xylosus* STF1 from petroleum contaminated soil. *J. Petrol. Sci. Eng.* 29, no. 3:334–341.

Santa Anna, L.M., Sebastin, G.V., Pereira Jr., N., Alves, T.L.M., Menezes, E.P., Freire, D.M.G. 2001. Production of biosurfactant from a new and processing strain of *Pseudomonas aeruginosa* PA1. *Appl. Biochem. Biotechnol.* 91–93, no. 1–9:459–467.

Santa Anna, L.M., Soriano, A.U., Gomes, A.C., Menezes, E.P., Gutarra, M.L.E., Freire, D.M.G., Pereira Jr., N. 2007. Use of biosurfactant in the removal of oil from contaminated sandy soil. *J. Chem. Technol. Biotechnol.* 82:687–691.

Santos, A.S., Sampaio, A.P.W., Vasquez, G.S., Santa Anna, L.M., Pereira Jr., N., Freire, D.M.G. 2002. Evaluation of different carbon and nitrogen sources in production of rhamnolipids by a strain of *Pseudomonas aeruginosa*. *Appl. Biochem. Biotechnol.* 98, no. 1:1025–1035.

Sarubbo, L.A., Marcal, M.C., Neves, M.L.C. 2001. Bioemulsifier production in batch culture using glucose as carbon source by *Candida lipolytica*. *Appl. Biochem. Biotechnol.* 95:59–67.

Sen, R. 1997. Response surface optimization of the critical media components for the production of surfactin. *J. Chem. Technol. Biotechnol.* 68:263–270.

Shafeeq, M., Kokub, D., Khalid, Z.M., Khan, A.M., Malik, K.A. 1989. Degradation of different hydrocarbons and production of biosurfactants by bacterial strain S8, isolated from coastal waters. *MIRCEN J. Appl. Microbiol. Biotechnol.* 5:505–510.

Shaligram, N.S., Singhal, R.S. 2010. Surfactin—A review on biosynthesis, fermentation, purification and applications. *Food Technol. Biotechnol.* 48:119–134.

Sineriz, F., Hommel, R.K., Kleber, H.-P. 2001. Production of biosurfactants. In *Encyclopedia of Life Support Systems*, Vol. 5. Oxford: Eolss Publishers.

Sobrinho, H.B.S., Rufino, R.D., Luna, J.M., Salgueiro, A.A., Campos-Takaki, G.M., Leite, L.F.C., Sarubbo, L.A. 2008. Utilization of two agroindustrial byproducts for the production of a surfactant by *Candida sphaerica* UCP0995. *Process Biochem.* 43:912–917.

Sudha, S., Kumanan, R., Muthusamy, K. 2010. Optimization of cultural conditions for the production of sophorolipids from *Candida Tropicalis*. *Der pharmacia Lettre* 2, no. 2:155–158.

Urum, K., Grigson, S., Pekdemir, T., Mcmenamy, S. 2006. A comparison of the efficiency of different surfactants for removal of crude oil from contaminated soils. *Chemosphere* 62, no. 9:1403–1410.

Van Dyke, M.I., Couture, P., Brauer, M., Lee, H., Trevors, J.T. 1993. *Pseudomonas aeruginosa* UG2 rhamnolipidbiosurfactants: Structural characterization and their use in removing hydrophobic compounds from soil. *Can. J. Microbiol.* 39:1071–1078.

Van Hamme, J.D., Singh, A., Ward, O.P. 2006. Physiological aspects. Part 1 in a series of papers devoted to surfactants in microbiology and biotechnology. *Biotechnol. Adv.* 24:604–620.

Vegilo, F., Beolchini, F., Toro, L. 1998. Kinetic modeling of copper biosorption by immobilized biomass. *Ind. Eng. Chem. Res.* 37, no. 3:1107–1111.

Venkata-Ramana, K., Karanth, N.G. 1989. Factors affecting biosurfactant production using *Pseudomonas aeruginosa* CFTR-6 under submerged conditions. *Chem. Technol. Biotechnol.* 45, no. 1:249–257.

Yeh, M.S., Wei, Y.H., Chang, J.S. 2005. Enhanced production of surfactin from *Bacillussubtilis* by addition of solid carriers. *Biotechnol. Progr.* 21:1329–1334.

Zinjarde, S.S., Pant, A. 2002. Emulsifier from tropical marine yeast, *Yarrowialipolytica* NCIM 3589. *J. Basic Microbiol* 42:67–73.

CHAPTER **11**

Biodegradable Polymers

Recent Perspectives

Arunachalam Bose Sathya, Raja Sivashankar,
Velmurugan Sivasubramanian, J. Kanimozhi,
and Arockiasamy Santhiagu

CONTENTS

Abstract

The environmental impact of perpetual fossil fuel plastic wastes is a looming universal concern since disposal techniques are restricted. Though synthetic plastics are one of the greatest inventions, the durability of these disposed plastics contributes to several adverse environmental conditions when they are introduced into the waste flow. The main disadvantage of fossil-fuel-based thermopolymers are their limited reserves, and disposal of these wastes has a negative environmental impact. A new generation of materials is needed in order to significantly reduce the environmental impact in terms of energy consumption and greenhouse effects. This encourages the production of biodegradable plastic or polymer, a plastic that is made partly or wholly from biological sources or living organisms. Biodegradable polymers have been recently developed from renewable resources, microbial biomass, and biotechnology. Potential application of biopolymers is included in the fields of medicine and pharmacy, food and packaging, and aerospace. Also, the enhanced biopolymeric material prevents mineral loudening and corrosion in wastewater treatment. In fact, the list of possible applications is almost infinite. Degradability of polymers using microbial (enzymatic) action in an ecofriendly method is one way to degrade these biopolymers.

11.1 INTRODUCTION

Synthetic plastics have emerged as the most important material in our daily life. Accumulation of these plastic wastes has become a major concern in waste management. Biodegradable plastics became an approach to solving this issue and has been prominent since 1970s. Biopolymers are made partly or wholly from polymers derived from biological sources such as sugar cane, potato starch, or the cellulose from trees, and are biodegradable with the action of microorganisms, heat, and moisture (Koushal et al., 2014). Biopolymers are sustainable, carbon neutral, and are always renewable. They are synthesized as intracellular carbon and energy reserves by a variety of microorganisms, especially when cultured in limiting nutritional conditions. The mechanical behavior of biodegradable materials depends on their chemical composition, production, storage and processing characteristics, aging, and application conditions (Willet, 1994). Bioplastics in India are still at a very nascent stage with only a few participants operating in this segment. As compared to the European market, where bioplastic products are commercially available, the Indian bioplastic industry has a long way to go in terms of production, raw materials, and technology. The market for biodegradable materials is growing significantly every year. A positive reaction can be expected from those applications, where biodegradability offers a clear advantage for customers and the environment, with typical examples being packaging, compost bags, agricultural films, and coatings (Yamamoto et al., 2005). Biopolymers have found wide acceptance in various industries, on account of their environmentally friendly properties. Numerous synthetic biodegradable polymers are available and still being developed for sustained and targeted drug delivery applications. Major advantages of bioplastics include:

- Potentially lower carbon footprint

- Lower manufacturing cost

- Do not use scarce crude oil

- Reduction in litter and improved compostability from using biodegradable bioplastics

- Improved acceptability to many households (Chen, 2014)

Bioplastics, currently accounting for less than half of one percent of all plastics manufactured, are growing rapidly because of the advantages that they have in many applications. As oil supplies tighten, these advantages will grow. However, use of fossil fuels should be reduced and research should focus on the use of renewable raw materials. The three drivers of growth—the importance of brand image, the value of joint composting, and the reduction of litter—will provide the impetus for continued growth in bioplastics across the world. This chapter deals with the classification of various biodegradable polymers, biodegradation, their application, and recent developments.

11.2 BIOPOLYMERS

Biopolymers are referred to as materials that are biodegradable, derived from both the renewable and nonrenewable resources. There are many sources of biodegradable plastics such as synthetic and natural polymers.

TABLE 11.1　Properties of Biopolymer and Conventional Plastics

Properties	Biopolymer (PHB)	Polypropylene
Crystallinity mass fraction	0.6 (0.8)	0.65
Tg	−5°C to 5°C (15°C)	15°C
Tm	175°C	174°C
Tensile strength	40 MPa	38 MPa
Percent elongation to break	2% to 5% (30% with annealing at 150°C)	400%

Natural polymers are available in large quantities from renewable sources, whereas synthetic polymers are produced from nonrenewable petroleum resources. Renewable sources are available indefinitely and include agricultural crops, microbes, and animals. Biopolymers possess similar physical and chemical properties to petroleum-derived thermoplastics, but they are totally biodegradable to CO_2 and H_2O (Vroman and Tighzert, 2009). Table 11.1 shows some of the properties of biopolymers and petroleum-derived polymers (Chen, 2014).

The direct production of biopolymers can be achieved by microorganisms (polyhydroxyalkanoates, PHA), by algae, by superior plants, or by several types of producers (e.g., cellulose is produced by superior plants and also by bacteria; chitosan is produced by crustacean and also by fungi). Biodegradation takes place through the action of enzymes and/or chemical deterioration associated with living organisms. The term biodegradable implies that it can be broken down into simpler substances by the activities of living organisms, and therefore is unlikely to persist in the environment (Gross and Kalra, 2002). There are many different standards used to measure biodegradability, with each country having its own. The requirements range from 90% to 60% decomposition of the product within 60 to 180 days of being placed in a standard composting environment (NIIR Board of Consultants and Engineers, 2006). Scope and acceptance of biopolymers is vast and versatile, finding a place in all human activities. They can be used as adhesives, adsorbents, lubricants, soil conditioners, paper coating, cosmetics, drug delivery vehicles, textiles, and high-strength structural materials (Kalia and Avérous, 2011). These polymers are costly compared to thermoplastics, and it is necessary to improve the availability and quality of these products to make them fully competitive.

11.3　CLASSIFICATION OF BIODEGRADABLE POLYMERS

Biopolymers are biodegradable, and some are also compostable. Biodegradable polymers are broken down into CO_2 and water by microorganisms. Compostable biopolymers can be put into an industrial composting process and will break down by 90% within six months.

There are primarily two classes of biopolymers: one that is obtained from living organisms and another that is produced from renewable resources but require polymerization (Chandra, 1998).

There are mainly three types of bioplastics in the commercial scale of production: (1) plastics derived from a fossil carbon source but biodegradable; (2) plastics derived from polymers converted from biomass and biodegradable; and (3) plastics derived from polymers converted from biomass but not biodegradable.

Biopolymers may also be divided into four categories based on their production (Figure 11.1).

Category 1—Polymers directly extracted/removed from biomass. Examples are polysaccharides such as starch and cellulose and proteins like casein and gluten.

Category 2—Polymers produced by microorganisms or genetically modified bacteria. To date, this group of biobased polymers consists mainly of the polyhydroxyalkanoates, but developments with bacterial cellulose are in progress.

Category 3—Polymers produced by classical chemical synthesis using renewable biobased monomers. A good example is polylactic acid, a biopolyester polymerized from lactic acid monomers.

Category 4—Polymers produced by conventional synthesis from synthetic monomers. Examples are aliphatic and aromatic polyesters.

11.4 BIOPOLYMERS FROM AGRO RESOURCES

Biopolymers similar to conventional polymers are produced by bacterial fermentation processes by synthesizing the building blocks (monomers) from renewable resources, including lignocellulosic biomass (starch and cellulose), fatty acids, and organic waste. Natural biobased polymers are the other class of polymers that are found naturally, such as proteins, nucleic acids, and polysaccharides (Babu et al., 2013). Agropolymers preserve the petrol resources replacing conventional polymers and have common characteristics such as hydrophilicity. They are mainly extracted from plants and are compostable. Since they are produced from renewable resources, they will be an interesting alternative to nondegradable

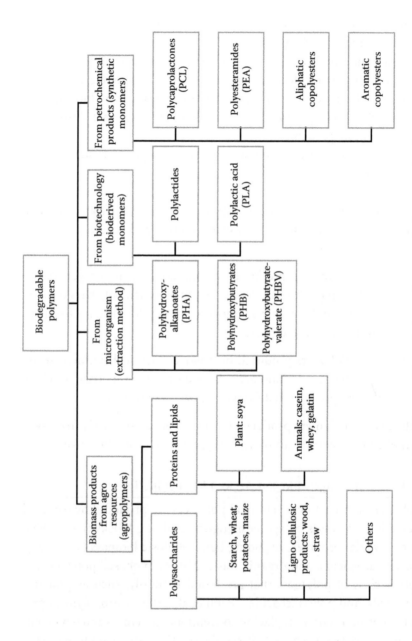

FIGURE 11.1   Classification of biodegradable polymers. (From Avérous, L., *Journal of Macromolecular Science. Polymer Reviews*, C44, no. 3:231–274, ISSN 1532–1797, 2004.)

polymers for short-life range applications. Agropolymers can find biomedical applications linked with intrinsic properties and could contribute to carbon footprint reduction in the future (Habibi and Lucia, 2012). They are classified into polysaccharides and proteins.

11.4.1 Polysaccharides

Polysaccharides and oligosaccharides are widely produced in nature. Polysaccharides are the most abundant macromolecules in the biosphere. These complex carbohydrates constituted of glycosidic bonds are often one of the main structural elements of plants and animals exoskeleton (Avérous and Pollet, 2012). Thus, animals and plants are the most important producers in terms of volume, whereas microorganisms produce much wider diversity. Also, in the small-scale structure, the proportion of different sugars and the type of linkage play an important role in the final properties of these biopolymers. Thus, applications of these poly- and oligosaccharides are also very different, from low- to high-added value products.

11.4.1.1 Polysaccharides from Plants

Plants are the most important producers of polysaccharides. Their main products are cellulose, hemicellulose, starch, inulin, and pectin. Cellulose and starch are both homopolymers, only composed by D-glucose units (Figure 11.2). The only difference between them is the type of linkage between the sugar units. Inulin is another homopolymer produced by plants. It is composed mainly of fructose units, even if a starting glucose moiety is present (Berezina and Maria, 2014).

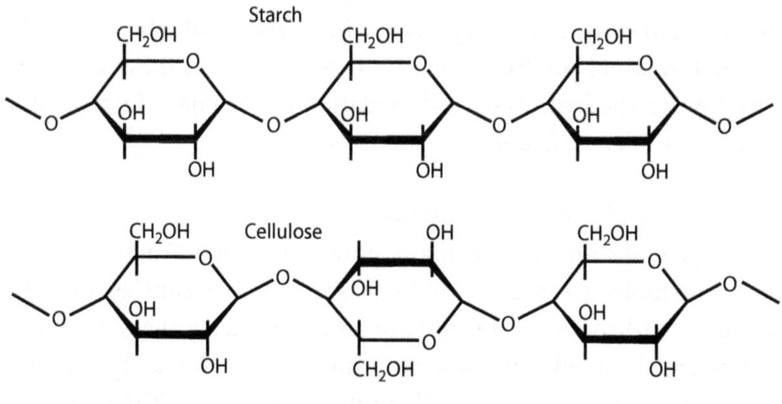

FIGURE 11.2 Structure of starch and cellulose.

Starch is extensively used since it is the main storage material and is a well-known hydrocolloid biopolymer that is abundantly available and one of the cheapest biodegradable polymers. It is a natural biopolymer found in roots, seeds, stem of plants such as corn, wheat, and potatoes (Dufresne et al., 2013). Starch, used by humans for centuries for food and feed (as well for its nutritional value as a thickener and emulsifier), has found application in the textile and paper industries and as a biodegradable packaging material. Food packaging and edible films are other two major applications of starch-based biodegradable polymers in the food industry. Starch can be used as filling agent in the granular state and as a component in synthetic polymer blends. At present and in the near future, different physical and chemical approaches are effective strategies for developing starch-based completely biodegradable polymers of appropriate biocompatibility, degradation rate, and physical properties for various applications. Starch is totally biodegradable in a wide variety of environments. It can be hydrolyzed into glucose by microorganism or enzymes, and then metabolized into carbon dioxide and water (Lu et al., 2009). It is worth noting that carbon dioxide will recycle into starch again by plants and sunshine. As starch is completely biodegradable and renewable, it will be an important component of biopolymer industries.

Biobased polymers can also be based on lipids, oils, and fatty acids produced from oil crops (e.g., oil palm, canola, soybean), chitin from insects and marine invertebrates, and lignin from biomass crops. Biomass crops, either woody species (e.g., pine, poplar, spruce, eucalyptus, willow) or grasses (e.g., sugarcane, sorghum, miscanthus) consist primarily of cell walls, which are a complex matrix in which cellulose microfibrils are embedded in a network of cross-linking hemicellulosic polysaccharides (Ten and Vermerris, 2013). Cellulose differs from synthetic polymers by virtue of its distinct polyfunctionality, its high chain stiffness, and its sensitivity toward the hydrolysis and oxidation of the chain-forming acetal groups (Klemm et al., 2005).

11.4.1.2 Polysaccharide from Animals

Chitin and chitosan are the main polysaccharides produced by animals. Those molecules are part of insects' and crustaceans' exoskeletons; even some mushrooms and fungi are able to produce them. Chitin and chitosan are aminoglucopyranans composed of N-acetylglucosamine (GlcNAc) and glucosamine (GlcN) units (Figure 11.3). The main applications of these polysaccharides have been, until now, based on their

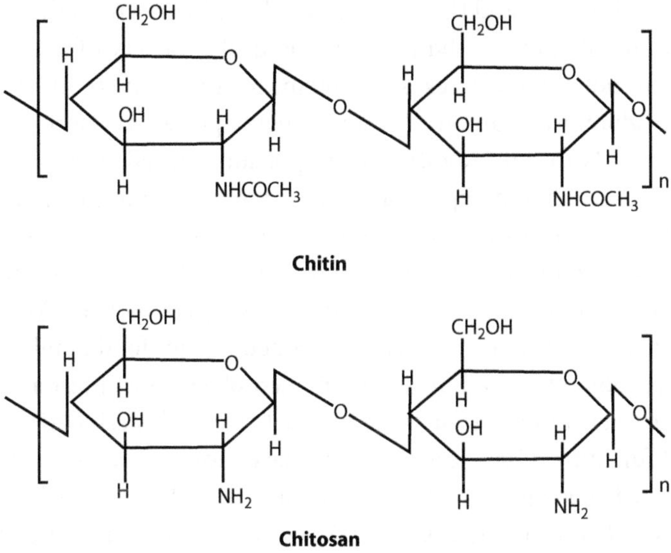

Chitin

Chitosan

FIGURE 11.3   Structure of chitin and chitosan.

antimicrobial properties as applied to either the food or the cosmetic industries (Kumar, 2000).

11.4.1.3 Polysaccharides from Microorganisms

Compared to higher plants and animals, microorganisms are characterized by the wide diversity of poly- and oligosaccharides they produce. Among the microorganisms, bacteria have the widest spread of possibilities. Polysaccharides produced by bacteria are mainly extracellular polysaccharides (EPS), also called exopolysaccharides, whereas those produced by algae are mainly cell wall and structural constituents (Valepyn et al., 2012). Some of the microbial polysaccharides are also produced by higher organisms, such as cellulose in plants or chitin and chitosan in animals. However, microbial production is often better controlled and offers the possibility of higher-added value applications. Microbial polysaccharides can also be synthesized using agricultural feedstocks. Cereals are agricultural raw materials that are very rich in biopolymers such as starch, protein, polysaccharides, and lipids.

11.4.2 Proteins

Proteins are agropolymers produced by plants, animals, and bacteria. Two main biosynthetic pathways for protein production have been identified so far: the ribosomal and the nonribosomal–multienzyme paths. The proteins are mainly heteropolymers composed of a variety of amino acids. In terms of potential sources, soy protein, corn protein (zein), and wheat proteins (gluten) are among the main plant proteins. Casein, collagen protein or gelatin, and keratin are important animal proteins. Lactate dehydrogenase, chymotrypsin, and fumarase constitute the main bacterial proteins. In all cases, the stability of the proteins and their sensitivity to moisture require strengthening by plasticization, compatibilization, cross-linking, or production of protein–nanoclay composites (Mohanty et al., 2005). These proteins could be developed for nonfood applications in the future. Some targeted areas are molded objects for interior use such as toys, leather imitation for office products (covers), design articles, and pieces of furniture. Caseins and gelatins have shown to be useful in a great number of fields such as adhesives, controlled releases and biomedical applications (Avérous and Pollet, 2012). Gelatin is commonly used for biomedical applications due to its biodegradability and biocompatibility in physiological environments in contact with living tissues.

11.5 BIOPOLYMERS FROM MICROORGANISMS

In response to problems associated with plastic waste and its effect on the environment, there has been considerable interest in the development and production of alternative, biodegradable plastics or bioplastics. The high cost associated with the production of bioplastics compared to the low cost of synthetic polymers has resulted in a loss of interest in bioplastic research (Varsha and Savitha, 2011). Their development became essential due to the depletion of petroleum reserves, high oil prices, and increased greenhouse gas emissions. Thus, bioplastics produced from microorganisms may provide a solution.

Microorganisms can be used for the biosynthesis of various biopolymers such as xanthan, dextrans, pullulan, glucans, gellan, alginate, cellulose, cyanophycin, poly(gammaglutamic acid), levan, hyaluronic acid, cellulose, organic acids, oligosaccharides, polysaccharides, and polyhydroxyalkanoates (Verma, 2012). Some microorganisms are particularly capable of converting biomass into biopolymers while employing a set of catalytic enzymes. Fermentation procedures, even though costly, are considered best for producing such polymers since they give good outcome within less time (Benerji et al., 2010). Fermentation carried out by the microorganisms may or may not rely on a separate polymerization step. This depends on the type of substrate that has to be degraded, the microbe that is involved, and the conditions that are provided during the process. The main advantage is that biodegradable polymers are completely degraded to water, carbon dioxide, and methane by anaerobic microorganisms (Santhanam and Sasidharan 2010).

11.5.1 Polyhydroxyalkanoates

Polyhydroxyalkanoates (PHAs) are among the most investigated biodegradable polymers in recent years. They are superior to other biodegradable polymers because of the large number of different monomer constituents that are incorporated. These polyesters have chemical and physical properties similar to conventional plastics, which add to their biodegradability and biocompatibility (Braunegg et al., 1998). PHAs are mainly produced from renewable resources by fermentation. A wide variety of prokaryotic organisms accumulate PHA from 30% to 80% of their cellular dry weight. PHAs are considered biodegradable and thus suitable for, e.g., short-term packaging, and also considered as biocompatible in contact with living tissues and can be used for biomedical applications (Avérous and Pollet, 2012).

PHAs are polyesters of hydroxyalkanoates with the general structural formula as shown in Figure 11.4, where n varies from 600 to 35,000 and if

R = hydrogen   Poly(3-hydroxypropionate)

R = methyl   Poly(3-hydroxybutyrate)

R = ethyl   Poly(3-hydroxyvalerate)

R = profile   Poly(3-hydroxyhexanoate)

R = pentyl   Poly(3-hydroxyoctanoate)

R = nonyl   Poly(3-hydroxydodecanpate)

PHAs are generally classified into short-chain-length PHAs (sCL-PHAs) and medium-chain-length PHAs (mCL-PHAs) by the different number of carbons in their repeating units. For instance, sCL-PHAs contain four

$$(-O-\underset{|}{\overset{R}{C}}H-CH_2-\underset{\|}{\overset{O}{C}}-)_n$$

FIGURE 11.4 Generic chemical structure of PHA. (From Lee, I.Y. et al., *FEMS Microbiol. Lett.* 131, 35–39, 1995.)

or five carbons in their repeating units, while mCL-PHAs contain six or more carbons in the repeating units.

Accumulation of PHA by microorganisms can be stimulated under unbalanced growth conditions, for example, when nutrients such as nitrogen, phosphorus, or sulfate become limited; when oxygen concentration is low; or when the C:N ratio of the feed substrate is higher (Lafferty et al., 1988). During starvation, PHA serves as a carbon and energy source and is rapidly oxidized thereby retarding the degradation of cellular components, combating the adverse conditions as in rhizosphere (Okon and Itzigsohn, 1992).

Besides functioning as a carbon and energy storage compound, other possible functions of PHA have been gaining interest. Studies on the ability of a microbial cell to take up genetic material from an external medium (known as competence) have led to the identification of another type of PHA that is a constituent of cytoplasm and cytoplasm membranes (Reusch and Sadoff, 1983).

PHA production in bacterial cells was initially described by Beijerinck and colleagues. PHA inclusions within the cells were first described as lipids by biochemists; however, in 1925 Lemoigne determined the inclusion bodies to be polyhydroxybutyrate (PHB). Lemoigne first reported the discovery of PHB as a component of the bacterium *Bacillus megaterium* in 1926. An improved fermentation process, a more efficient recovery/purification process, and the use of inexpensive carbon sources for the production of PHB have also been found to substantially reduce the cost of production (Lemos et al., 2006). Table 11.2 represents some PHA-accumulating bacteria.

TABLE 11.2   Polyhydroxyalkanoate (PHA)-Accumulating Genera of Prokaryotic Microorganisms

Microorganisms That Accumulate PHA				
Acidovorax	Bacillus	Erwinia	Klebsiella	Ralstonia
Acinetobacter	Brachymonas	Escherichia	Leptothrix	Spirulina
Actinobacillus	Clostridium	Ferrobacillus	Methanomonas	Synechococcus
Anabenna	Cupriavidus	Gloeocapsa	Nitromonas	Thiobacillus
Aquaspirillum	Cyanobacteria	Haemophilus	Nostoc	Vibrio
Azospirillum	Delfia	Halobacterium	Pseudomonas	Xanthobacter

Source: Koller, M., A. Atlic, M. Dias, A. Reiterer, and G. Braunegg, 2010, Microbial PHA production from waste raw materials. In *Plastics from Bacteria: Natural Functions and Applications*, edited by G.-Q. Chen and A. Steinbüchel, 85–119, Münster: Springer.

Microbial PHA is a potential renewable biopolymer with properties closely resembling some common petrochemical plastics. Because of the vast range of structurally different monomers that can be polymerized by microbes, a wide range of material properties can be achieved (Kunasundari and Sudesh, 2011). PHA production starts in response to stress imposed on cells, usually by nitrogen or phosphorus limitation, in the presence of an abundant carbon source. Under these conditions (PHA accumulation phase), the cells do not grow or divide but instead divert their metabolites toward the biosynthesis of hydoxyalkyl-CoA (HA-CoA). HA-CoA is an activated monomeric precursor that is polymerized by the enzymatic action of PHA synthase to form a PHA polyester. Being insoluble in water, PHA begins to form amorphous and nearly spherical granules that gradually fill the cells and force them to expand.

The most widely produced PHA is PHB, which is water insoluble and relatively resistant to hydrolytic degradation. PHB is soluble in chloroform and other chlorinated hydrocarbons. PHB is suitable for medical applications like bone plates, nails, screws, and in the treatment of osteomyelitis. PHB has its melting point at 175°C and glass transition temperature of 15°C (Avérous and Pollet, 2012). The biosynthesis of PHB is initiated by the condensation of two acetyl-CoA molecules by α ketothiolase (PhaA) to form acetoacetyl-CoA. Subsequently, NADPH-dependent acetoacetyl-CoA reductase (PhaB) catalyzes the reduction of acetoacetyl-CoA to the (R)-isomer of 3-hydroxybutyryl-CoA, which is then polymerized into P(3HB). Figure 11.5 explains the metabolic pathway of PHB synthesis.

11.6 BIOPOLYMERS FROM BIOTECHNOLOGY

An environmentally conscious alternative for petroleum-derived plastics is to design/synthesize polymers that are biodegradable. The presence of hydrolyzable or oxidizable linkage on the polymer main chain, the presence of suitable substituents, correct stereoconfiguration, balance of hydrophobicity and hydrophilicity, and conformation flexibility contribute to the biodegradation of hydrolysable polymers, which proceeds in a diffuse manner, with the amorphous regions degrading prior to the degradation of the crystalline and cross-linked regions. In this context, the biotechnological approaches are being increasingly recognized as a key to developing better biodegradables and low-cost biopolymers. Fully biodegradable synthetic polymers, such as polylactic acid (PLA), polycaprolactone (PCL), and polyhydroxybutyrate-valerate (PHBV), have been commercially available since 1990. However, these synthetic polymers

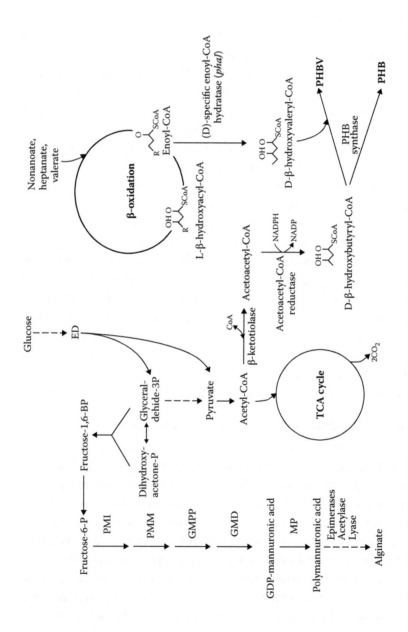

FIGURE 11.5 The most studied metabolic pathway for the biosynthesis of polyhydroxybutyrate (PHB). (Adapted from Avérous, L., and E. Pollet, eds., *Environmental Silicate Nano-Biocomposites*, Green Energy and Technology, London, Springer, 2012.)

from natural resources are usually more expensive than petroleum-based polymers (Pandey et al., 2010). For decades, scientists have been fascinated with microbial production of PHB (Byrom, 1987; Madison and Huisman, 1999). As an alternative approach, microorganisms have been engineered to produce hydroxyalkanoates (HAs) (Chen and Wu, 2005; Ren et al., 2010). Besides wild-type strains, recombinant strains are also being developed. Various types of recombinant *Escherichia coli* strains are able to synthesize PHA to high intracellular levels and some are amenable to genetically mediated lysis systems to facilitate the release of the PHA granules (Khanna et al., 2005).

11.6.1 Polylactic Acid

Polylactic acid (polylactide) is the most widely used biodegradable aliphatic polyester with the basic constitutional unit lactic acid. The monomer lactic acid is the hydroxyl carboxylic acid, which can be obtained via bacterial fermentation from corn (starch) or sugars obtained from renewable resources. Although other renewable resources can be used, corn has the advantage of providing a high-quality feedstock for fermentation, which results in a high-purity lactic acid, that is required for an efficient synthetic process. L-lactic acid or D-lactic acid is obtained depending on the microbial strain used during the fermentation process. It is difficult to obtain high molecular weight PLA via polycondensation reaction because of water formation during the reaction. A low-cost continuous process can be developed for the production of PLA (Erwin et al., 2007). In this process, low molecular weight prepolymer lactide dimers are formed during a condensation process. In the second step, the prepolymers are converted into high molecular weight PLA via ring opening polymerization with selected catalysts. Depending on the ratio and stereochemical nature of the monomer (L or D), various types of PLA and PLA copolymers can be obtained. The final properties of PLA produced are highly dependent on the ratio of the D and L forms of the lactic acid.

The chemical reactions, leading to the formation of a cyclic dimer, the lactide, as an intermediate step to the production of PLA, could lead to long macromolecular chains with L- and D-lactic acid monomers. This mechanism of ring-opening polymerization (ROP) from the lactide explains the formation of two enantiomers. This ROP route (Figure 11.6) has the advantage of reaching high molecular weight and of allowing the control of PLA final properties by adjusting the proportions and sequencing of L- and D-lactic acid units. Properties of PLA depend on their molecular

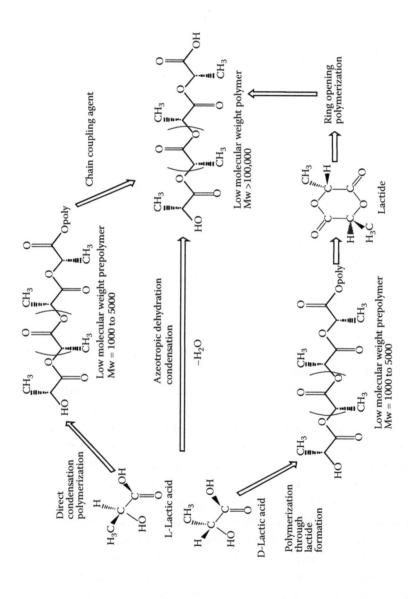

FIGURE 11.6   Synthesis methods for obtaining high molecular weight PLA. (Adapted from Avérous, L., and E. Pollet, eds., *Environmental Silicate Nano-Biocomposites*, Green Energy and Technology, London, Springer, 2012.)

characteristics as well as ordered structures such as crystalline thickness, crystallinity, spherulite size, morphology, and degree of chain orientation (Avérous and Pollet, 2012).

11.7 BIOPOLYMERS FROM SYNTHETIC MONOMERS

A large number of biodegradable polyesters are based on petroleum resources, obtained chemically from synthetic monomers. According to the chemical structures, we can distinguish polycaprolactone, aliphatic copolyesters, and aromatic copolyesters. All these polyesters are soft at room temperature. These are polymers that do not occur in nature as such, and, therefore, when they end up in the natural environment, they represent a durable foreign object, since they cannot be incorporated into the natural cycle (Okada, 2002). It is estimated that the discarded plastic bottle remains in the natural environment for 450 years. Among these polymers, degradability is otherwise achieved with the integration of hydrolytically unstable bonds into the polymer (e.g., ester-, amide groups).

11.7.1 Aliphatic Copolyesters

Aliphatic polyesters made from dimethylesters, and diols are expected to be some of the most economically competitive biodegradable polymers. Aliphatic polyesters are attracting increasing attention to solve "white pollution" concerns caused by traditional nonbiodegradable polymers and also for their use as specialty polymers in applications that mainly involve the biomedical field. In fact, polyester amides derived from naturally occurring amino acids offer great possibilities as biodegradable materials for biomedical applications. The biodegradability of these polymers depends mainly on their chemical structure and especially on the hydrolyzable ester bond in the main chain, which is susceptible to microbial attack. The main disadvantage of these polymers is their poor mechanical properties as compared with those of commodity plastics, which restricts their use in many applications. The lack of sufficient mechanical properties can be mainly attributed to their chemical structure and their relatively low molecular weight (Tserki et al., 2006). In order to obtain biodegradable polyesters with useful mechanical properties, many scientists have worked on the synthesis of copolyesters composed of aliphatic and aromatic units. Another possible way to improve the mechanical properties of aliphatic polyesters without a loss of biodegradability could be the incorporation of a chain extender, resulting in increased molecular weight.

Polyalkylene dicarboxylates can be easily prepared by thermal polycondensation of diols and dicarboxylic acids or either their diesters or dichlorides. Nevertheless, this synthesis has problems associated with the use of solvents, and high vacuum and temperature conditions required to favor condensation reactions and the removal of water or alcohol by-products. Polycondensation of diols with dicarboxylic acids can also be performed by employing nonspecific lipases, but in this case final cost is also increased due to the expensive enzyme and even for the subsequent separation process that is required to eliminate the enzyme. ROP has several advantages over polycondensation, but the low availability and high manufacturing costs of cyclic ester oligomers (CEOs) at the industrial scale has prevented its general application for the synthesis of poly(alkylene dicarboxylate) (Díaz et al., 2014).

Biodegradable aliphatic polyesters such as polybutylene succinate (PBSu), poly(butylene succinate-co-butylene adipate) (PBSA) copolymer, and polyethylene succinate (PESu) were invented in 1990 and produced successfully through a polycondensation reaction of glycols with aliphatic dicarboxylic acids and their derivatives (Avérous and Pollet, 2012).

$$\left[\!\!\begin{array}{c}O\\\|\\C\end{array}\!\!-(CH_2)_2-\begin{array}{c}O\\\|\\C\end{array}\!\!-O-(CH_2)_4-O\right]_x\left[\begin{array}{c}O\\\|\\C\end{array}\!\!-(CH_2)_4-\begin{array}{c}O\\\|\\C\end{array}\!\!-O-(CH_2)_4-O\right]_y$$

Chemical structure of poly(butylene succinate adipate) (PBSA).

11.7.2 Aromatic Polyesters

Compared to aliphatic copolyesters, aromatic copolyesters have excellent physical and mechanical properties but appear to be very resistant to bacterial and fungal attack under environmental conditions. Aromatic polyesters are often based on terephthalic acid. An increase of terephthalic acid content tends to decrease the degradation rate (Muller et al., 1998.). However, recent investigations indicate that the aromatic ester groups can also be degraded under specific conditions. Polymerization time should be optimum for thermal polyesterification reaction because degradation appears to be significant when samples are maintained at a high temperature for long periods. Poly(butylene adipate-co-terephthalate) (PBAT) is an aliphatic–aromatic copolymer that is prepared by melt polycondensation of 1, 4-butanediol (BDO), dimethylterephthalate (DMT), and adipic acid with tetrabutylorthotitanate (TBOT) as the catalyst.

Chemical structure of poly(butylene adipate-co-terephthalate) (PBAT).

11.8 BIODEGRADATION

Biodegradation is a term used in ecology to indicate the biochemical processes in which organic substances produced directly or indirectly from photosynthesis are broken down and transformed back into the inorganic state. Combustion of synthetic plastics releases hazardous pollutants and greenhouse gases such as sulfur, nitrogen oxide, and carbon dioxide. The phenomenon of biodegradation is very important for the environment, which must get rid of waste and residues in order to make space for new life. The biodegradation is carried out by the decomposers, i.e., microorganisms (fungi, bacteria, protozoa), that grow on the dead organic matter, i.e., on the refuse produced by the ecosystem.

This event occurs in two steps. The first one is the fragmentation of the polymers into lower molecular mass species by means of either abiotic reactions (e.g., oxidation, photodegradation, hydrolysis) or biotic reactions (e.g., degradations by microorganisms). This is followed by bioassimilation of the polymer fragments by microorganisms and their mineralization. Biodegradability depends not only on the origin of the polymer but also on its chemical structure and the environmental degrading conditions. The mechanical behavior of biodegradable materials depends on their chemical composition, the production, the storage and processing characteristics, the aging, and the application conditions (Vroman and Tighzert, 2009). Regarding biodegradation in compost, adequate conditions are only found in industrial units with a high temperature (above 50°C) and a high relative humidity (RH) to promote chain hydrolysis.

11.9 APPLICATIONS OF BIODEGRADABLE POLYMERS

Biodegradable polymers are of significant interest to a variety of fields including medicine, agriculture, and packaging. One of the most active areas of research in biodegradable polymers is in controlled drug delivery and release. Biopolymers that may be employed in packaging continue to receive more attention than those designated for any other application. The renewable and biodegradable characteristics of biopolymers are what render them appealing for innovative uses in packaging. The biopolymers used in medical applications must be compatible with the tissue they are found in, and may or may not be expected to break down after a given time period, as reported by researchers working in tissue engineering who are attempting to develop organs from polymeric materials that are fit for transplantation into humans. The plastics would require injections with growth factors in order to encourage cell and blood vessel growth in the new organ (Kumar et al., 2011).

11.9.1 Packaging

Biodegradable polymers are often used to reduce the volume of waste in packaging materials. Packaging and bottles are the major biopolymer applications and serve an array of industries due to their various properties and performance advantages. Bioplastics are also used in waste collection bags, carrier bags, mulch-film, and food-service ware such as cutlery. One of the most commonly used polymers for packaging purposes is polylactic acid (PLA). PLA is used for a variety of films, wrappings, and containers (including bottles and cups). The new food-packaging systems have been developed as a response to trends in consumer preferences toward mildly preserved, fresh, tasty, healthier, and convenient food products with prolonged shelf life. The novel packaging technologies can also be used to compensate for shortcomings in the packaging design, for instance, in order to control the oxygen, water, or CO_2 levels in the package headspace.

11.9.2 Medicinal Applications

Biodegradable polymers have innumerable uses in the biomedical field, particularly in the fields of tissue engineering and drug delivery. In order for a biodegradable polymer to be used as a therapeutic, it must meet several criteria:

1. Be nontoxic in order to eliminate foreign body response

2. The time it takes for the polymer to degrade must be proportional to the time required for therapy

3. The products resulting from biodegredation are not cytotoxic and are readily eliminated from the body

4. The material must be easily processed in order to tailor the mechanical properties for the required task

5. Be easily sterilized

6. Have acceptable shelf life

Injectable polymers based on urethane and urethane/acrylate have shown great promise in developing delivery systems for tissue-engineered products and therapies. The main application of collagen films in ophthalmology is as drug delivery systems for slow release of incorporated drugs. It was also used for tissue engineering including skin replacement, bone substitutes, and artificial blood vessels and valves (Lee et al., 2001).

Being natural, biodegradable, and biocompatible properties, biopolymers have a wide range of applications in different fields such as membranes, medicine, drug delivery, hydrogels, water treatment, adhesives, food packaging, fuel cells, and surface conditioners. Biopolymers and their derivatives are suitable for tissue engineering applications (Verma, 2012).

11.10 RECENT ADVANCES

Synthetic plastics are resistant to degradation, and consequently their disposal is fueling an international drive for the development of biodegradable polymers. As the development of these materials continues, industry must find novel applications for them. Material usage and final mode of biodegradation are dependent on the composition and processing method employed. An integrated waste management system may be necessary in order to efficiently use, recycle, and dispose of biopolymer materials. In spite of the plethora of applications, the large-scale introduction of biopolymers into the market has often been forestalled by high production costs mainly due to complex or inefficient downstream processing. With ongoing research the number of possible applications has increased rapidly, ranging from use as food additives and biomedical agents to biodegradable plastics from renewable resources (Chen, 2014). Nowadays, biobased polymers are commonly found in many applications from commodity to high-tech applications due to advancement in biotechnologies and public awareness. However, despite these advancements, there are still some drawbacks that prevent the wider commercialization of biobased polymers in many applications. This is mainly due to performance and price when compared with their conventional counterparts, which remains a significant challenge for biobased polymers. Efforts to create more economical production strategies have surged in recent years with the utilization of mixed microbial cultures (MMCs) (Mohan and Reddy, 2013). The application of MMCs avoids the cost of equipment sterilization and the use of expensive refined substrates, as they can be operated in open systems and utilize cheap industrial fermentable by-products (Reis et al., 2011).

Algae-based plastics have been a recent trend in the era of bioplastics compared to traditional methods of utilizing feedstocks of corn and potatoes as plastics. Although algae-based plastics are in their infancy, once they are commercialized they are likely to find applications in a wide range of industries. Algae bioplastics mainly evolved as a by-product of algae biofuel production, where companies were exploring alternative sources of revenues along with those from biofuels. In addition, the use of algae opens the possibility of utilizing carbon, neutralizing greenhouse gas emissions from factories or power plants. Continued research and development in bioplastics is creating high-quality products for a wide variety of industries.

11.11 CONCLUSION

Plastic waste is a major environmental and public health problem in India, particularly in urban areas. Plastic pollution can be reduced by using fewer plastics products and switching to alternatives. Each year, an estimated 500 billion to 1 trillion plastic bags are consumed worldwide. That comes out to over 1 million per minute. Billions end up as litter each year or in landfills. As oil runs out and the use of fossil fuels becomes increasingly expensive, the need for replacement of sources of raw material for the manufacture of vital plastics becomes increasingly urgent. In addition, the use of carbon-based sources of energy for use in plastics manufacturing adds greenhouse gases to the atmosphere, impeding the world's attempts to cut CO_2 emissions. A possible solution to the problem is the prospect of biodegradable polymers.

Biodegradable polymers offer the following advantages. First, they are made from renewable resources and thus do not face the problem of exhaustion. Second, they are biodegradable, which implies that the polymer after use will ultimately turn into compost. A variety of novel biomaterials have been developed by modifying biopolymers or by synthesizing bioinspired macromolecules. The designs of such intelligent biomaterials are green and environmental friendly. Bioplastics can provide excellent biodegradability, helping the world deal with the increasing problems of litter, particularly in the world's rivers and seas. Biodegradable polymers have received much more attention in the last decades due to their potential applications in the fields related to environmental protection and the maintenance of physical health.

The use of biotechnology can play a key role in conducting feasibility and sustainability studies in bioplastics. Fermentation and genetic engineering can take the lead in using novel techniques to make bioplastics commercially viable. To improve the properties of biodegradable polymers, lot of methods have been developed, such as random and block copolymerization or grafting. These methods improve both the biodegradation rate and the mechanical properties of the final products. Physical blending is another route to prepare biodegradable materials with different morphologies and physical characteristics. There is a need to increase public awareness through litter education as an important supporting element and other initiatives that may be undertaken to reduce plastic waste and its impact.

REFERENCES

Avérous, L. 2004. Biodegradable multiphase systems based on plasticized starch: A review. *Journal of Macromolecular Science. Polymer Reviews*, C44, no. 3: 231–274, ISSN 1532–1797.

Avérous, L., and E. Pollet, eds. 2012. *Environmental Silicate Nano-Biocomposites*. Green Energy and Technology. London: Springer.

Babu, R.P., K. O'Connor, and R. Seeram. 2013. Current progress on bio-based polymers and their future trends. *Prog. Biomater.* 2, no. 1:8.

Benerji, D.S.N., Ayyanna, C., Rajini, K., Rao, B.S., Banerjee, D.R.N. et al. 2010. Studies on physico-chemical and nutritional parameters for the production of ethanol from mahua flower (madhuca indica) using saccharomyces cerevisiae—3090 through submerged fermentation (smf). *J. Microbial. Biochem. Technol.* 2:46–50.

Berezina, N., and S. Maria. 2014. Bio-based polymers and materials. In *Renewable Resources for Biorefineries*, 1–28.

Braunegg, G., Lefebvre, G., and Genser, K.F. 1998. Polyhydroxyalkanoates, biopolymers from renewable resources: Physiological and engineering aspects. *J. Biotechnol.* 65:127–161.

Byrom, D. 1987. Polymer synthesis by microorganisms: Technology and economics. *Trends Biotechnol.* 5:246–250.

Chandra, R. 1998. Biodegradable polymers. *Polym. Sci.* 23, no. 7:1273–1335.

Chen, Y.J. 2014. Bioplastics and their role in achieving global sustainability. *J. Chem. Pharm. Res.* 6, no. 1:226–231.

Chen, G., and Wu, Q. 2005. Microbial production and applications of chiral hydroxyalkanoates. *Appl. Microbiol. Biotechnol.* 67:592–599.

Díaz, A., R. Katsarava, and J. Puiggalí. 2014. Synthesis, properties and applications of biodegradable polymers derived from diols and dicarboxylic acids: From polyesters to poly (ester amide) s. *Int. J. Mol. Sci.* 15, no. 5:7064–7123.

Dufresne, A., S. Thomas, and L.A. Pothan. 2013. *Biopolymer Nanocomposites: Processing, Properties, and Applications*. Hoboken, NJ: John Wiley & Sons.

Erwin, T.H., David, A.G., Jeffrey, J.K., Robert, J.W., and Ryan, P.O. 2007. The eco-profiles for current and near-future NatureWorks—Polylactide (PLA) production. *Industrial Biotechnology* 3:58–81.

Gross, R.A., and B. Kalra. 2002. Biodegradable polymers for the environment. *Science* 297, no. 5582:803–807.

Habibi, Y., and L.A. Lucia. 2012. *Polysaccharide Building Blocks: A Sustainable Approach to the Development of Renewable Biomaterials*. Hoboken, NJ: John Wiley & Sons.

Kalia, S., and L. Avérous. 2011. *Biopolymers: Biomedical and Environmental Applications*. Hoboken, NJ: John Wiley & Sons.

Khanna, S., and Srivastava, A. 2005. Recent advances in microbial polyhydroxyalkanoates. *Process Biochem.* 40:607–619.

Klemm, D., Heublein, B., Fink, H.P., and Bohn, A. 2005. Cellulose: Fascinating biopolymer and sustainable raw material. *Angew. Chem.-Int. Edit.* 44, no. 22:3358–3393.

Koller, M., A. Atlic, M. Dias, A. Reiterer, and G. Braunegg. 2010. Microbial PHA production from waste raw materials. In *Plastics from Bacteria: Natural Functions and Applications*, edited by G.-Q. Chen and A. Steinbüchel, 85–119. Münster: Springer.

Koushal, V., R. Sharma, M. Sharma, R. Sharma, and V. Sharma. 2014. Plastics: Issues challenges and remediation. *Int. J. Waste Resour.* 4, no. 1:1–6.

Kumar, A., Karthick, K., and Arumugam, K.P. 2011. Biodegradable Polymers and Its Applications. *International Journal of Bioscience, Biochemistry and Bioinformatics* 1, no. 3:1–4. doi:10.7763/IJBBB.2011.V1.32.

Kumar, M.N.V.R. 2000. A review of chitin and chitosan applications. *Reactive Funct. Polymers* 46:1–27.

Kunasundari, B., and K. Sudesh. 2011. Isolation and recovery of microbial polyhydroxyalkanoates. *Express Polym. Lett.* 5, no. 7:620–634.

Lafferty, R.M., Korsatko, B., and Korsatko, W. 1988. Microbial production of poly-β-hydroxybutyric acid. In: Rehm H. J., Reed, G., editors. *Biotechnology Special microbial Processes* 136–176.

Lee, H.C., Anuj, S., and Lee, Y. 2001. Biomedical applications of collagen. *Inter. J. Pharma.* 221:1–22.

Lee, I.Y., Kim, M.K., Chang, H.N., and Park, Y.H. 1995. Regulation of poly β hydroxybutyrate biosynthesis by nicotinamide nucleotides in Alcaligenes eutrophus. *FEMS Microbiol. Lett.* 131, 35–39.

Lemos, P.C., Serafim, L.S., and Reis, M.A.M. 2006. Synthesis of polyhydroxyalkanoates from different short-chain fatty acids by mixed cultures submitted to aerobic dynamic feeding. *J. Biotechnol.* 122:226–238.

Lu, D.R., Xiao, C.M., and Xu, S.J. 2009. Starch-based completely biodegradable polymer materials. *Express Polymer Lett.* 3, no. 6:366–375. doi:10.3144/expresspolymlett.2009.46.

Madison, L.L., and Huisman, G.W. 1999. Metabolic engineering of poly(3 hydroxyalkanoates): From DNA to plastic. *Microbiol. Mol. Biol. Rev.* 63, no. 1:21–53.

Mohan, S.V., and Reddy, M.V. 2013. Optimization of critical factors to enhance polyhydroxyalkanoates (PHA) synthesis by mixed culture using Taguchi design of experimental methodology. *Bioresour. Technol.* 128, 409–416.

Mohanty, A.K., W. Liu, P. Tummula, L.T. Drzal, M. Manjusri, and R. Narayan. 2005. Soy protein-based plastics, blends, and composites. In *Natural Fibers, Biopolymers, and Biocomposites*, edited by A.K. Mohanty, M. Misra, and L.T. Drzal, 699–725. Boca Raton, FL: Taylor & Francis.

Muller, R.-J., Witt, U., Rantze, E., and Deckwer, W.-D. 1998. Architecture of bio-degradable copolyesters containing aromatic constituents. *Polym. Degrad. Stab.* 59, no. 1–3:203–208.

NIIR Board of Consultants and Engineers. 2006.

Okada, M. 2002. Chemical syntheses of biodegradable polymers. *Prog. Polym. Sci. (Oxford)* 27, no. 1:87–133.

Okon, Y., and Itzigsohn, R. 1992, Poly-β-hydroxybutyrate metabolism in Azospirillum brasilense and the ecological role of PHB in the rhizosphere. *FEMS Microbiol. Rev.* 103:131–140.

Pandey, P., Kumar, B., and Tiwari, D. 2010. Environmental considerations concerning the release of genetically modified organisms. *ProEnvironment Promediu* 3, no. 6:381–384.

Reis, M., Albuquerque, M., Villano, M., and Majone, M. 2011. Mixed culture processes for polyhydroxyalkanoate production from agro-industrial surplus/wastes as feedstocks. In: Moo-Young, M. (Ed.), Comprehensive Biotechnology. Academic Press, Burlington, 669–683.

Ren, Q., Ruth, K., Thony-Meyer, L., and Zinn, M. 2010. Enatiomerically pure hydroxycarboxylic acids: Current approaches and future perspectives. *Appl. Microbiol. Biotechnol.* 87:41–52.

Reusch, R.N., and Sadoff, H.L. 1983. d-(2)-poly-b-hydroxybutyrate in membranes of genetically competent bacteria. *J. Bacteriol.* 156, no. 2:778–788.

Santhanam, A. and Sasidharan, S. 2010. Microbial production of polyhydroxy alkanote (PHA) from *Alcaligens* spp. and *Pseudomonas oleovorans* using different carbon sources. *Afri. J. Biotechnol.* 9:3144–3150.

Ten, E., and W. Vermerris. 2013. Functionalized polymers from lignocellulosic biomass: State of the art. *Polymers (Basel)* 5, no. 2:600–642.

Tserki, V., P. Matzinos, E. Pavlidou, D. Vachliotis, and C. Panayiotou. 2006. Biodegradable aliphatic polyesters. Part I. Properties and biodegradation of poly(butylene succinate-co-butylene adipate). *Polym. Degrad. Stab.* 91, no. 2:367–376.

Valepyn, E., N. Berezina, and M. Paquot. 2012. Optimization of production and preliminary characterization of new exopolysaccharides from *Gluconacetobacter hansenii* LMG1524. *Adv. Microbiol.* 2, no. 4:488–496.

Varsha, Y.M., and R. Savitha. 2011. Overview on polyhydroxyalkanoates: A promising biopolymer. *Microbial Biochem. Technol.* 3, no. 5:99–105.

Verma, M.L. 2012. Microbial biosynthesis of biopolymers and applications in the bio-pharamaceutical, biomedical and food industries. *Crit. Rev. Biotechnol.* 27:1–19.

Vroman, I., and L. Tighzert. 2009. Biodegradable polymers. *Materials* 2, no. 2:307–344.

Willett, J.L. 1994. Mechanical properties of LDPE/granular starch composites. *J. Appl. Polym. Sci.* 54:1685–1695.

Yamamoto, M., U. Witt, D.I.G. Skupin, D. Beimborn, and R.J. Muller. 2005. Biodegradable aliphatic-aromatic polyesters. *Biopolym. Online* 299–305.

CHAPTER 12

Microbial Fuel Cell Technology

A Sustainable Energy Producer from a Wide Variety of Waste Sources

D. Vidhyeswari, A. Surendhar, M. Jerold,

S. Bhuvaneshwari, and Velmurugan Sivasubramanian

CONTENTS

Abstract

Increasing demand for power production and need for low impact on the climatic change or environmental sustainability has led to the rediscovery of one of the oldest methods of energy recovery from wastewater using microbes, referred to as microbial fuel cell technology. Limited nonrenewable resources on the earth and their effects on the environment also urge sustainability. The rate of waste generation from various sources should not exceed the assimilative capacity of the environment (sustainable waste disposal). Microbial fuel cells are bioelectrochemical systems designed to convert the chemical energy contained in organic matter into electrical energy utilizing the catalytic (metabolic) activity of microorganisms. Microbial fuel cell systems operate on the fundamental mechanisms of microbiology (e.g., biofilm growth on anode particles), biochemistry (e.g., chemical oxygen demand reduction inside biofilms), and electrochemistry (e.g., electron generation and transfer on anodes and cathodes), which make it truly challenging to achieve effective power generation simultaneously with effective wastewater treatment. Microbial fuel cells appear promising for wastewater treatment as well as metal recovery by bioelectrocatalysis, because metal ions can be reduced and deposited by bacteria, algae, yeasts, and fungi. This chapter discusses in detail the working principle, design, and construction of different types of biofilm utilized in the fuel cell; a performance evaluation of microbial fuel cell systems and their application; and modeling studies.

12.1 INTRODUCTION

Power generation is a main area of focus for researchers and industrialists around the world. A major part of power production was dependent on energy from nonrenewable energy sources, such as fossil fuels, nuclear fuels, and earth minerals (Makarieva et al., 2008). Demand for these sources and the pollution caused by them have led to alternative resources. Renewable energy sources, such as water, wind, and solar, have been utilized to generate power, but they have their own limitations and advantages. In recent years, clean and green energy, a new track of fuel cell systems has been utilized for power generation. Basic knowledge about fuel cells and their possibility to emerge as microbial fuel cells (MFCs) are discussed in this chapter.

12.1.1 Fuel Cell and Its Components

A fuel cell is an energy conversion device that converts the chemical energy of a fuel directly into electricity without any intermediate thermal or mechanical processes. In a fuel cell, the reaction occurs electrochemically, and the energy is released as a combination of low-voltage DC electrical energy and heat. The electrical energy can be used to do useful work directly while the heat is either wasted or used for other purposes. However, unlike batteries, the reductant and oxidant in fuel cells must be continuously replenished to allow continuous operation (EG&G Technical Services, 2004).

The basic components of fuel cells are an anodic chamber, cathodic chamber, electrolyte, proton exchange membrane, and an external circuit

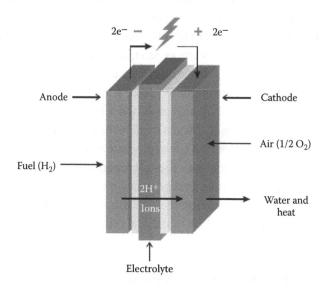

FIGURE 12.1 Typical hydrogen fuel cell. (From Hydrogen fuel cell engines and related technologies, Module 4: Fuel cell engine technology, Energy Technology Training Center, College of the Desert, Palm Desert, CA, 2001.)

to collect the current, as shown in Figure 12.1 (EG&G Technical Services, 2004). Based on this basic design, many fuel cells are developed and classified based on the type of electrolyte used. These fuel cells are fed with various sources, including hydrogen, methanol, and formate, usually smaller molecules. Limitations of these fuel cells from using larger molecules such as glucose, propanol due to incomplete oxidation process, and higher operating temperature led to modifications in the fuel cell (Scott, 2014) with microorganisms or enzymes as catalysts in the electrochemical reaction, generally called microbial fuel cells and enzyme fuel cells, respectively.

12.1.2 Wastewater Treatment along with Simultaneous Power Generation

Microbial fuel cells have the capability to oxidize the organic matter present in the wastewater and also act as a sustainable energy producer. Fornero et al. (2010) had treated municipal, food, and animal wastewater in MFCs and reported that MFCs produce low power density but their efficiency to treat wastewater was particularly attractive. G. Zhang et al. (2015) used dairy manure as fuel in an MFC system and noticed a stable power generation for 110 days. J. Choi and Ahn (2015) also found that treating food waste leachate from biohydrogen fermentation with MFCs produced a maximum voltage of 0.56 V and power density of 1540 mW/m². Catal et al. (2009), Z. Li et al. (2008), A. Singh and Yakhmi (2014), Luo et al. (2014), C. Choi et al. (2014), and C. Choi and Hu (2013) investigated MFCs with the recovery of various heavy metals such as selenite, chromium, copper, nickel, cadmium, gold, and silver, from wastewater as well as power generation. Recovery percentage of chromium IV, gold, selenite, and silver was found to be greater than 99%, and other metals had a reasonable value. Thus the result from the aforementioned works depicts the capability of an MFC as a sustainable power generator and effective wastewater treating system.

This chapter mainly deals with the microbial fuel cell and its working mechanism to treat wastewater and generation of power. Contents include the components of MFC, reaction mechanism involved, type of materials adapted, performance evaluation of the fuel cell, varying applications, and modeling and simulation studies.

12.2 BRIEF HISTORY OF MICROBIAL FUEL CELL TECHNOLOGY

Even though the interest in fuel cell technology started to increase in 1980s, its history dates back to the nineteenth century. In 1839, the first

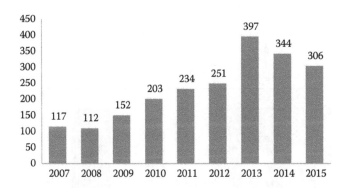

FIGURE 12.2   Number of publications per year as per the Science Direct database.

fuel cell was fabricated. Later Potter (1911) developed the concept of the MFC using living cultures of *Escherichia coli* and *Saccharomyces* with platinum electrodes. Since then work related to MFCs was untouched for many years until Allen and Bennetto in the 1980s started their work with the use of electron mediators for electron transfer from the microorganisms to the anode surface. Cost of the mediators used in the fuel cell, toxicity, its reaction with the anodic chamber, and, most important, the ability of some microbes themselves to transfer the electron to the anode surface with the help of cytochromes initiated the development of the MFC without a mediator for electron transfer (Gil et al., 2003). After the discovery of the mediator-less MFC, a lot of work was carried out in the 1990s to figure out the microbes with the highest efficiency to transfer electrons to the anode surface. Recently, most of the research work has focused on increasing the energy efficiency of the fuel cell by replacing the materials used with anode, cathode, proton exchange membrane (PEM), and microbial species (Antolini, 2015; He et al., 2015; Janicek et al., 2015; W. Li et al., 2011; Schechter et al., 2014). Santoro et al. (2012) and S. Choi and Chae (2012) have also reported that there was a considerable increase in the power production when the fuel cells are connected as a series in large numbers. Research and publications in the field of renewable energy and microbial fuel cell has gradually increased since 2007 (Figure 12.2).

12.3 MICROBIAL FUEL CELL SYSTEM

12.3.1 Working Principle

Microbial fuel cells are bioelectrochemical systems designed to convert the chemical energy contained in organic matter into electrical energy utilizing the catalytic (metabolic) activity of microorganisms. Consider

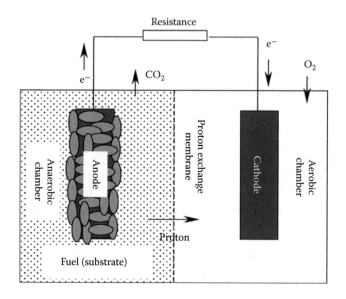

FIGURE 12.3 Schematic diagram of a typical two-chamber microbial fuel cell. (From Du, Z. et al., *Biotechnology Advances*, 25: 464–482, 2015.)

a dual-chambered MFC system with an anode, cathode, and a proton exchange membrane as shown in Figure 12.3. When introducing fuel (wastewater) into the MFC anode chamber, microorganisms immobilized in the anode surface area oxidize the organic matters or heavy metals present in the wastewater into electrons, protons, and carbon dioxide. Electrons flow from the anode to the cathode through an external wire completing the circuit. Protons migrate to the cathode chamber through the proton exchange membrane held between the anode and cathode chamber. In the cathode chamber, protons and electrons migrated from the anode chamber combine with the oxygen supplied in the chamber producing water molecules.

12.3.2 Classification of Microbial Fuel Cell Systems

Microbial fuel cells are classified into two categories based on electron transportation from the bacterial cell to the electrode, as shown in Figure 12.4 which are as follows: mediator microbial fuel cells and mediator-less microbial fuel cells. MFCs are electrochemically inactive. They require mediators to transfer electrons from microorganisms to the anode. Commonly used mediators are thionine, methyl viologen, methyl blue, humic acid, neutral red, anthraquinone-2,6-disulfonate (AQDS), and

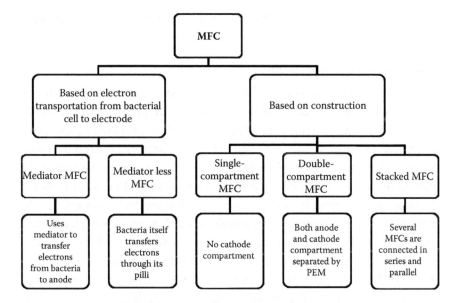

FIGURE 12.4   Classification of microbial fuel cell systems.

ethylene diamine tetra acetic acid. These types of MFCs are termed medi-ator MFCs. Even though they have advantages, usage of these mediators cause corrosion to the anode, which can become toxic to microbes. Hence microbes with special redox chemicals can be used to transfer electrons directly to the anode through its pili. These types of MFCs are termed as Mediator less MFCs.

MFCs can be further classified into three categories based on their construction.

Single-compartment MFC—It contains an anode compartment and the cathode is directly exposed to air (i.e., there is no cathode compart-ment), and may not contain proton exchange membranes.

Double-compartment MFC—It consists of both an anode and cathode in individual compartments, are separated by PEM, and are connected through an external circuit.

Stacked MFC—In stacked MFCs, several MFCs are connected in a series and parallel to increase the power generation (Rahimnejad et al., 2015; Tao et al., 2015).

12.4 COMPONENTS OF MICROBIAL FUEL CELL SYSTEMS

Irrespective of the type, MFCs are composed of an anode, cathode, pro-ton exchange membrane, and the biofilm. Brief notes on electrochemical and microbiological reaction that take place inside the MFC are discussed individually.

12.4.1 Anode Compartment

The anode chamber is one of the key components of an MFC. All the essential conditions to degrade the wastewater are provided in the anode chamber. This compartment is filled with substrate, mediator (optional), and microorganisms, and the anode electrode acts as electron acceptor (Rahimnejad et al., 2015). Microbes are adsorbed to the anode surface to oxidize the organic matters present in wastewater and produce protons, electrons, and carbon dioxide. These generated electrons are utilized to reduce electron acceptors in the cathode when they pass through external circuit. Anaerobic conditions are very important for the substrate oxidation by the microbes through anaerobic respiration (Heilmann and Logan 2006). If the anode chamber is exposed to the aerobic environment via leak or diffusion from the cathodic chamber, organic matters present in the chamber would be oxidized by the oxygen directly producing a lower amount of heat instead of power generation (Zhou et al., 2014). The general reaction in the anode chamber is given as

$$\text{Biodegradable organics} \rightarrow CO_2 + H^+ + e^- \qquad (12.1)$$

Anodic materials must have certain properties for effective functioning, for example, conductivity, biocompatibility, chemically stability, microbial adhesion performance, and electrochemical reversibility (D. Singh et al., 2010; Zheng et al., 2015). The most commonly used electrode material is carbon. It is available in various forms such as rods, granular structures, and fibrous material. Some of the carbon-based materials used are carbon cloth, carbon brush, carbon paper, graphite fiber brush, graphite rod, reticulated vitreous carbon (RVC), and carbon felt (Santoro et al., 2012; D. Singh et al., 2010). Use of composite materials as electrodes has been extensively studied to enhance its effect on the MFC performance. Antolini (2015) tested MFCs using composite metal–carbon, metal–polymer, polymer–carbon, polymer–polymer, and carbon–carbon materials, and found the characteristics improved with respect to the sole component. A binder-free carbon black/stainless steel mesh composite electrode showed higher biocatalytic activity, higher current density, robustness, low-cost, and was more suitable for up-scaling than carbon felt (Zheng et al., 2015).

12.4.1.1 Anode Reaction

The microbes in the anodic chamber oxidize substrates such as glucose, acetate, and some refractory organics. For example, glucose is oxidized as in Equation 12.2 to generate electrons, protons, and carbon dioxide (Pham et al., 2006):

$$C_6H_{12}O_6 + 6H_2O \rightarrow 6CO_2 + 24H^+ + 24e^- \tag{12.2}$$

12.4.2 Cathode Compartment

Protons produced in the anode chamber as per reaction 12.2 migrate through the proton exchange membrane into the cathode chamber. This completes the electrical circuit. The oxygen supplied as the oxidizing agent in the cathode chamber reacts with the positive ions to form water. The basic reactions are as follows:

$$H_2 \rightarrow 2H^+ + 2e^- \tag{12.3}$$

$$O_2 + 4H^+ + 4e^- \rightarrow 2H_2O \tag{12.4}$$

Similar to the anode materials, cathode materials should have high performance and stability. Performance of the cathode is dependent on the concentration and species of the oxidant (electron acceptor), proton availability, catalyst performance, and electrode structure and its catalytic ability. Carbon-based electrodes are usually employed as cathode electrodes. Recently, a composite metal electrode with higher surface area and biocatalytic activity has been used. Platinum, palladium, and gold are a few catalysts having high catalytic activity for oxygen reduction used in the cathode. Their usage is limited by cost, availability, and chemical fouling. Using ferricyanide, $K_3[Fe(CN)_6]$, as the electron acceptor in the cathode chamber increased the power density due to the availability of a good electron acceptor at high concentrations. Ferricyanide increased power by 1.5 to 1.8 times compared to a Pt-catalyst cathode and dissolved oxygen (Heilmann and Logan 2006). Phthalocyanines, porphyrins, metal complexes, and metal dioxides have also been considered as alternatives to platinum. Among them, MnO_2 and Fe- and Co-chelates are the most studied. In most cases, the performance of MFCs with these catalysts was comparable or higher than that of cells with platinum (Antolini, 2015).

12.4.2.1 Cathode Reaction

The cathode reaction has a major impact on MFC performance. The electrons are coming from the anode via the external circuit, the protons are coming from the anodic chamber via the PEM, and the electron acceptors will react with the help of catalysts on the cathode (Pham et al., 2006).

$$24H^+ + 24e^- + 6O_2 \rightarrow 12H_2O \qquad (12.5)$$

12.4.3 Proton Exchange Membranes (PEMs)

PEMs are generally known as separators, which physically separate anodic and cathodic chamber. The importance of PEMs in MFCs is based on a few characteristic features exhibited by them, for example, the ability to separate the two chambers without mixing, permitting the flow of protons from the anode chamber to the cathode chamber, and being a barrier for the diffusion of oxygen from the cathode chamber and substrate (wastewater) from the anode chamber. Even though PEM performs these features, it also produces a few negative characteristics with respect to MFC performance. For example, PEMs not only allow the protons but also other cations present in the anode chamber (such as sodium, potassium, calcium, and magnesium), they increase the pH in the cathode chamber, and create higher internal resistance in the chamber.

12.4.3.1 PEM Classification

PEMs can be mainly classified into three categories according to their filtration characteristics. These are ion exchange membranes (IEMs), size-selective separators, and salt bridge. IEMs can further be classified into cation exchange membranes (CEMs), anion exchange membranes (AEMs), and bipolar membranes (BPMs). Size-selective separators include microfiltration membranes (MFMs) and ultrafiltration membranes (UFMs). Apart from these types, there are glass fibers, porous fabrics, and other coarse-pore filter materials. Separator electrode assemblies (SEAs) are recently developed separators to increase MFC performance (W. Li et al., 2011). The advantages and drawbacks of various separators are given in Table 12.1.

12.4.3.2 PEM Materials

Nafion, a DuPont product developed in 1970, is the commonly used ion exchange membrane in most MFCs. However, its response to biofouling

TABLE 12.1   Advantages and Drawbacks of Various Separators

Separator	Advantages	Drawbacks
None	High proton transfer rate, high power density, low costs, simple configuration	Serious oxygen permeation, high substrate loss, cathode fouling and deactivation, high internal resistance caused by large electrode spacing
Ion exchange membranes	Effective isolation of anodic and cathodic solution/cathode, low oxygen permeation and substrate loss	Constrained proton transfer, pH splitting, membrane fouling, high costs
Microporous filtration membranes	High proton transfer and low pH accumulation, moderate costs	High oxygen permeation, high internal resistance
Coarse-pore filters	High proton transfer and low pH accumulation, low costs	High substrate loss caused by biofilm on filter, inferior durability
Salt bridge	Simple configuration, low costs	High internal resistance

Source: Adapted from Li, W. et al., *Bioresource Technology* 102, no. 1: 244–252, 2011.

and cost hinder the application. A comparative study on power generation tested Nafion 112, SPEEK, and Nafion 117. Results were higher for Nafion 117 with 179.7 mW/m^2, while it was 126.1 for SPEEK and 19.7 for Nafion 112 (Ghasemi et al., 2015). Ultrex, bipolar membranes, dialyzed membrane, polystyrene, divinylbenzene with sulfuric acid group, glass wool, nanoporous filters and microfiltration membranes, SPEEK, CMI 7000, Hyflon, Zirfon are some of the commonly used PEM (Antolini, 2015; W. Li et al., 2011; Rahimnejad et al., 2015; D. Singh et al., 2010).

12.4.4 Biofilm

Biofilm is defined as any group of microorganisms in which cells stick to each other on a surface and are enclosed in a self-developed polymeric matrix of a primarily polysaccharide material (Hall-Stoodley et al., 2004). Biofilm is considered as the heart of the MFC due to its unique contributions toward wastewater treatment and power generation solely dependent upon its activity. As mentioned earlier, microbes oxidize organic matter to produce protons, electrons, and carbon dioxide, which form the base for the whole system. Composition and thickness of the biofilm play a major role in the electron kinetics within the biofilm, which in turn retard the electron transfer mechanism. The formation of biofilm on the anode surface can be characterized by SEM analysis as shown in

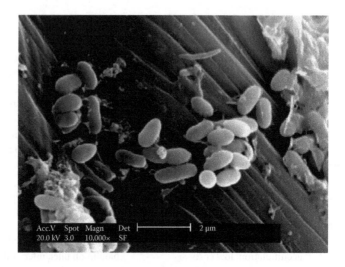

FIGURE 12.5   SEM image of electron-transfer bacteria attached to the surface of the anode. (From Li, Z. et al., *Process Biochemistry* 43: 1352–1358, 2008.)

Figure 12.5 (Zheng et al., 2015). Based on the electrogenic activity (ability to transfer electrons to the electrode), microorganisms can be divided into two types:

Nonelectrogenic—*Proteus mirabilis, Saccharomyces cerevisiae, Shewanella oneidensis, Streptococcus lactis, and Micrococcus luteus* are nonelectrogenic microorganisms that functions with the help of mediators for the electron transfer to the anode surface (Rahimnejad et al., 2015).

Electrogenic—*Shewanella putrefaciens, Clostridium butyricum, Desulfuromonas acetoxidans, Geobacter metallireducens, Geobacter sulfurreducens, Rhodoferax ferrireducens, Pseudomonas aeruginosa, Desulfobulbus propionicus, Geothrix fermentans, Shewanella oneidensis, Escherichia coli, Rhodopseudomonas palustris, Ochrobactrum anthropi, Desulfovibrio desulfuricans, Acidiphilium* sp., *Klebsiella pneumonia, Thermincola* sp., and *Cupriavidus basilensis* are microbes with electrogenic activity that do not need the support of mediators for the transfer of electrons.

A few families *Geobacteraceae, Desulfuromonaceae, Alteromonadaceae, Enterobacteriaceae, Pasteurellaceae, Clostridiaceae, Aeromonadaceae,* and *Comamonadaceae,* and *Escherichia coli* K12, *Clostridium beijerinckii, Proteus vulgaris, Aeromonas hydrophila, Hansenula anomala, Enterococcus faecium, Erwinia dissolvens, Desulfovibrio vulgaris, Shewanella putrefaciens* IR-1, *Shewanella putrefaciens* MR-1, *Shewanella putrefaciens* SR-1, *Aeromonas hydrophila* PA 3, and *Clostridium* sp. EG 3 were also tested in MFC systems (Zhou et al., 2014).

12.5 PERFORMANCE EVALUATION OF MICROBIAL FUEL CELLS

Performance of MFCs can be determined using certain factors such as cell potential, power, power density, columbic efficiency, and energy efficiency. Electricity is generated in an MFC only if the overall reaction is thermodynamically favorable. The reaction can be evaluated in terms of Gibbs free energy expressed in units of joules (J), which is a measure of the maximal work that can be derived from the reaction calculated as

$$\Delta G_r = \Delta G_r^{\circ} + RT\ln(\Pi) \tag{12.6}$$

where ΔG_r (J) is the Gibbs free energy for the specific conditions; ΔG_r° (J) is the Gibbs free energy under standard conditions usually defined as 298.15 K, 1 bar pressure, and 1 M concentration for all species; R (8.31447 J mol^{-1} K^{-1}) is the universal gas constant; T (K) is the absolute temperature; and Π (dimensionless) is the reaction quotient calculated as the activities of the products divided by those of the reactants. For the standard reaction, Gibbs free energy is calculated from tabulated energies of formation for organic compounds in water available from many sources (Alberty et al., 2011).

For MFC calculations, it is more convenient to evaluate the reaction in terms of overall cell electromotive force (emf). E_{emf} (V) is defined as the potential difference between the cathode and anode. This is related to the work W (J), produced by the cell

$$W = E_{emf} Q = -\Delta G_r \tag{12.7}$$

where $Q = nF$ is the charge transferred in the reaction, expressed in coulomb, which is determined by the number of electrons exchanged in the reaction; n is the number of electrons per reaction mol; and F is Faraday's constant (9.64853×10^4 C/mol). Combining these two equations, we have

$$E_{emf} = -\Delta G_r/nF \tag{12.8}$$

If all reactions are evaluated at standard conditions, $\Pi = 1$, then

$$E_{emf}^\circ = -\Delta G_r^\circ/nF \tag{12.9}$$

where E_{emf}° (V) is the standard cell electromotive force. We can therefore use the preceding equations to express the overall reaction in terms of the potentials as

$$E_{emf} = E_{emf}^\circ - (RT/nf)\ln(\Pi) \tag{12.10}$$

In general, measured voltage will be lower than the theoretical voltage calculated. This may be due to various potential losses, such as ohmic losses, activation losses, bacterial metabolic losses, and concentration losses.

12.5.1 Standard Electrode Potentials

The reactions occurring in the MFC can be analyzed in terms of the half-cell reactions or the separate reactions occurring at the anode and the cathode. According to the IUPAC convention, standard potentials (at 298 K, 1 bar, 1 M) are reported as a reduction potential, i.e., the reaction as consuming electrons. For example, if acetate is oxidized by bacteria at the anode, we write the reaction as

$$2HCO_3^- + 9H^+ + 8e^- \rightarrow CH_3COO^- + 4H_2O \tag{12.11}$$

The standard potentials are reported relative to the normal hydrogen electrode (NHE), which has a potential of zero at standard conditions. To obtain the theoretical anode potential, E_{An}, under specific conditions, we use Equation 12.5, with the activities of the different species assumed to be equal to their concentrations. For acetate oxidation, we have

$$E_{An} = E^\circ_{An} - (RT/8F) \ln([CH_3COO^-]/[HCO_3^-]2[H^+]^9)$$ (12.12)

For theoretical cathode potential, E_{cat}, if we consider the case where oxygen is used as the electron acceptor for the reaction, we can write

$$O_2 + 4H^+ + 4e^- \rightarrow 2H_2O$$ (12.13)

$$E_{cat} = E^\circ_{cat} - (RT/4F) \ln(1/pO_2[H^+]^4)$$ (12.14)

The cell emf is calculated as

$$E_{emf} = E_{cat} - E_{an}$$ (12.15)

Power is calculated as

$$P = IE_{cell}$$ (12.16)

Normally the voltage is measured across a fixed external resistor (R_{ext}), while the current is calculated from Ohm's law (I) (E_{cell}/R_{ext}).

Thus, power is usually calculated as

$$p = \frac{E^2_{cell}}{R_{ext}}$$ (12.17)

12.5.2 Power Density

To perform engineering calculations for size and costs of reactors, and as a useful comparison to chemical fuel cells, the power is normalized to the reactor volume as

$$p_v = \frac{E^2_{cell}}{vR_{ext}}$$ (12.18)

The columbic efficiency (CE) is calculated using

$$CE = \frac{M \int_0^t I\,dt}{FbV_{\text{anode}}\Delta COD} \qquad (12.19)$$

12.5.3 Energy Efficiency

The overall energy efficiency, ε_E, is calculated as the ratio of power produced by the cell over a time interval t to the heat of combustion of the organic substrate added in that time.

$$\varepsilon_E = \frac{\int_0^t E_{\text{cell}} I\,dt}{\Delta HM_{\text{added}}} \qquad (12.20)$$

12.6 MODELING AND SIMULATION

Modeling of MFCs is one of the large areas that needs to be explored. Very few works based on modeling were studied and they were limited to specific functions of the system. Standard models are yet to be developed to compare the experimental data with them. MFC systems with a complex interdisciplinary approach to treat wastewater as well as power generation have lot of room for modeling where experimental work can be avoided thereby reducing the cost and time.

Based on the literature, Hernández-Fernández et al. (2015) classified the models into anode-based models, cathode-based models, and specific models. The developed models were based on the common equations: Monod, Bulter-Volmer, Tafel, Nernst, Nernst-Planck equations, Fick's law, the Maxwell-Stefan diffusion equation, and Ohm's law. Development of model equations and simulation of maximum power produced in gold recovery using MFCs was studied, where power was predicted using the internal resistance measured from the system (C. Choi and Hu, 2013). Similarly model equations can be derived for growth kinetics of microorganisms, proton generation in the anode chamber, and COD reduction. X. Zhang and Halme (1995) were the first to develop models considering anode compartment as a rate-limiting factor in one dimension. Piciorenau et al. (1998) developed models based on two and three dimensions in the anode chamber. Marcus and Sutin (1985) and Merkey and Chopp (2012) also developed models for the anode, and Pinto et al. (2010) developed models for the cathode compartment. Models considering both the anode and cathode in a dual-chamber MFC were developed by Oliveira et al. (2013). It was observed that very few areas have been explored in the modeling. Utilization of the commercially available, user-friendly software such as Fluent and COMSOL for modeling would be an effective tool for identification of new materials for anodes and cathodes and PEMs, and also for calculating the power generation based on various parameters in the system.

12.7 APPLICATIONS

Power generation from biobased materials, wastewater treatment, biohydrogen production, and biosensors are the major applications of MFC systems. Among these, wastewater treatment with simultaneous power generation is the most widely used application. Much research has been carried out with various kinds of wastewater, starting from domestic wastewater to highly toxic industrial effluent. A reasonable percentage of chemical oxygen demand (COD) reduction and recovery of heavy metals are found to be possible with MFCs and lags in power generation, producing very low power density. Research focusing on wastewater treatment was discussed in Section 12.1.2. A biosensor for glucose has been developed utilizing a glucose oxidase-based anode and cytochrome c cathode to generate electrical current (Katz et al., 2001). MFCs can power many electronic devices requiring lower power densities for their operations. Small-scale MFCs powering drug delivery systems had also been reported (Mano et al., 2003). Hence MFCs tuned with novel technologies could lead to many diversified applications.

12.8 CONCLUSION

This chapter presents an overview of the MFC system, its working mechanism in COD reduction, and power generation in a single system. The mechanisms involved in the anode and cathode chambers were summarized. Reviews on recently developed anode, cathode, and PEM materials were illustrated with their advantages and drawbacks. Biofilm has a special place in the MFCs due its biocatalytic activity on organic matter present in a system. Different types of microbes utilized in MFC systems were classified based on electrogenic activity. Basic performance evaluations of MFCs were also discussed. Current status of the modeling studies and its scope in the future were briefly mentioned. To conclude, this gave an overall idea of MFC basics and areas where improvements could be made to increase the efficiency of the system.

REFERENCES

Alberty, R. A., Cornish-Bowden, A., Goldberg, R. N., Hammes, G. G., Tipton, K., and Westerhoff, H. V. 2011. Recommendations for terminology and databases for biochemical thermodynamics. *Biophysical Chemistry* 155, no. 2–3: 89–103.

Antolini, E. 2015. Biosensors and bioelectronics composite materials for polymer electrolyte membrane microbial fuel cells. *Biosensors and Bioelectronics* 69: 54–70.

Catal, T., Bermek, H., and Liu, H. 2009. Removal of selenite from wastewater using microbial fuel cells. *Biotechnology Letters* 31, no. 8: 1211–1216.

Choi, C., and Hu, N. 2013. The modeling of gold recovery from tetrachloroaurate wastewater using a microbial fuel cell. *Bioresource Technology* 133: 589–598.

Choi, C., Hu, N., and Lim, B. 2014. Cadmium recovery by coupling double microbial fuel cells. *Bioresource Technology* 170: 361–369.

Choi, J., and Ahn, Y. 2015. Enhanced bioelectricity harvesting in microbial fuel cells treating food waste leachate produced from biohydrogen fermentation. *Bioresource Technology* 183: 53–60.

Choi, S., and Chae, J. 2012. An array of microliter-sized microbial fuel cells generating 100 W of power. *Sensors & Actuators: A. Physical* 177: 10–15.

Du, Z., Li, H., and Gu, T. 2015. A state of the art review on microbial fuel cells: A promising technology for wastewater treatment and bioenergy. *Biotechnology Advances* 25: 464–482.

EG&G Technical Services. 2004. *Fuel Cell Handbook* (7th ed.). Morgantown, WV: U.S. Department of Energy.

Fornero, J. J., Rosenbaum, M., and Angenent, T. 2010. Electric power generation from municipal, food and animal wastewaters using microbial fuel cells. *Electroanalysis* 22, no. 7–8: 832–843.

Ghasemi, M., Halakoo, E., Sedighi, M., Alam, J., and Sadeqzadeh, M. 2015. Performance comparison of three common proton exchange membranes for sustainable bioenergy production in microbial fuel cell. *Procedia CIRP* 26: 162–166.

Gil, G. C., Chang, I. S., Kim, B. H., Kim, M., Jang, J. K., Park, H. S., and Kim, H. J. 2003. Operational parameters affecting the performannce of a mediator-less microbial fuel cell. *Biosensors & Bioelectronics* 18, no. 4: 327–334.

Hall-Stoodley, L., Costerton, J. W., and Stoodley, P. 2004. Bacterial biofilms: From the natural environment to infectious diseases. *Nature Reviews Microbiology* 2, no. 2: 95–108.

He, C., Mu, Z., Yang, H., Wang, Y., Mu, Y., and Yu, H. 2015. Electron acceptors for energy generation in microbial fuel cells fed with wastewaters: A mini-review. *Chemosphere* 140: 12–17.

Heilmann, J., and Logan, B. E. 2006. Production of electricity from proteins using a microbial fuel cell. *Water Environment Research* 78, no. 2: 531–537.

Hernández-Fernández, F. J., Pérez de los Ríos, A., Salar-García, M. J., Ortiz-Martínez, V. M., Lozano-Blanco, L. J., Godínez, C., and Quesada-Medina, J. 2015. Recent progress and perspectives in microbial fuel cells for bioenergy generation and wastewater treatment. *Fuel Processing Technology* 138: 284–297.

Hydrogen fuel cell engines and related technologies. 2001. Module 4: Fuel cell engine technology. Energy Technology Training Center, College of the Desert, Palm Desert, CA, USA.

Janicek, A., Fan, Y., and Liu, H. 2015. Performance and stability of different cathode base materials for use in microbial fuel cells. *Journal of Power Sources* 280: 159–165.

Katz, E., Bückmann, A. F., and Willner, I. 2001. Self-powered enzyme-based biosensors. *Journal of the American Chemical Society* 123, no. 43: 10752–10753.

Li, W., Sheng, G., Liu, X., and Yu, H. 2011. Recent advances in the separators for microbial fuel cells. *Bioresource Technology* 102, no. 1: 244–252.

Li, Z., Zhang, X., and Lei, L. 2008. Electricity production during the treatment of real electroplating wastewater containing Cr6+ using microbial fuel cell. *Process Biochemistry* 43: 1352–1358.

Luo, H., Liu, G., Zhang, R., Bai, Y., Fu, S., and Hou, Y. 2014. Heavy metal recovery combined with H_2 production from artificial acid mine drainage using the microbial electrolysis cell. *Journal of Hazardous Materials* 270: 153–159.

Makarieva, A. M., Gorshkov, V. G., and Li, B. L. 2008. Energy budget of the biosphere and civilization: Rethinking environmental security of global renewable and non-renewable resources. *Ecological Complexity* 5, no. 4: 281–288.

Mano, N., Mao, F., and Heller, A. 2003. Characteristics of a miniature compartmentless glucose-O_2 biofuel cell and its operation in a living plant. *Journal of the American Chemical Society* 125, no. 21: 6588–6594.

Marcus, R. A., and Sutin, N. 1985. Electron transfers in chemistry and biology. *Biochimica et Biophysica Acta–Reviews on Bioenergetics* 811, no. 3: 265–322.

Merkey, B. V., and Chopp, D. L. 2012. The performance of a microbial fuel cell depends strongly on anode geometry: A multidimensional modeling study. *Bulletin of Mathematical Biology* 74, no. 4: 834–857.

Oliveira, V. B., Simões, M., Melo, L. F., and Pinto, A. M. F. R. 2013. Overview on the developments of microbial fuel cells. *Biochemical Engineering Journal* 73: 53–64.

Pham, T. H., Rabaey, K., Aelterman, P., Clauwaert, P., De Schamphelaire, L., Boon, N., and Verstraete, W. 2006. Microbial fuel cells in relation to conventional anaerobic digestion technology. *Engineering in Life Sciences* 6, no. 3: 285–292.

Piciorenau, C., Loosdrecht, M., and Heijnen, J. 1998. Mathematical modeling of biofilm strucutre with a hybrid differential-discrete cellular automata approach. *Biotechnology and Bioengineering* 58, no. 1: 101–116.

Pinto, R. P., Srinivasan, B., Manuel, M. F., and Tartakovsky, B. 2010. A two-population bio-electrochemical model of a microbial fuel cell. *Bioresource Technology* 101, no. 14: 5256–5265.

Potter, M. C. 1911. Electrical effects accompanying the decomposition of organic compounds. *Proceedings of the Royal Society B: Biological Sciences* 84, no. 571: 260–276.

Rahimnejad, M., Adhami, A., Darvari, S., Zirepour, A., and Oh, S.-E. 2015. Microbial fuel cell as new technology for bioelectricity generation: A review. *Alexandria Engineering Journal* 54, no. 3: 745–756.

Santoro, C., Lei, Y., Li, B., and Cristiani, P. 2012. Power generation from wastewater using single chamber microbial fuel cells (MFCs) with platinum-free cathodes and pre-colonized anodes. *Biochemical Engineering Journal* 62: 8–16.

Schechter, M., Schechter, A., Rozenfeld, S., Efrat, E., and Cahan, R. 2014. Anode biofilm. In *Technology and Applications of Microbial fuel cells*, edited by C.-T. Wang, chap. 4.

Scott, K. 2014. *Membranes for Clean and Renewable Power Applications*. Cambridge, UK: Woodhead Publishing Limited.

Singh, A., and Yakhmi, M. J. V. 2014. Microbial fuel cells to recover heavy metals. *Environmental Chemistry Letters* 12, no. 4: 483–494.

Singh, D., Pratap, D., Baranwal, Y., Kumar, B., and Chaudhary, R. K. 2010. Microbial fuel cells: A green technology for power generation. *Annals of Biological Research* 1, no. 3: 128–138.

Tao, Q., Zhou, S., Luo, J., and Yuan, J. 2015. Nutrient removal and electricity production from wastewater using microbial fuel cell technique. *Desalination* 365: 92–98.

Zhang, G., Zhao, Q., Jiao, Y., and Lee, D. 2015. Long-term operation of manure-microbial fuel cell. *Bioresource Technology* 180: 365–369.

Zhang, X., and Halme, A. 1995. Modelling of a microbial fuel cell process. *Biotechnology letters* 17, no. 8: 809–814.

Zheng, S., Yang, F., Chen, S., Liu, L., Xiong, Q., and Yu, T. 2015. Binder-free carbon black/stainless steel mesh composite electrode for high-performance anode in microbial fuel cells. *Journal of Power Sources* 284: 252–257.

Zhou, M., Yang, J., Wang, H., Jin, T., Hassett, D. J., and Gu, T. 2014. Bioelectrochemistry of microbial fuel cells and their potential applications in bioenergy. In *Bioenergy Research: Advances and Applications*, edited by V. G. Gupta, M. Tuohy, C. P. Kubicek, J. Saddler, and F. Xu, 131–152. Waltham, MA: Elsevier.

Recent Advancements in Biodiesel from Nonedible Oils Using Packed Bed Membrane Reactors

Uma Krishnakumar
and Velmurugan Sivasubramanian

CONTENTS

Abstract

It is well known that energy consumption is rapidly increasing due to population growth, a higher standard of living, and increased production. Significant amounts of energy resources are being consumed by the transportation sector, leading to the fast depletion of fossil fuels and environmental pollution. Biodiesel is one of the technically and economically feasible options to tackle the aforesaid problems. Biodiesel is mainly produced from edible oils. However, it is believed that the extensive use of edible oils for biodiesel production may lead to food shortages in most developing countries. Membrane technology has attracted the interest of researchers for its ability to provide high purity and quality biodiesel fuel, in addition to its remarkable biodiesel yield. Membranes have numerous useful properties such as resistance to mechanical, chemical, and thermal stress; high available surface area per unit volume; high selectivity; and the ability to control the contact between reactants and catalysts. This study includes the recent advancements in the field of membrane technology for the production of biodiesel.

13.1 INTRODUCTION

Today, with increasingly expensive and less secure petroleum supplies, and mounting evidence of the link between vehicle emissions, global warming, and climate change, biofuels are getting much more attention. They have particular importance as technologies for reducing pollutants and greenhouse gas emissions from motor vehicles. Corn, soy, and other biomass crops such as sawgrass and miscanthus absorb carbon dioxide from the atmosphere while emitting less carbon dioxide from the tailpipe during combustion. *Biofuel* is a generic term that is used to refer to liquid or gaseous fuels that are produced from a biological source. The term *liquid biofuel* is more commonly used to refer to specific types of biofuels used as fossil fuel substitutes. Biofuels can be further classified based on the type of biomass used for their production and by the degree to which they are refined. The most common type of liquid biofuel is straight vegetable oil (SVO), which is directly extracted from the oil seeds. Biodiesel is made from the transesterification of suitable vegetable oils. Biodiesel is very similar to fossil-based diesel fuel and can be used in almost any type of diesel engine without modification and has a long shelf life. Bioethanol is another important liquid biofuel that can be produced from a wide range of biomass (plant) materials using a complex chemical process. In general, ethanol is mixed with gasoline in varying concentrations in order to reduce emissions and increase efficiency.

First-generation biofuels are created largely from feedstocks that have traditionally been used as food, for example, ethanol from corn and biodiesel from vegetable oil and animal fats. However, the extensive use of edible oils might lead to some negative impacts such as starvation and higher food prices in developing countries. However, many researchers agree that nonedible oils are suitable alternatives to edible oils for biodiesel production. Hence, the recent focus is to find nonedible oil feedstocks for biodiesel production (Balat 2011).

13.2 NONEDIBLE FEEDSTOCKS FOR BIODIESEL PRODUCTION

Numerous nonedible feedstocks have been experimented with for biodiesel production. Advancements from such experimentations led to the establishment of waste-to-wealth biodiesel production. Cheap and readily available raw materials such as used cooking oil and yellow grease are used for producing biodiesel. These efforts help in reducing the environmental impact associated with dumping in landfills as well as saving the cost of paying for such dumping. Another notable success is the use of jatropha or the "miracle plant" in many developing countries. The fact that it can be cultivated almost anywhere with minimal irrigation and minimal care made it suitable for peasant farmers. Sustained high yields were obtained throughout its average life cycle of 30 to 50 years. Castor plantation is also intercropped with jatropha to improve the economic viability of jatropha within the first 2 to 3 years. Another oil crop that is used to improve soil quality is the nitrogen-fixing *Pongamia pinnata*. It produces seeds with significant oil contents.

In order to reduce production costs and to avoid the "food versus fuel" conflict, nonedible oils are used as the major sources for biodiesel production. Compared to edible oils, nonedible oils are affordable and readily available. They are obtained from *Jatropha curcas* (jatropha), *Pongamia pinnata* (karanja/honge), *Calophyllum inophyllum* (nagchampa), *Hevea brasiliensis* (rubber seed tree), *Azadirachta indica* (neem), *Madhuca indica* and *Madhuca longifolia* (mahua), *Ceiba pentandra* (silk cotton tree), *Simmondsia chinensis* (jojoba), *Euphorbia tirucalli*, babassu tree, and microalgae. Among the 75 plant species that have more than 29% oil in their seed/kernel, palm, *Jatropha curcas*, and *Pongamia pinnata* (Karanja) were found to be the most suitable for biodiesel production.

TABLE 13.1 Major Oils Used for Producing Biodiesel

Group	Source of Oil
Major oils	Coconut (copra), corn (maize), cottonseed, canola (a variety of rapeseed), olive, peanut (groundnut), safflower, sesame, soybean, sunflower
Nut oils	Almond, cashew, hazelnut, macadamia, pecan, pistachio, walnut
Other edible oils	Amaranth, apricot, argan, artichoke, avocado, babassu, bay laurel, beech nut, ben, Borneo tallow nut, carob pod (algaroba), cohune, coriander seed, false flax, grape seed, hemp, kapok seed, lallemantia, lemon seed, macauba fruit (*Acrocomia sclerocarpa*), meadowfoam seed, mustard, okra seed (hibiscus seed), perilla seed, pequi (*Caryocar brasiliensis* seed), pine nut, poppy seed, prune kernel, quinoa, ramtil (*Guizotia abyssinica* seed or Niger pea), rice bran, tallow, tea (camellia), thistle (*Silybum marianum* seed), wheat germ
Inedible oils	Algae, babassu tree, copaiba, honge, jatropha or ratanjyote, jojoba, karanja or honge, mahua, milk bush, nagchampa, neem, petroleum nut, rubber seed tree, silk cotton tree, tall
Other oils	Castor, radish, and tung

Many European countries utilize rapeseed. During World War II, oil from *Jatropha* seeds was used as blends with and substituted for diesel. It has been reported that biodiesel produced from palm and Jatropha have physical properties in the right balance; conferring it with adequate oxidation stability and cold performance. The major oils used for producing biodiesel are presented in Table 13.1 (Chhetri et al. 2008).

13.3 BIODIESEL PRODUCTION TECHNOLOGIES

Vegetable oils are not suitable for direct use in engines due to their high viscosity, low volatility, and polyunsaturated characteristics. Methods such as pyrolysis, dilution, microemulsion, and transesterification are being used to solve these issues in order to make vegetable oils compatible for engine use (Sani et al. 2012). Simple processes like dilution and microemulsion cannot be used successfully due to the high viscosity and low volatility of the vegetable oils in use for the biodiesel production (Lin et al. 2011). Transesterification is the most commonly used method for biodiesel production. Transesterification is the reaction of a triglyceride (fat/oil) with an alcohol to form esters and glycerol. A triglyceride has a glycerin molecule as its base with three long-chain fatty acids attached. In most production, methanol or ethanol is the alcohol used (methanol produces methyl esters, ethanol produces ethyl esters) and is base catalyzed by either potassium or sodium hydroxide. Figure 13.1 shows different processes employed for biodiesel production (Yunus Khan et al. 2013).

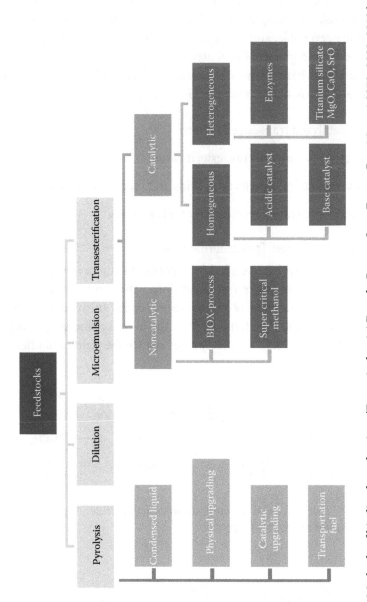

FIGURE 13.1   Methods of biodiesel production. (From Atabani AE et al., *Renew. Sust. Energy Rev.* 16 no. 4:2070–2093, 2012.)

13.3.1 Transesterification

Transesterification consists of a number of consecutive, reversible reactions. Triglycerides are converted stepwise to diglycerides, monoglycerides, and finally glycerol, liberating a mole of ester in each step (Figure 13.2). The reactions are reversible, and the equilibrium lies toward the production of fatty acid ester and glycerol (Venkateswarulu et al. 2013).

There are four basic transesterification routes to biodiesel production from oils and fats (Canakci and Van Gerpen 2003):

1. Base-catalyzed transesterification

2. Direct-acid catalyzed transesterification

3. Conversion of oil into its fatty acids and then into biodiesel

4. Noncatalytic transesterification of oils and fats

13.3.2 Acid-Catalyzed Transesterification

Carboxylic acids can be esterified by alcohols in the presence of a suitable acidic catalyst in water free conditions. The most frequently cited reagent for the preparation of methyl esters is 5% anhydrous hydrogen chloride in methanol. Free fatty acids in the oil are reduced by using an acid catalyst (Figure 13.3). Free fatty acids reduce the yield of biodiesel by

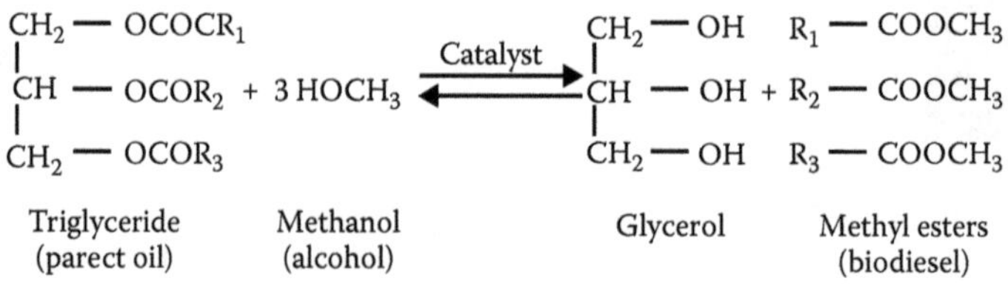

FIGURE 13.2 Mechanism of decomposition of triglycerides. (From Kumar A et al., *J. Biosci. Bioeng.* 67:77–82, 2002.)

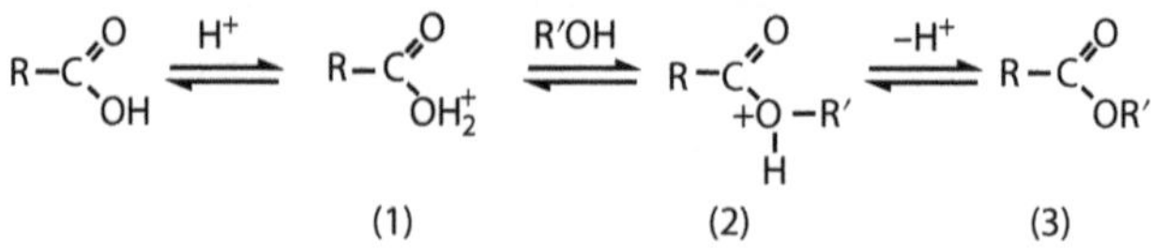

FIGURE 13.3 Mechanism of acid catalyzed transesterification.

saponification, hence acid-catalyzed transesterification has an advantage of more yields. However, the reaction is so slow that one run can take more than a day.

13.3.3 Base-Catalyzed Transesterification

Esters in the presence of a base such as an alcoholate anion form an anionic intermediate that can dissolve back to the original ester or form the new ester. The most useful basic transesterifying agents are sodium or potassium methoxide in anhydrous methanol (Ramdas et al. 2005). The main advantage of base-catalyzed transesterification over an acid one is that it is fast and can be conducted at low temperatures (303–308 K) and pressure (0.1 MPa).

13.3.4 Basic Heterogeneous Catalysts

Heterogeneous catalysts are environmentally benign, noncorrosive, easily separable from the liquid products, and have higher activity and longer catalyst lifetime. Homogeneous catalysts sodium hydroxide (NaOH), potassium hydroxide (KOH) and sulfuric acid, phosphoric acid, sulfonic acid, and hydrochloric acid are highly reproducible, with a definite stoichiometry and structure. The removal of a conventional homogeneous catalyst after reaction is technically difficult and large volumes of wastewater are produced to separate and clean the catalyst and the products. Therefore, conventional homogeneous catalysts are expected to be replaced with environmentally friendly heterogeneous catalysts (Table 13.2). Heterogeneous catalysts can easily be separated from reaction mixtures and reduce environmental pollutants. Also heterogeneous catalysts are known for their activity, stability, economic benefits, ease of recovery, and resistance to

TABLE 13.2 Homogeneous versus Heterogeneous Process

Homogeneous Process	Heterogeneous Process
Reaction is very fast	Relatively slow process
Reaction is 100% complete	Conversion is relatively poor
Catalyst dissolved in the reaction mixture	Catalyst does not dissolved in reaction mixture
Purification is difficult	Purification is much easier
Purification by water wash	Biodiesel purification by distillation
Glycerin is crude need further purification	Product and glycerin received is pure
Catalyst cannot be recycled	Catalyst can be recycled
Process is cheaper	Process is presently expensive

poisoning. They can be removed from the reaction mixture by the coarse filtration technique (Demirbas 2007; Han et al. 2005).

13.3.5 Disadvantages of Existing Methods

Due to the low cost of methanol and the high rates of base-catalyzed transesterification, most commercial biodiesel is made by alkali-catalyzed reaction of vegetable oils with methanol in stirred tank reactors. There are some technical problems related to this process, such as the limit of reaction rate by mass transfer between the oil and alcohol because they are immiscible and hence there is a requirement of high energy mixing. Because of this, multiple downstream processing steps are necessary to obtain a refined biodiesel. This also involves high-strength wastewater and its treatment. Transesterification itself is a reversible reaction and in thermodynamic equilibrium, and hence there is an upper limit to raw materials conversion in the absence of product removal. At present, the "separation" process accounts for 60% to 80% of the overall production cost (Ragauskas et al. 2006). Conventional separation techniques for liquid mixtures include distillation, low-temperature crystallization, adsorption, and extraction. Among them, distillation is the dominant refinery process. However, due to the energy intensive nature, negative environmental impact, and complicated operation procedure, these techniques are generally not economical and practical to stand alone for the entire bioalcohol separation process. New biorefinery concepts and hybrid technologies must be developed. Some of these technical challenges can be overcome by using a membrane reactor.

A membrane reactor is a device for simultaneously carrying out a reaction and membrane-based separation in the same physical enclosure. Due to the immiscibility of lipids feedstock and alcohol, lipids form droplets that are excluded from passing through the membrane pores. The microporous inorganic membrane selectively permeates free fatty acid methyl ester (FAME), alcohol, and glycerol, while retaining the emulsified oil droplets, thus increasing the conversion of equilibrium limited reactions.

13.4 MEMBRANE REACTOR TECHNOLOGY

Membrane separation technologies—microfiltration (MF), ultrafiltration (UF), nanofiltration (NF), and reverse osmosis (RO)—are proving to be effective ways to achieve optimum yields and reduce energy costs for biofuel production. In particular, membrane filtration technology shows promise to improve "second-generation" cellulosic ethanol processes.

Membrane use is rising in biodiesel processes where membranes facilitate water reuse, particularly in areas where water is scarce. Membrane technology shows promise for concentrating and purifying organic acids, commonly used as the base for a variety of new biodegradable plastics. In addition, so-called integrated biorefineries are using a variety of membrane technologies, particularly those that facilitate continuous, rather than batch, fermentation. Membranes are poised to play a significant role in this important advancement. In the next few years, a number of cellulosic biorefinery pilot and demonstration plants will be placed online using membrane technology in parts of the production process.

As pointed out earlier, the crude biodiesel product obtained from both conventional and membrane reactors comes with impurities such as glycerol, soap, residual catalysts, and excess alcohol that need to be removed. Purification of crude biodiesel from these impurities is necessary so as to make biodiesel suitable for diesel engine consumption and other applications. The presence of these impurities in biodiesel fuel could cause severe engine problems and damage. The removal of these impurities via conventional separation and purification techniques poses severe difficulties, such as a huge amount of water usage, high energy consumption, time wasting, and treatment of wastewater. This has led to the recent application of organic/inorganic separative membranes for the refining of crude biodiesel. Membrane biodiesel separation processes seem to provide high-quality biodiesel fuel. A large portion of membrane separation processes are carried out under moderate temperature and pressure conditions and their scale-up is less cumbersome. Furthermore, membranes are generally most preferred in refining processes for the following reasons: low energy consumption; safety; simple operation; elimination of wastewater treatment; easy change of scale; higher mechanical, thermal, and chemical stability; and resistance to corrosion.

Development of a membrane reactor and its successful application in producing biodiesel has renewed the strong interest to develop alternative renewable and sustainable fuel to replace petrodiesel fuel. Membrane reactors can serve different purposes such as intensifying the contact between reactants and catalysts, selectively removing the products from the reaction mixture, and control the addition of reactants to the reaction mixture. These reactors can be employed to avoid the equilibrium conversion limits of conventional reactors. Besides, the reactors can efficiently improve the maximum achievable conversion of reversible reactions and the general reaction pathways. Furthermore, membrane reactors provide the potential of higher selectivity and yields in many different processes as well as being safe and more environmentally friendly. Dubé et al. (2007) developed a novel membrane reactor that enabled both acid- or base-catalyzed transesterification of canola oil as well as separation of unre-acted canola oil from reaction products. The membrane reactor consisted of a membrane pore size of 0.05 m, inside and outside diameters of the membrane were 6 and 8 mm, a length of carbon membrane tube of 1200 mm, and a surface area of 0.022 m². The membrane reactor was charged with canola oil (100 g) and sealed. Following a circulation time of 10 min, the reactor was operated continuously at a pressure of 138 kPa with a feed (mixture of methanol and acid) pump flow rate of 6.1 mL/min. The heat exchanger was switched on to achieve temperatures of reaction (60°C, 65°C, and 70°C), which was monitored by a thermocouple. Therefore after starting the heat exchanger, a stable reaction (±0.1°C) time was achieved with 30 min for 60°C, 40 min for 65°C, and 45 min for 70°C. The experiments were all conducted for 6 h. The authors remarked that an additional experiment was also performed to study the effects of a methanol/acid catalyst feed flow rate on conversion for both acid- and base-catalyzed trans-esterifications. The flow rates were 2.5, 3.2, and 6.1 mL/min. The schematic diagram of the membrane reactor is depicted in Figure 13.4 (Atadashi et al. 2011).

Similarly, Canakci and Van Gerpen (2003) transesterified a number of vegetable oils including canola, soybean, palm, and yellow grease lipids via a membrane reactor. The authors noted that despite the wide range of feed-stocks used, the membrane reactor presented a moderately consistent performance at one set of operating conditions and enabled the production of high-quality biodiesel fuel, which was confirmed by gas chromatography (GC) analysis based on the ASTM D6584 standard. The biodiesel from all the feedstocks met the ASTM D6751 standard. Also, the glycerol content of biodiesel produced using a membrane reactor was significantly lower than that produced via a conventional batch reaction. Another study conducted by Canakci and Van Gerpen (2003) compared three different recycling ratios for production of biodiesel via membrane reactor: 100%, 75%, and 50% by volume; for instance, 75% recycling entails that every 0.75 L of polar phase was mixed with 0.25 L methanol with 1 wt% (by weight of oil) NaOH catalyst and pumped into the reactor circulating loop at a feed rate of 3 L/h, while the feed rate of canola oil was also kept at 3 L/h. The authors noted that the catalysts and glycerol were also recycled. Furthermore, to main-tain a biodiesel-rich nonpolar phase containing 85 wt%, the permeate was

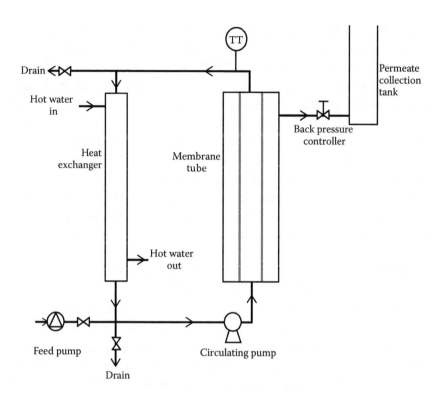

FIGURE 13.4   Schematic diagram of a biodiesel membrane reactor.

consistently removed as well as during the methanol/glycerol polar phase. Consequently, at maximum recycle ratio, the FAME concentration ranged from 85.7 to 92.4 wt% in the biodiesel-rich nonpolar phase. In addition, the overall molar ratio of methanol:oil in the reaction system significantly decreased to 10:1 while maintaining a FAME production rate of 0.04 kg/min. Also in the biodiesel-rich nonpolar phase, no triglycerides (TGs), mono-glycerides (MGs), or glycerol were observed. The authors noted that despite the samples not being water washed prior to analysis, high purity biodiesel product free of nonsaponifiable materials was produced. As a result, a high purity biodiesel product was produced. However, the yield obtained via homogeneous catalyst in a membrane reactor was below the EN13213 stan-dard. Therefore, to circumvent the problems associated with use of homoge-neous catalyst, using a membrane reactor still poses difficulties.

Baroutian et al. (2011) developed a novel continuous packed bed mem-brane reactor (a tubular ceramic [TiO$_2$/Al$_2$O$_3$] membrane) to produce biodiesel fuel using solid alkaline catalyst (potassium hydroxide catalyst supported on activated carbon). The membrane reactor comprised of a length, inner diameter, outer diameter, and pore size of the membrane of 40 cm, 1.60 cm, 2.54 cm, and 0.05 m, respectively. The filtration surface area for the entire membrane was 0.0201 m². The temperature and pressure were monitored via pressure gauges and a temperature indicator. The authors noted that during transesterification the membrane reactor was able to block the triglycerides, but biodiesel and by-product glycerol alongside methanol passed through the membrane pore size due to their smaller molecular sizes. As discussed earlier, the ability to block triglycerides provided high-quality biodiesel fuel. Conversion of 94% was obtained at a 70°C reaction temperature, 157.04 g catalyst per unit volume of reactor, and 0.21 cm/s cross-flow circulation velocity. The characteristics of the product under the optimum condition were within the ASTM standard.

13.5 REVIEW OF RECENT REPORTS ON BIOFUEL USING MEMBRANE TECHNOLOGY

Scientists in Japan conducted the transesterification reactions of triolein with ethanol using various ion-exchange resin catalysts to produce ethyl oleate as a biodiesel. The effects of the resin's structural factors and the operating factors on the reaction rate were investigated. The possibility of a continuous biodiesel production was studied by constructing an expanded bed reactor packed with active resin. The anion exchange resins exhibited much higher catalytic activities than the cation-exchange resin. The anion-exchange resin with a lower cross-linking density and a smaller particle size gave a high reaction rate as well as a high conversion. By combining the three-step regeneration method, the resin could be repeatedly used for batch transesterification without any loss in catalytic activity. A continuous transesterification reaction was carried out using an expanded bed reactor packed with the most active resin. The reactor system permitted the continuous production of ethyl oleate with a high conversion (Shibasaki-Kitakawa et al. 2007).

The enzymatic production of biodiesel by methanolysis of cottonseed oil was studied by scientists in Argentina using immobilized *Candida antarctica* lipase as a catalyst in *t*-butanol solvent. Methyl ester production and triacylglycerol disappearance were followed by high-performance liquid chromatography (HPLC). It was found, using a batch system, that enzyme inhibition caused by undissolved methanol was eliminated by adding *t*-butanol to the reaction medium, which also gave a noticeable increase of reaction rate and ester yield. The effect of *t*-butanol, methanol concentration, and temperature on this system was determined. A methanolysis yield of 97% was observed after 24 hours at 50°C with a reaction mixture containing 32.5% *t*-butanol, 13.5% methanol, 54% oil, and 0.017 g enzyme/g oil. With the same mixture, a 95% ester yield was obtained using a one-step fixed-bed continuous reactor with a flow rate of 9.6 ml/hour/g enzyme.

Experiments with the continuous reactor over 500 hours did not show any appreciable decrease in ester yields (Royon et al. 2007).

Researchers at Kocaeli University in Turkey conducted a study with the objective to present the availability and the properties of restaurant waste oils and rendered animal fat as low-cost feedstocks for biodiesel production, and determined the level of these contaminants in feedstock samples from a rendering plant. Waste restaurant oils and animal fats have relatively high levels of saturation. Levels of free fatty acids varied from 0.7% to 41.8% and moisture from 0.01% to 55.38%. These wide ranges indicate that an efficient process for converting waste grease and animal fats must tolerate a wide range of feedstock properties (Canakci 2007).

The immiscibility of canola oil in methanol provides a mass-transfer challenge in the early stages of the transesterification of canola oil in the production of FAME or biodiesel. To overcome, or rather exploit, this situation, a two-phase membrane reactor was developed by researchers at the University of Ottawa, Canada, to produce FAME from canola oil and methanol. The transesterification of canola oil was performed via both acid and base catalysis. Runs were performed in the membrane reactor in semibatch mode at 60°C, 65°C, and 70°C and at different catalyst concentrations and feed flow rates. Increases in temperature, catalyst concentration, and feedstock (methanol/oil) flow rate significantly increased the conversion of oil to biodiesel. The novel reactor enabled the separation of reaction products (FAME/glycerol in methanol) from the original canola oil feed. The two-phase membrane reactor was particularly useful in removing unreacted canola oil from the FAME product yielding high purity biodiesel and shifting the reaction equilibrium to the product side (Dubé et al. 2007).

13.6 CONCLUSION

Membrane reactors were observed to restrict the passage of unreacted oils into the biodiesel product mixture, hence providing high-quality biodiesel fuel. Application of alkaline catalysts in the membrane system for the production of biodiesel was found to produce soap, whereas acid catalysts avoided soap formation but slow reaction rates were observed. Membrane biodiesel production processes were reported to produce products whose properties met ASTM standard specification. To achieve high biodiesel yield via membrane reactor operation, development of cheap and very active heterogeneous catalysts is necessary. Membranes need to be maximally harnessed for their unique characteristics of high available surface area per unit volume, high selectivity/conversion, capacity to control contact between reactants and catalysts (which is rare with conventional systems), ability to recover important products, treat effluents, and minimize their harm to the atmosphere. More research should be conducted to study the fouling effects and stability of the membranes for the production and refining of biodiesel.

REFERENCES

Atabani AE et al. 2012. A comprehensive review on biodiesel as an alternative energy resource and its characteristics. *Renew. Sust. Energy Rev.* 16 no. 4:2070–2093.

Atadashi IM et al. 2011. Biodiesel separation and purification: A review. *Renew. Sust. Energy Rev.* 56:39–53.

Balat M. 2011. Potential alternatives to edible oils for biodiesel production—A review of current work. *Energy Convers. Manage.* 52 no. 2:1379–1392.

Baroutian et al. 2011. A packed bed membrane reactor for production of biodiesel using activated carbon supported catalyst. *Bioresource Technology* 102 no. 2: 1095–1102.

Canakci M, Van Gerpen J. 2003. A pilot plant to produce biodiesel from high free fatty acid feed stocks. *Am. Soc. Agri. Eng.* 46 no. 4:945–954.

Canakci M. 2007. The potential of restaurant waste lipids as biodiesel feedstocks, *Bioresour. Technol.* 98 no. 1:183–190.

Chhetri AB et al. 2008. Non-edible plant oils as new sources for biodiesel production. *Int. J. Mol. Sci.* 9 no. 2:169–180.

Demirbas A. 2007. Biodiesel from sunflower oil in supercritical methanol with calcium oxide. *Energy Convers. Manag.* 48:937–941.

Dubé MA et al. 2007. Biodiesel production using a membrane reactor. *Bioresour. Technol.* 98 no. 3:639–647.

Han H et al. 2005. Preparation of biodiesel from soybean oil using supercritical methanol and co-solvent. *Fuel* 84:347–351.

Kumar A et al. 2002. Acid-base transesterification of oil with high free fatty acid content. *J. Biosci. Bioeng.* 67:77–82.

Lin L et al. 2011. Opportunities and challenges for biodiesel fuel. *Appl. Energy* 88 no. 4:1020–1031.

Ragauskas AJ et al. 2006. The path forward for biofuels and biomaterials. *Science* 311:484–489.

Ramdas AS et al. 2005. Biodiesel production from high free fatty acid rubber seed oil. *Fuel* 84 no. 4:335–340.

Royon D et al. 2007. Enzymatic production of biodiesel from cotton seed oil using *t*-butanol as a solvent. *Bioresour. Technol.* 98 no. 3:648–653.

Sani YM et al. 2012. Biodiesel feedstock and production technologies: Successes, challenges and prospects. In *Biodiesel: Feedstocks, Production and Applications*, edited by Z Fang, 77–101. doi: 10.5772/52790.

Shibasaki-Kitakawa N et al. 2007. Biodiesel production using anionic ion-exchange resin as heterogeneous catalyst. *Bioresour. Technol.* 98 no. 2:416–421.

Venkateswarulu TC et al. 2013. *Int. J. Chem. Tech. Res.* 6 no. 4:2568–2576.

Yunus Khan TM et al. 2013. Recent scenario and technologies to utilize non-edible oils for biodiesel production. *Renew. Sust. Energy Rev.* 37:840–851.

CHAPTER 14

Production of Ecofriendly Biodiesel from Marine Sources

Sahadevan Renganathan,
Ramachandran Kasirajan, and Theresa Veeranan

CONTENTS

Abstract

Drastic changes in climatic conditions due to increased CO_2 levels and rising costs of fossil fuels has led to a search for biofuels from marine sources. Biodiesel are fuels that can be economically viable alternatives for petrodiesel fuels. Most marine sources, such as micro- and macroalgae consume CO_2, which is one of the toxic pollutants increasing global warming. This increase in the temperature is melting ice caps in the polar regions and causing abnormal changes in climatic conditions. Hence, for an ecofriendly remedial solution, cultivation of marine sources that are potential consumers of CO_2 and in turn extraction of biodiesel from them decreases the release of CO_2 into the environment. This chapter focuses on marine sources such as micro- and macroalgae as sources and biodiesel production methodologies.

14.1 INTRODUCTION

The creation of energy from renewable and dissipate materials is an ideal substitute to conservative agricultural feedstocks (Ramachandran et al. 2011). This chapter describes an approach to segregate oil from marine algae and to convert the extracted oil into biodiesel. The amount of oil in the algae source varies from 0.5% to 70% depending on its types. This technique yields 10% to 65% of oil depending on the algae species. The obtained biodiesel was achieved as 100% conversion of oil to biodiesel.

Biodiesel is an alternative biofuel produced by chemically catalyzing the reactant of vegetable oils or animal fats with short-chain alcohols such as methanol, ethanol, or butanol (Meher et al. 2006). Biodiesel is obtained from vegetable oils and animal fats, so it is an important material for fighting environmental degradation because of its liquid nature, ecofriendly nature, and easy portability (Balat 2007). On the other hand, a global challenge has now emerged because biodiesel fuel is derived mainly from soybean oil or other seeds and nuts, and using food to produce fuel is not rational considering the increase in world population.

14.2 BENEFITS OF BIODIESEL

Biodiesel is highly recommended by the U.S. Department of Energy as a perfect fuel for buses and trucks (U.S. Department of Energy 2003). Its renewability is the most attractive property of biodiesel. Biodiesel made from an inexhaustible supply (e.g., plant or animal materials) gives hope for the future fuel. Biodiesel is expected to build up a consolidated and sustainable fuel chain around the world. Environmental-friendliness is the second benefit of biodiesel. The application of biodiesel can greatly reduce greenhouse gas emissions by 78%, hydrocarbon emissions by 56%, and carcinogenic properties by 94% compared to conventional diesel fuel.

The presence of oxygen in biodiesel (~10%) improves combustion and reduces carbon monoxide and hydrocarbon emissions. Since nearly all the carbon in biodiesel is derived from atmospheric CO_2, which is released again on combustion, the life cycle of biodiesel is estimated to be a near carbon-neutral process. Unlike petrodiesel, biodiesel is free of sulfur impurities, so firing of biodiesel energy also efficiently eradicates sulfur oxide and sulfate emissions, which are major contributors to acid rain. Major benefits of biofuels are listed next.

Energy security

- Domestically distributed
- Supply reliability
- Reducing use of fossil fuels
- Reducing the dependency on imported petroleum
- Renewable
- Fuel diversity

Environmental impact

- Reduction of greenhouse gasses
- Reduction of air pollution
- Higher combustion efficiency
- Easily biodegradable
- Carbon neutral

Economic impact

- Sustainability
- Increased number of rural manufacturing jobs
- Increased farmer income
- Agricultural development

Furthermore, the use of biodiesel yields significant reductions in carbon monoxide and soot particles compared to petrodiesel. The third highlight of biodiesel is that it can be used in late-model diesel engines without any modifications. Biodiesel is actually good for diesel engines. It provides better lubricant than petrodiesel and has excellent solvent properties. Higher lubricant prolongs engine life and reduces the frequency of replacement of engine parts. Conventional diesel leaves deposits inside fuel lines, storage tanks, and fuel delivery systems over a period. Biodiesel dissolves this sediment while adding no deposits of its own, resulting in cleaner and more trouble-free fuel handling systems. A higher cetane number and higher flash point of biodiesel leads to better and safer performance. Other advantages of biodiesel include biodegradable and nontoxic properties. Biodiesel is safe to handle and transport because it is as biodegradable as sugar, ten times less toxic than table salt, and burns at relatively high temperature. Biodiesel actually degrades about four times faster than petrodiesel when accidentally released into the environment (Suganya and Renganathan 2013).

14.3 SOURCES FOR BIODIESEL

Biodiesel production processes with nonedible sources and their oil yields are listed next (Ramachandran et al. 2013).

14.3.1 Plant Seed and Crops

Linseed oil (*Linum usitatissimum*): 37%–42%

Karanja oil (*P. pinnata*): 27%–39%

Jojoba oil (*Simmondsia chinensis*): 44%

Jatropha oil (*J. curcas*): 45%

Hemp oil (*Cannabis sativa*): 33%

Cuphea oil (*Cuphea viscosissima*): 25%–43%

Castor oil (*Ricinus communis*): 45%–50%

Camelina oil (*C. sativa*): 38%–40%

Borage oil (*Borago officinalis*): 20%

Babassu oil (*Attalea speciosa*): 60%–70%

Types and availability of the feedstock and production method contribute to the overall cost of biodiesel production (Ng et al. 2010). Miscellaneous sources for biodiesel making are based on convenience, characteristics such as durability and cold flow property, and cost (Sharma and Singh 2009). Edible oils such as rapeseed and soybean are used for biodiesel invention in the United States and Europe. Asian countries of Malaysia and Philippines utilize edible oil such as palm oil and coconut oil, respectively, for biodiesel. The raw materials oppressed commercially by some developed countries are edible oils such as palm, soybean, coconut, rapeseed, and sunflower oils. The proportion of raw resources utilized for world commercial biodiesel invention is approximately 84% of rapeseed oil, 13% of sunflower oil, 1% of palm oil, and 2% of soybean and others.

The main raw resource used in India is jatropha as nonedible oil (Sohpal et al. 2011).

Microalgae and macroalgae oils have also been tried by several researchers as a medium of feedstock for biodiesel due to their superior biomass invention, faster escalation, and photosynthetic efficiency as compared to other energy sources. Nowadays, many countries are encouraging exploration to increase the escalation rate of algal as a biomass and oil content for biodiesel invention. Balezentiene et al. (2013) studied that line of attack for environmentally friendly bioenergy techniques and gave practical examples of approaches based on practical experience or technical knowledge. The most significant approaches gain maximum environmental and energetic assistance from bioenergy. The following section lists the different algal raw materials commonly used for biodiesel invention and their oil yields (Chisti 2007; Demirbas 2009a).

14.3.2 Algal Sources

Anabaena cylindrical: 4%–7%

Botryococcus bruanii: 25%–75%

Chlamydomonas rheinhardii: 21%

Chlorella pyrenoidosa: 2%

Chlorella sp.: 28%–32%

Chlorella vulgaris: 14%–22%

Crypthecodinium cohni: 20%

Cylindrotheca sp.: 16%–37%

Dunaliella bioculata: 8%

Dunaliella salina: 6%

Euglena gracilis: 14%–20%

Isochrysis galbana: 30%–41%

Nannochloropsis sp.: 31%–68%

Nitzschia sp.: 45%–47%

Pavlova lutheri: 35.5%

Phaeodactylum tricornutu: 20%–30%

Porphyridium cruentum: 9%–14%

Prymnesium parvum: 22%–38%

Scenedesmus dimorphus: 16%–40%

Scenedesmus obliquus: 12%–14%

Scenedesmus quadricauda: 1.9%

Schizochytrium sp.: 50%–77%

Spirogyra sp.: 11%–21%

Spirulina maxima: 6%–7%

Spirulina platensis: 4%–9%

Synechoccus sp.: 11%

Tetraselmis maculate: 3%

Tetraselmis sueica: 15%–23%

14.4 MICROALGAE

Microalgae have long been renowned as a potentially superior resource for biofuel invention because of their high oil content and rapid biomass invention (Wen and Johnson 2009). Microalgae as a feedstock for biodiesel has been comprehensively reviewed in recent years (Khan et al. 2009). Algae biomass can play a vital role in solving the difficulty between the production of food and that of biofuel in the near future. Microalgae emerged to be the only source of renewable biodiesel that is capable of meeting the demand for transportation fuels. The development of microalgae does not necessitate much land as compared to that of terraneous plants (Huang et al. 2010). Microalgae are photosynthetic microorganisms that convert water, sunlight, and CO_2 to algal biomass (Chisti 2008).

Microalgae include green algae (*chlorophyceae*), diatoms (*bacillariophyceae*), blue-green algae (*cyanophyceae*), and golden-brown algae (*chrysophyceae*) (Rengel 2008). Microalgae are unexploited sources, with more than 25,000 species of which only 15 and more are in use. According to Benemann (2008) total world mercantile microalgal biomass production is about 10,000 tons per annum. Microalgae have superior growth rates and productivity when compared to conventional agricultural crops, forestry, and other aquatic plants. Microalgae do not require as much land area as other biodiesel feedstocks of agricultural origin, up to 49 or 132 times less when compared to rapeseed or soybean crops, for a 30% (w/w) of oil substance in algae biomass.

Algae have the potential to dwarf all the other biodiesel feedstocks due to their efficiency in photosynthesizing solar energy into chemical energy (Mateos 2007). Plants can collect a maximum of 5% of the solar energy over a three-month period. The best algal biomass yields under tropical conditions are about 50 tons per hectare per year (Pimentel et al. 2008). The yields reported or claimed for microalgae outperform traditional energy crops. Currently, suspend-based open ponds and enclosed photobioreactors are commonly used for algal-biofuel production. Nevertheless, about 98% of commercial algae biomass production is currently with open ponds, even for high value nutritional products, which sell for prices over a hundred- and even a thousandfold higher than allowable for biofuels. Open ponds are a very efficient and cost-effective method of cultivating algae, but they can become contaminated with unwanted species very quickly (Benemann 2008; Schenk et al. 2008).

Sustained open pond production has been successful for a limited number of algae including *Chlorella*, *Spirulina*, and *Dunaliella* (Reijnders and Huijbregts 2009). Commercially operated open ponds yields cover a vast range from 3.5 to 10–30 mg dry weight per hectare per year, with the highest productivities achieved under tropical or subtropical conditions. The extreme conditions in the raceway ponds are not favorable to maximizing yield (Reijnders 2009). Microalgae production in closed photobioreactors is highly expensive. Closed systems are much more expensive than ponds. However, closed systems require much less light and agricultural land to grow the algae. Photobioreactors offer a closed culture environment, which is protected from direct fallout, relatively safe from invading microorganisms, where temperatures are controlled with an enhanced CO_2 fixation that is bubbled through the culture medium (Patil et al. 2008). Maximal volumetric productivities can be higher in a closed photobioreactor than in an open pond because the surface-to-volume ratio can be higher. But in large scale, both open pond and closed photobioreactors are limited by the amount of incident sunlight on the earth's surface and cannot exceed a maximum of 100 g/m^2 per day (Pienkos and Darzins 2009).

Biofuel production using microalgal farming offers many advantages as follows:

- Microalgal cultivation consumes less water than land crops.

- Microalgal farming could be potentially more cost effective than conventional farming.

- Nitrous oxide release could be minimized when microalgae are used for biofuel production.

- The high growth rate of microalgae makes it possible to satisfy the massive demand on biofuels using limited land resources without causing potential biomass deficit.

- The tolerance of microalgae to high CO_2 content in gas streams allows high efficiency CO_2 mitigation.

A major problem with the culture of algae in ponds or tanks is the harvesting of the algae. The large water content of harvested algal biomass means its drying would be an energy-consuming process (Li et al. 2008). This problem was observed at the University of Florida when algae were being cultured in managed ponds for the production of nutrients for hogs. After 2 years with a lack of success, the algal nutrient culture was abandoned (Pimentel 2008).

Many researchers reported that microalgae might be better for higher biodiesel production. However, microalgal biodiesel technology, while promising, has not yet been developed to the point of full commercial-scale production, but working feasibility studies have been conducted. Specially bred mustard varieties can produce reasonably high oil yields and have the added benefit that the meal left over after the oil has been pressed out can act as an effective and biodegradable pesticide (Demirbas 2009d; Sharif Hossain and Salleh 2008).

Microalgae contain lipids and fatty acids as membrane components, storage products, metabolites, and sources of energy. Algae present an exciting possibility as a feedstock for biodiesel, especially when you realize that oil was originally formed from algae. Algae contain anywhere between 2% and 40% of lipids/oils by weight (Demirbas 2009a). The oil contents of some microalgae were given earlier. The lipid and fatty acid contents of microalgae vary in accordance with culture conditions. Algal oil contains saturated and monounsaturated fatty acids. The fatty acids were determined in the algal oil in the following proportions: oleic acid (36%), palmitic acid (15%), stearic acid (11%), isoleic acid 17:0 (8.4%), and linoleic acid (7.4%). The high proportion of saturated and monounsaturated fatty acids in this alga is considered optimal from a fuel quality standpoint, in that fuel polymerization during combustion would be substantially less than what would occur with polyunsaturated fatty acid-derived fuel (Demirbas 2009b). The main problem associated with algal use on land is caused by the land utilization due to higher population growth and expensive processing.

14.5 MACROALGAE

One of the contending candidates for biodiesel production is macroalgae. Macroalgae are a renewable source of oil that could meet the global demand for transport fuels (Miao and Wu 2006). This seaweed is selected due to its numerous advantages. Some species of macroalgae are nonedible and highly toxic to costal environments. The occurrence of algal blooms greatly disturbs ecosystems by modifying food chains and faunal neighborhood structure. The accumulation of algal biomass relocates natural neighborhoods of seagrasses and higher plants (Taylor et al. 1995). Correspondingly, macroalgal blooms cause changes in the main biogeochemical cycles of carbon, nitrogen, phosphorus, and sulfur (Viaroli et al. 2001). Therefore, it is very necessary to clean up the toxic macroalgae from the environment.

Less information is available about the production of biodiesel from marine macroalgae and compared with plant algae, but there are plenty of marine algae resources available in the ocean. Marine macroalgae used for biodiesel production include:

Ulva lactuca

Enteromorpha compressa

Caulerpa peltata

Caulerpa scalpelliformis (under research)

Ulva faciata (under research)

Chaetomorpha linum

Pterocladiella capillacea

Gracilariopsis longissima

Fucus vesiculosus

Chorda filum

Laminaria digitata

Fucus serratus

Laminaria hyperborean

Macrocystis

Pyrifera

Fucus spiralis

Saccorhiza polyschides

Sargassum muticum

Codium tomentosum

Ulva rigida

Enteromorpha intestinalis

Ascophyllum nodosum

Pelvetia canaliculata

Enteromorpha prolifera

Sargassum spp.

14.6 BIODIESEL PRODUCTION METHODS

A number of methods are currently available and have been adopted for the production of biodiesel fuel. There are four primary ways to produce biodiesel: direct and blending exploit of raw oils (Adams et al. 1983; Anon. 1982; Engler et al. 1983), microemulsions (Schwab et al. 1988), thermal cracking (Chang and Wan 1947; Pioch et al. 1993), and transesterification (Ma and Hanna 1999).

14.6.1 Direct and Blending Exploit

In 1893, Rudolf Diesel successfully demonstrated his innovation about the diesel engine by running it on peanut oil (a biomass fuel), and for the first time the possibility of utilizing a biomass fuel for engines was established.

As early as the 1980s, Bartholomew (1981) proposed the usage of plant and vegetable oils as fuels (Bartholomew 1981). Vegetable oils can be used by blending with diesel fuel. Later the functional ratio of vegetable oil to diesel in the fuel blend was raised to 20% and 50%. Nevertheless, the long-term use of this blending in a modern diesel engine becomes impractical because of the decrease in the power output and thermal efficiency by carbon deposits (Ma and Hanna 1999; Srivastava and Prasad 2000). However, a total substitution of diesel by 100% vegetable oil is not applicable because pure vegetable oil as a fuel has two severe problems: oil deterioration and incomplete combustion in the diesel engine. Several other problems include high viscosity, acid composition, FFA content, as well as gum formation due to oxidation and polymerization during storage and lubricating oil thickening (Peterson et al. 1983; Srivastava and Prasad 2000).

Advantages

    Liquid nature portability

    Heat content (80% of diesel fuel) readily available

Disadvantages

    High viscosity

    Lower volatility

    Reactivity of unsaturated hydrocarbon chains

Trouble in engine

    Coking and trumpet formation

    Carbon deposits

    Sticking

    Thickening and gelling of the lubricating oil

14.6.2 Microemulsions

To solve the problem of the high viscosity of vegetable oils, microemulsions with solvents, such as methanol, ethanol and 1-butanol, have been studied. A microemulsion is defined as the combination of two normal immiscible liquids and one or more ionic or nonionic amphiphiles (Schwab et al. 1988; Ziejewski et al. 1984). Microemulsions are elucidated as a colloidal equilibrium dispersion of optically isotropic fluid having microstructures with dimensions generally in the range of 1 to 150 nm. Microemulsions achieved good spray characteristics, with explosive vaporization, which improved the combustion characteristics (Ma and Hanna 1999; Pryde 1984). But irregular injector needle sticking, heavy carbon deposits, incomplete combustion and an increase of lubricating oil viscosity were reported as disadvantages of microemulsions (Ma and Hanna 1999).

Advantages

Better spray patterns during combustion

Lower fuel viscosities

Disadvantages

Lower cetane number

Lower energy content

Trouble in engine

Irregular injector needle sticking

Incomplete combustion cause heavy carbon deposits

Increase lubricating oil viscosity

14.6.3 Thermal Cracking

Pyrolysis, also referred to as thermal cracking, is the breakdown of a large molecule into smaller molecules by means of heating in the absence of air or oxygen (Sonntag 1979). The triglyceride (TG) molecule is relatively large with and average molar mass of 900 g/mol. By treating it under temperatures of 450°C~850°C or by heating with the aid of a Lewis acid catalyst, a TG molecule decomposes into small molecules, such as, alkanes, alkenes, aromatics, and fatty acids (Schwab et al. 1988). The experimental setup required for pyrolysis is expensive. However, the products are chemically similar to diesel oil. The removal of oxygen during the thermal process reduces the benefits of an oxygenated fuel, eliminates the environmental benefits, and usually produces a fuel closer to gasoline than diesel (Ma and Hanna 1999). Pyrolyzate has a lower viscosity, flash point, and pour point than diesel fuel and has equivalent calorific values. However, it is not considered a promising approach to biodiesel production from virgin oils.

Since the feedstock cost of biodiesel is already much higher than petroleum, by applying the expensive pyrolysis facilities and energy-intensive processes, biodiesel has no way to beat petrodiesel on price.

Advantages

Chemically similar to petroleum-derived gasoline and diesel fuel

Disadvantages

Energy intensive and hence higher cost

14.6.4 Transesterification

Transesterification is a reaction in which the glycerol backbone of the triglycerides is replaced by methanol, which esterifies fatty acids, the side chains of TG, into methyl ester (ME). The general equation for transesterification is shown in Figure 14.1. For manufacturing biodiesel, transesterification is performed to lower the viscosity of vegetable oils. Specifically, a TG molecule (the primary compound in vegetable oils) reacts with a low molecular weight alcohol, yielding a mono alkyl ester and a by-product glycerin, which is used in the pharmaceutical and cosmetic industries. The transesterification reaction proceeds in three steps:

1. TG reacts with methanol in the presence of a catalyst to produce diglycerides (DG).

2. DG reacts with methanol to generate monoglycerides (MG).

3. Finally, MG reacts with methanol to produce ME and glycerol. One mole of ME is generated per mole of methanol reacted at each step; in all three moles of ME are produced.

The major problem associated with the use of algal oils as fuels for diesel engines is caused by high fuel viscosity in compression ignition. Due to their high viscosity (about 10–20 times higher than diesel fuel) and low volatility, they do not burn completely and form deposits in the fuel injector of diesel engines (Demirbas 2009c). The transesterification of marine algal oils will greatly reduce the original viscosity and increase the fluidity. The physical and fuel properties of biodiesel from marine algal oil, in general, (e.g., density, viscosity, flash point, heating value) are comparable to those of diesel fuel (Amin 2009). Xu et al. (2006) investigated the fuel characteristics

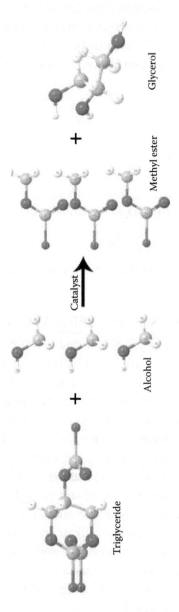

FIGURE 14.1   Transesterification reaction.

of biodiesel fuel from the acidic transesterification of heterotrophic marine algal oil with methanol. The biodiesel was characterized by a high heating value of 41 MJ/kg, a density of 0.864 kg/L, and a viscosity of 5.2 mm²/s at 313 K. Vijayaraghavan and Hemanathan (2009) reported fuel characteristics of biodiesel derived from freshwater algae as density, 0.801 kg/L; ash content, 0.21%; flash point, 371 K; pour point, 259 K; cetane number, 52; minimum gross calorific value, 40 MJ/kg; and water content, 0.02 vol.%.

Advantages

Renewability

Higher cetane number

Lower emission

Higher combustion efficiency

Disadvantages

Disposal of by-products (glycerol and waste water)

14.7 CATALYSTS FOR TRANSESTERIFICATION

In general, there are three categories of catalysts used for biodiesel production: alkalis, acids, and enzymes (Canakci and Van Gerpen 1999; Nelson et al. 1996; Shimada et al. 1999). Enzyme catalysts have become more attractive, since they can avoid soap formation and the purification process is simple to accomplish. However, they are less often used commercially because of their longer reaction time and higher cost. To reduce the cost, some researchers developed new biocatalysts in recent years. An example are the so-called whole cell biocatalysts, which are immobilized within biomass support particles. An advantage is that no purification is necessary for using these biocatalysts (Devanesan et al. 2007; Kaieda et al. 1999). Compared to enzyme catalysts, alkali and acid catalysts are more commonly used in biodiesel production (Ma and Hanna 1999).

Alkali and acid catalysts include homogeneous and heterogeneous catalysts. H_2SO_4, hydrochloric acid, and sulfonic acid are usually preferred as acid catalysts. However, these have been less studied (Canakci and Van Gerpen 1999; Kildiran et al. 1996; Siler-Marinkovic and Tomasevic 1998). H_2SO_4 gives a very high yield in esters, but the reaction occurs very slowly. It normally takes more than a day to complete. If an excess of alcohol is used in the experiment, a better conversion of TG is obtained. However, recovering glycerol becomes more difficult. The optimal relationship between the alcohol and raw material should be determined for better conversion. Acid-catalyzed transesterification is shown in Figure 14.2 (Stavarache et al. 2006; Talebian-Kiakalaieh et al. 2013).

The use of liquid acid catalysts has advantages over base catalysts with respect to biodiesel synthesis utilizing low-cost feedstocks. The performance of acid catalysts is insensitive to the presence of free fatty acid (FFA) in the feedstocks, allowing simultaneous esterification and transesterification.

Recently, a number of studies have increasingly paid attention to the use of liquid acids for biodiesel formation from high FFA content feedstocks, and they have been considered as a commercially viable alternative route to alkali catalysis.

Due to the fact that the reaction rate of a liquid-acid catalyzed transesterification was 4000 times slower than using alkali catalysts (Fukuda et al. 2001) and their stronger corrosive nature than liquid bases, the use of liquid acids is never preferred for industrial processes. Alkali catalysts are preferred in commercialized biodiesel plants due to the high speed (in seconds or minutes) (Zheng et al. 2006) and mild heating conditions (ambient temperature). Alkali homogeneous catalysts such as sodium hydroxide and potassium hydroxide are frequently used, and alkali-catalyzed transesterification is most commonly used for commercial production of biodiesel (Dias et al. 2008; Dizge et al. 2009). These materials are considered the most economical because the alkali-catalyzed transesterification process is carried out under mild conditions with no intermediate steps (low temperature and pressure environment), and the conversion rate is high with minimal side reactions and reaction time.

The base-catalyzed transesterification reaction is presented in Figure 14.3. However, the alkali homogeneous catalysts are highly hygroscopic and absorb water from air during storage. They also form water when dissolved in the alcohol reactant and affect the yield. Therefore, they should be handled properly.

However, the standard biodiesel production suffers from the presence of water and FFA in feedstocks. Water favors the formation of FFA by hydrolysis of TG. The formation of FFA in the presence of basic homogeneous catalysts gives rise to soap, creating serious problems for product separation and ultimately hindering catalytic activity. As a result, highly refined vegetable oils are required for the process; otherwise, the

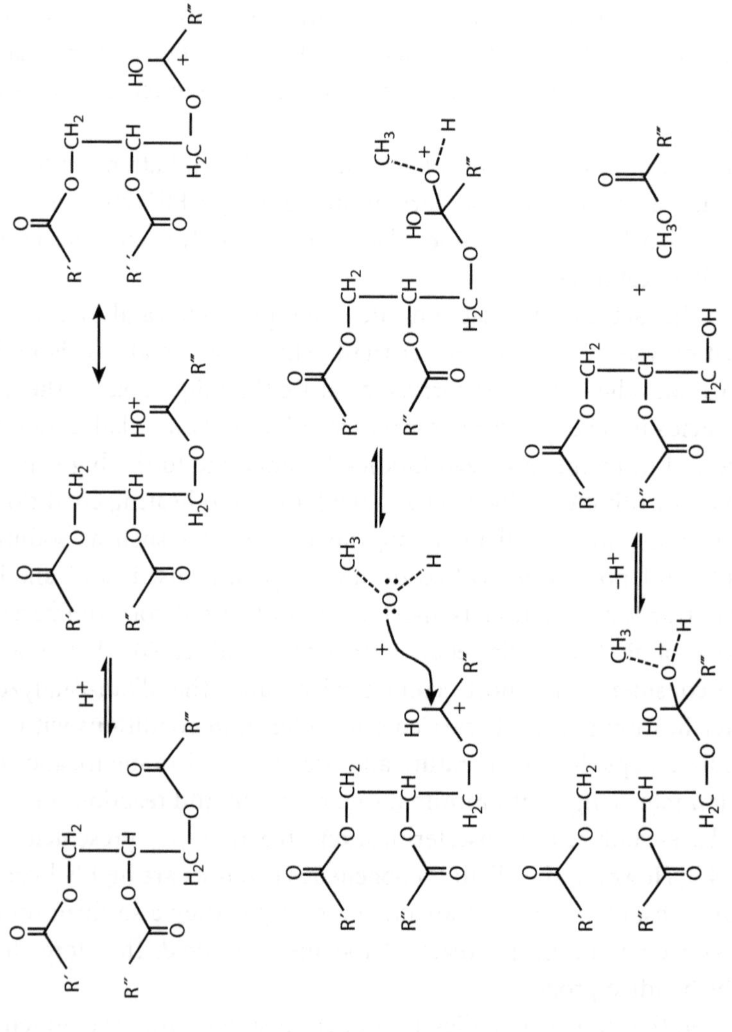

FIGURE 14.2 Acid-catalyzed transesterification reaction mechanism.

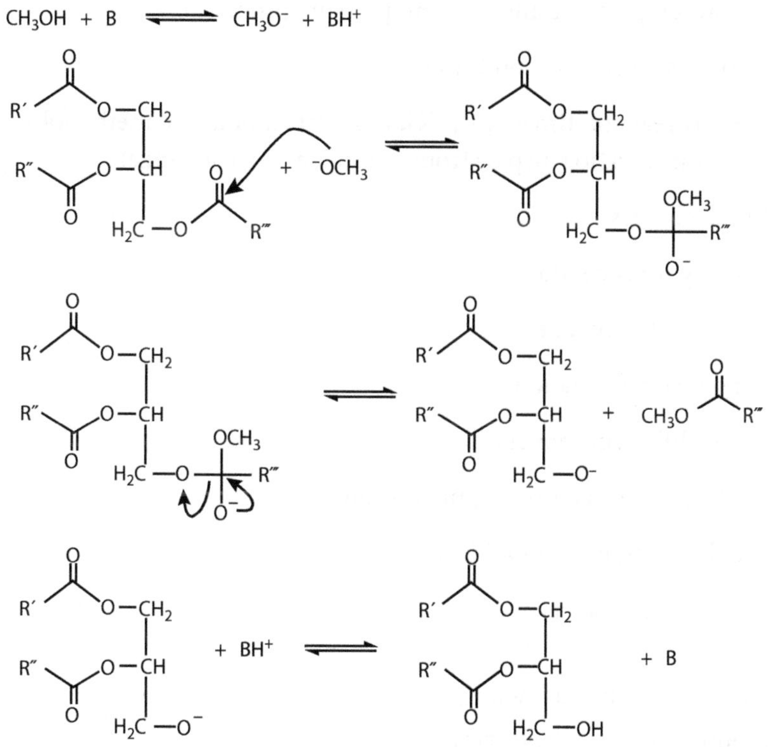

FIGURE 14.3   Mechanism of base-catalyzed transesterification reaction.

pretreatment steps are necessary for the feedstocks to reduce the acid and water concentrations below an optimum threshold limit, i.e., FFA <1 wt% and water <0.5 wt% (Zhang et al. 2003). The catalyst types, advantages, and disadvantages are listed next.

14.7.1 Homogeneous Alkali Catalyst

NaOH (sodium hydroxide) and KOH (potassium hydroxide).

Advantages

Favorable kinetics

Good quality biodiesel

High catalytic activity

High conversion rate of TG

Low cost

Low temperature and ambient pressure process

Modest operation conditions

Short reaction times with TG, moderate amount of methanol-to-oil molar ratio (surplus from the stoichiometric range)

Disadvantages

Anhydrous conditions

Difficult to recycle

Emulsion formation

Low FFA requirement

More wastewater from purification

Side reaction of saponification

Waste disposable

14.7.2 Homogeneous Acid Catalyst

Concentrated H_2SO_4 (sulfuric acid).

Advantages

Avoid soap formation

Catalyzed esterification and transesterification simultaneously

Effective pretreatment for high FFA feedstock

Short reaction times with FFA

Disadvantages

Difficult to recycle

Equipment corrosion

Higher reaction temperature

Long reaction times

More waste from neutralization

Weak catalytic activity

14.7.3 Heterogeneous Base Catalyst

CaO, $CaZrO_3$, $CaTiO_3$, CaO/CeO_2, $Ca_2Fe_2O_5$, $CaMnO_3$, KOH/NaY, KOH/Al_2O_3, Al2O3/KI, alumina/silica supported K_2CO_3, and ETS-10 zeolite.

Advantages

Easy separation

Environmentally benign

Fewer disposal problems

Higher selectivity

Longer catalyst lifetimes

Noncorrosive

Recyclable

Disadvantages

Anhydrous conditions

Diffusion limitations

High cost

High molar ratio of alcohol-to-oil requirement

High reaction temperature and pressure

Low FFA requirement

More wastewater from purification

14.7.4 Heterogeneous Acid Catalyst

TiO_2/SO_2, ZnO/I_2, ZrO_2/SO_2, carbohydrate-derived catalyst, niobic acid, carbon-based solid acid catalyst, vanadyl phosphate, sulfated zirconia, Nafion-NR50, and Amberlyst-15.

Advantages

Catalyze esterification and transesterification simultaneously

Diffusion limitations

Ecofriendly

High cost

Low acid site concentrations

Low microporosity

Recyclable

Disadvantages

Requires high temperatures

High methanol-to-oil molar ratio

14.7.5 Enzyme Catalyst

Rhizomucor mieher lipase and *Candida Antarctica* fraction B lipase.

Advantages

Avoid soap formation

Easier purification

Low methanol-to-oil molar ratio

Low temperature and ambient pressure reaction

No by-product

Nonpolluting

Simple product recovery

Treats FFA and TG

Essentially pure product

Disadvantages

Enzyme inactivation (water)

Expensive denaturation

Long reaction time

14.8 EXTRACTION OF LIPID AND TRANSESTERIFICATION PROCESS

The overall downstream processes for biodiesel production from algae are shown in Figure 14.4 (Halim et al. 2012). The feedstock must have a moisture content of less than 0.05 wt%, a free fatty acid content of less than 0.5 wt%,

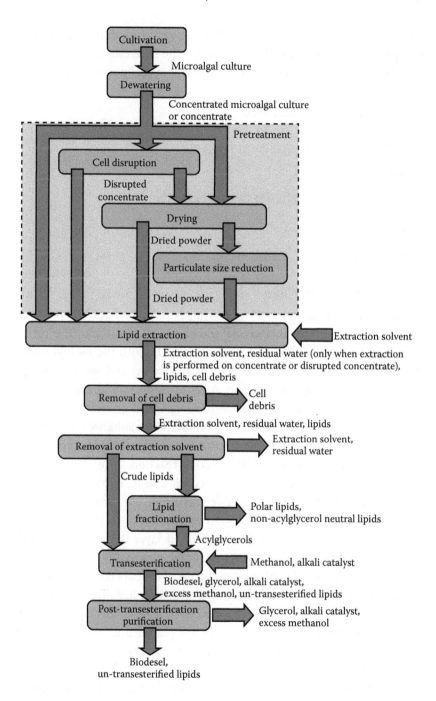

FIGURE 14.4   Process flow diagram for downstream processing steps needed to produce biodiesel from algae biomass.

a phosphorous content of less than 10 ppm, and a combined calcium and magnesium content of less than 5 ppm for the transesterification process. Each feedstock was transesterified using the same production procedure.

14.8.1 Raw Materials Required

Bio-oil (triglyceride)

Methanol or ethanol

Sodium hydroxide (NaOH) as a base catalyst

Hydrochloric acid (0.2N)

Deionized water

14.8.2 Standard Procedure

To the reactor are added: 700 grams feedstock; methanol, 17.6 wt% of feedstock; and sodium methoxide, 2.64 wt% of feedstock. The reactor temperature is set to 65°C and remains at 65°C until the methyl ester is removed from the reactor. The mixer is turned on and set to 1200 rpm for 15 minutes for the first reaction. After 15 minutes in the first reaction, the mixer is turned off and the methyl esters and glycerin settle for 15 minutes. The glycerin is removed, and 4.4 wt% of methanol and 0.66 wt% of sodium methoxide are added for the second reaction. The mixer is set to 600 rpm for 15 minutes for the second reaction. After 15 minutes in the second reaction, the mixer is turned off and 0.2N hydrochloric acid solution (13 wt% of feedstock) is added. The mixer is turned back on and neutralization occurs for 2 minutes. After the 2 minutes, the methyl esters settle for 15 minutes. The hydrochloric acid layer is removed from the reaction vessel and then the methyl esters are removed.

When two batches of methyl ester are made, they are combined together into the same flask before proceeding to the next step. The methyl esters are transferred into a flask that is heated to 70°C with stirring and the use of a vacuum pump. The methyl esters are placed under vacuum for 30 minutes to remove the methanol. After the methanol has been removed, the methyl esters are poured into a separating funnel. Deionized water, 10 wt%, at 70°C is added to the separating funnel, and the mixture is shaken vigorously for 2 minutes. The methyl esters are then settled for 15 minutes at room temperature. After 15 minutes, the water phase is removed. The methyl esters are then transferred to a flask to be dried.

The methyl esters are dried using a hotplate and a vacuum pump. The methyl esters are heated to 110°C under vacuum for 1 hour. After the methyl esters have been dried and cooled to room temperature, diatomaceous earth (5 wt% of methyl ester) is added to the methyl ester. The methyl esters and diatomaceous earth are stirred and chilled at 15°C for 30 minutes. The slurry is removed and filtered through a filter press equipped with a 0.7 μm filter paper. The filtered methyl esters are then filtered again through a 0.7 μm glass fiber filter paper to remove all the diatomaceous earth. This transesterification procedure may be differing due to the presence of free fatty acids in the feedstock (more than 1%).

14.9 BIODIESEL CHARACTERISTICS

14.9.1 Cloud Point (CP)

Low temperature operability of biodiesel fuel is an important aspect from the engine performance standpoint in cold weather conditions. There are several tests that are commonly used to determine the low temperature operability of biodiesel. Cloud point (CP) is one of these tests and is included as a standard in ASTM D6751. The cloud point is the temperature at which crystals first appear in the fuel when cooled. ASTM D6751 requires the producer to report the cloud point of the biodiesel sold, but it does not set a range, as the desired cloud point is determined by the intended use of the fuel.

14.9.2 Cold Filter Plugging Point (CFPP)

Cold filter plugging point (CFPP) refers to the temperature at which the test filter starts to plug due to fuel components that have started to gel or crystallize. The CFPP is a commonly used indicator of low temperature operability of fuels. As with other low temperature properties, the CFPP of biodiesel also depends on the feedstock used for production of methyl esters. ASTM D6751 does not include the CFPP test as a standard. Similar to cloud point, the CFPP of biodiesel also varies with the fatty acid distribution; with a lower fraction of saturated fatty acids resulting in a lower CFPP, and a higher fraction of saturated fatty acids resulting in a higher CFPP. Usually, the CFPP of a fuel is lower than its cloud point.

14.9.3 Cold Soak Filtration

Cold soak filtration is the newest biodiesel requirement set in ASTM D6751. The cold soak filtration test is done to determine if crystals form at low temperatures and do not redissolve when the biodiesel returns to a higher temperature.

14.9.4 Fatty Acid Profile

Biodiesel is defined by ASTM D6751 as a mixture of mono alkyl esters of long-chain fatty acids derived from vegetable oils or animal fats. These mono alkyl esters are the predominant chemical species present in B100 biodiesel. About 90% of the structure of mono alkyl esters in biodiesel is made of long-chain fatty acids. The structure and composition of these long-chain fatty acid components have been associated with trends in cetane number, heat of combustion, cold flow properties, oxidation stability, viscosity, and lubricity. The fatty acid profile (FAP) is a list of fatty acids and their amounts in biodiesel.

14.9.5 Relative Density

Relative density is the density of the component compared to the density of water. Relative density is a measure of weight per unit volume. The relative density of biodiesel is needed to make mass-to-volume conversions, calculate flow and viscosity properties, and is used to judge the homogeneity of biodiesel tanks.

14.9.6 Kinematic Viscosity

Kinematic viscosity in biodiesel was determined using ASTM D445. Biodiesel kinematic viscosities are all lower that those presented by their respective oils or fats. This is an expected finding since biodiesel molecules are single, long-chain fatty esters with higher mobility than the bigger and bulkier triglyceride molecules. The same trends found for kinematic viscosity in the feedstocks are found in the biodiesels.

14.9.7 Sulfated Ash

ASTM D874 measures sulfated ash that may come from abrasive solids, soluble metallic soaps, and unremoved catalysts. The biodiesel is ignited, burned, and then treated with sulfuric acid to determine the percentage of sulfated ash present in the biodiesel. The maximum ASTM limit for sulfated ash is 0.020% mass and the majority of the evaluated biodiesels fell under the maximum limit. Biodiesels do not present high concentrations of calcium, magnesium, phosphorous, or sulfur, which are some common elements that compose sulfated ash, therefore, the source for these high sulfated ash results is unknown.

14.9.8 Carbon Residue

The carbon residue test indicates the extent of deposits that result from the combustion of a fuel. Carbon residue, which is formed by decomposition and subsequent pyrolysis of the fuel components, can clog the fuel injectors. ASTM D6751 includes carbon residue as a standard for biodiesel. The maximum allowable carbon residue for biodiesel is 0.050% by mass. The carbon residue for biodiesel was measured according to ASTM D524, Standard Test Method.

14.9.9 Water and Sediment

Water and sediment testing is done using 100 mL of biodiesel and centrifuging it at 1870 rpm for 11 minutes. If the water and sediment level is below 0.005% volume (vol), the result is reported as <0.005% vol. Water and sediment tests were done as per ASTM D2709, Standard Test Method.

14.9.10 Free and Total Glycerin

Free and total glycerin is a measurement of how much triglyceride remains unconverted into methyl esters. Total glycerin is calculated from the amount of free glycerin, monoglycerides, diglycerides, and triglycerides. Free and total glycerin was run in accordance with ASTM D6584, Standard Test Method.

14.9.11 Flash Point

The flash point is the lowest temperature at which fuel emits enough vapors to ignite. Biodiesel has a high flash point; usually more than 150°C, whereas conventional diesel fuel has a flash point of 55°C–66°C. If methanol, with its flash point of 12°C is present in the biodiesel, the flash point can be lowered considerably. To ensure that the methanol has been adequately stripped from the biodiesel, the Pensky-Martens closed cup flash point test was adopted. The flash points were measured using ASTM D93, Standard Test Method.

14.9.12 Copper Corrosion

Copper corrosion is tested using ASTM D130, Standard Test Method. The copper corrosion test measures corrosion-forming tendencies of fuel when used with copper, brass, or bronze parts. The presence of acids or sulfur can tarnish copper.

14.9.13 Phosphorous, Calcium, and Magnesium

The specifications from ASTM D6751 state that in biodiesel, the phosphorous must be less than 10 ppm, and calcium and magnesium combined must be less than 5 ppm. Phosphorous was determined using ASTM D4951, Standard Test Method for Determination of Additive Elements in Lubricating Oils by Inductively Coupled Plasma Atomic Emission Spectrometry. Calcium and magnesium were determined using EN Standard 14538, Fat and Oil Derivatives—Fatty Acid Methyl Ester (FAME)—Determination of Ca, K, Mg, and Na Content by Optical Emission Spectral Analysis with Inductively Coupled Plasma.

14.9.14 Total Acid Number (TAN)

The total acid number (TAN) determination is an important test to assess the quality of a particular biodiesel. It can indicate the degree of hydrolysis of the methyl ester, a particularly important aspect when considering storage and transportation as large quantities of free fatty acids can cause corrosion in tanks. The TAN determination in the biodiesel samples was performed following ASTM D664, Standard Test Method.

14.10 CONCLUSION

As verified at this point, marine algal biodiesel is technically practicable. It is the only renewable biodiesel that can potentially entirely displace liquid fuels derived from petroleum sources. The economics of producing marine algal biodiesel need to develop substantially to make it aggressive with petrodiesel. Producing low-cost marine algal biodiesel requires first and foremost improvements to algal biology through genetic and metabolic engineering. Use of the marine algae in biorefinery perception and advances will further lower the cost of biodiesel production.

REFERENCES

Adams, C., Peters, J., Rand, M., Schroer, B., and Ziemke, M. 1983. Investigation of soybean oil as a diesel fuel extender: Endurance tests. *J. Am. Oil Chem. Soc.* 60:1574–1579.

Amin, S. 2009. Review on biofuel oil and gas production processes from microalgae. *Energy Convers. Manage.* 50:1834–1840.

Anon. 1982. Filtered used frying fat powers diesel fleet. *J. Am. Oil Chem. Soc.* 59: 780A–781A.

Balat, M. 2007. Production of biodiesel from vegetable oils: A survey. *Energy Sources Part A* 29:895–913.

Balezentiene, L., Streimikiene, D., and Balezentis, T. 2013. Fuzzy decision support methodology for sustainable energy crop selection. *Renew. Sust. Energy. Rev.* 17, no. 1: 83–93.

Bartholomew, D. 1981. Vegetable oil fuel. *J. Am. Oil Chem. Soc.* 58:286–288.

Benemann, J.R. 2008. Opportunities and challenges in algae biofuels production. Paper presented at the Algae World Conference, Singapore, November 17–18.

Canakci, M., and Van Gerpen, J. 1999. Biodiesel production via acid catalysis. *Trans. ASAE* 42:1203–1210.

Chang, C.C., and Wan, S.W. 1947. China's motor fuels from tung oil. *Ind. Eng. Chem.* 39:1543–1548.

Chisti, Y. 2007. Biodiesel from microalgae. *Biotechnol. Adv.* 25:294–306.

Chisti, Y. 2008. Biodiesel from microalgae beats bioethanol. *Trends Biotechnol.* 26:126–131.

Demirbas, A. 2009a. Inexpensive oil and fats feedstocks for production of biodiesel. *Energy Educ. Sci. Technol. Part A* 23:1–13.

Demirbas, A. 2009b. Potential resources of non-edible oils for biodiesel. *Energy Sources Part B: Econ. Plann. Policy* 4:310–324.

Demirbas, A. 2009c. Production of biodiesel from algae oils. *Energy Sources, Part A: Recovery Utilization Environ. Effects* 31:163–168.

Demirbas, A. 2009d. Progress and recent trends in biodiesel fuels. *Energy Convers. Manage.* 50:14–34.

Devanesan, M.G., Viruthagiri, T., and Sugumar, N. 2007. Transesterification of *jatropha* oil using immobilized monas fluorescens. *Afr. J. Biotechnol.* 6:2497–2501.

Dias, J.M., Alvim-Ferraz, M.C.M., and Almeida, M.F. 2008. Comparison of the performance of different homogeneous alkali catalysts during transesterification of waste and virgin oils and evaluation of biodiesel quality. *Fuel* 87:3572–3578.

Dizge, N., Aydiner, C., Imer, D.Y., Bayramoglu, M., Tanriseven, A., and Keskinler, B. 2009. Biodiesel production from sunflower, soybean, and waste cooking oils by transesterification using lipase immobilized onto a novel microporous polymer. *Bioresour. Technol.* 100:1983–1991.

Engler, C., Johnson, L., Lepori, W., and Yarbrough, C. 1983. Effects of processing and chemical characteristics of plant oils on performance of an indirect-injection diesel engine. *J. Am. Oil Chem. Soc.* 60:1592–1596.

Fukuda, H., Kondo, A., and Noda, H. 2001. Review: Biodiesel fuel production by transesterification of oils. *J. Biosci. Bioeng.* 92:405–416.

Halim, R., Danquah, M.K., and Webley, P.A. 2012. Extraction of oil from microalgae for biodiesel production: A review. *Biotechnol. Adv.* 30:709–732.

Huang, G., Chen, F., Wei, D., Zhang, X., and Chen, G. 2010. Biodiesel production by microalgal biotechnology. *Appl. Energy* 87:38–46.

Kaieda, M., Samukawa, T., Matsumoto, T., Ban, K., Kondo, A., Shimada, Y., and Fukuda, H. 1999. Biodiesel fuel production from plant oil catalyzed by *Rhizopus oryzae* lipase in a water-containing system without an organic solvent. *J. Biosci. Bioeng.* 88:627–631.

Khan, S.A., Mir, Z., Hussain, R., Prasad, S., and Banerjee, U.C. 2009. Prospects of biodiesel production from microalgae in India. *Renew. Sust. Energy. Rev.* 13:2361–2372.

Kildiran, G., Ozgul-Yucel, S., and Turkay, S. 1996. In-situ alcoholysis of soybean oil. *J. Am. Oil Chem. Soc.* 73:225–228.

Li, Y., Horsman, M., Wu, N., Lan, C.Q., and Dubois-Calero, N. 2008. Biofuels from microalgae. *Biotechnol. Progr.* 24:815–820.

Ma, F., and Hanna, M.A. 1999. Biodiesel production: A review. *Bioresour. Technol.* 70:1–15.

Mateos, A. 2007. Biodiesel: Production process and main feedstocks. *Energia Spring* 1, no. 1.

Meher, L.C., Vidya Sagar, D., and Naik, S.N. 2006. Technical aspects of biodiesel production by transesterification—A review. *Renew. Sust. Energy Rev.* 10:248–268.

Miao, X., and Wu, Q. 2006. Biodiesel production from heterotrophic microalgal oil. *Bioresour. Technol.* 97:841–846.

Nelson, L., Foglia, T., and Marmer, W. 1996. Lipase-catalyzed production of biodiesel. *J. Am. Oil Chem. Soc.* 73:1191–1195.

Ng, J.H., Ng, H., and Gan, S. 2010. Recent trends in policies, socio economy and future directions of the biodiesel industry. *Clean Technol. Environ. Policy* 12:213–238.

Patil, V., Tran, K.O., and Giselrod, H.R. 2008. Towards sustainable production of biofuels from microalgae. *Int. J. Molecular Sci.* 9:1188–1195.

Peterson, C., Auld, D., and Korus, R. 1983. Winter rape oil fuel for diesel engines: Recovery and utilization. *J. Am. Oil Chem. Soc.* 60:1579–1587.

Pienkos, P.T., and Darzins, A. 2009. The promise and challenges of microalgal-derived biofuels. *Biofuels, Bioproducts Biorefining* 3:431–440.

Pimentel, D. 2008. *Biofuels, Solar and Wind as Renewable Energy Systems: Benefits and Risks.* Dordrecht: Springer-Verlag.

Pimentel, D., Marklein, A., Toth, M.A., Karpoff, M., Paul, G.S., and McCormack, R. 2008. Biofuel impacts on world food supply: Use of fossil fuel, land and water resources. *Energies* 1:41–78.

Pioch, D., Lozano, P., Rasoanantoandro, M.C., Graille, J., Geneste, P., and Guida, A. 1993. Biofuels from catalytic cracking of tropical vegetable oils. *Oleagineux* 48:289–291.

Pryde, E.H. 1984. Vegetable oils as fuel alternatives—Symposium overview. *J. Am. Oil Chem. Soc.* 61:1609–1610.

Ramachandran, K., Sivakumar, P., Suganya, T., and Renganathan, S. 2011. Production of biodiesel from mixed waste vegetable oil using an aluminium hydrogen sulphate as a heterogeneous acid catalyst. *Bioresour. Technol.* 102:7289–7293.

Ramachandran, K., Suganya, T., Nagendra Gandhi, N., and Renganathan, S. 2013. Recent developments for biodiesel production by ultrasonic assist transesterification using different heterogeneous catalyst: A review. *Renew. Sust. Energy Rev.* 22:410–418.

Reijnders, L. 2009. Microalgal and terrestrial transport biofuels to displace fossil fuels. *Energies* 2:48–56.

Reijnders, L., and Huijbregts, M.A.J. 2009. *Biofuels for Road Transport: A Seed to Wheel Perspective.* London: Springer-Verlag.

Rengel, A. 2008. Promising technologies for biodiesel production from algae growth systems. Paper presented at the Eighth European Symposium of the International Farming Systems Association (IFSA), Clermont-Ferrand, France, July 6–10.

Schenk, P.M., Thomas-Hall, S.R., Stephens, E., Marx, U.C., Marx, J.H., and Posten, C. 2008. Second generation biofuels: High efficiency microalgae for biodiesel production. *Bioenergy Res.* 1:20–43.

Schwab, A.W., Dykstra, G.J., Selke, E., Sorenson, S.C., and Pryde, E.H. 1988. Diesel fuel from thermal decomposition of soybean oil. *J. Am. Oil Chem. Soc.* 65:1781–1786.

Sharif Hossain, A.B.M., and Salleh, A. 2008. Biodiesel fuel production from algae as renewable energy. *Am. J. Biochem. Biotechnol.* 4:250–254.

Sharma, Y.C., and Singh, B. 2009. Development of biodiesel: Current scenario. *Renew. Sust. Energy. Rev.* 13:1646–1651.

Shimada, Y., Watanabe, Y., Samukawa, T., Sugihara, A., Noda, H., and Fukuda, H. 1999. Conversion of vegetable oil to biodiesel using immobilized *Candida antarctica* lipase. *J. Am. Oil Chem. Soc.* 76:789–793.

Siler-Marinkovic, S., and Tomasevic, A. 1998. Transesterification of sunflower oil in-situ. *Fuel* 77:1389–1391.

Sohpal, V.K., Singh, A., and Dey, A. 2011. Fuzzy modelling to evaluate the effect of temperature on batch transesterification of *Jatropha curcas* for biodiesel production. *Bull. Chem. Reaction Eng. Catalysis* 6, no. 1:31–38.

Sonntag, N.O.V. 1979. Reactions of fats and fatty acids. In *Bailey's Industrial Oil and Fat Products*, vol. 1, 4th ed., edited by D. Swern, 138. Somerset, NJ: John Wiley & Sons.

Srivastava, A., and Prasad, R. 2000. Triglycerides-based diesel fuels. *Renew. Sust. Energ. Rev.* 4:111–133.

Stavarache, C., Vinatoru, M., and Maeda, Y. 2006. Ultrasonic versus silent methylation of vegetable oils. *Ultrason. Sonochem.* 13:401–407.

Suganya, T., and Renganathan, S. 2013. Studies on production of biodiesel from marine macroalgae using homogeneous catalysts. Ph.D. thesis, Anna University.

Talebian-Kiakalaieh, A., Amin, N.A.S., and Mazaheri, H. 2013. A review on novel processes of biodiesel production from waste cooking oil. *Appl. Energ.* 104:683–710.

Taylor, D., Nixon, S., Granger, S., and Buckley, B. 1995. Nutrient limitation and the eutrophication of coastal lagoons. *Mar. Ecol. Prog. Ser.* 127:235–244.

U.S. Department of Energy. 2003. Energy Efficiency and Renewable Energy, August 2003. http://www1.eere.energy.gov/vehiclesandfuels/pdfs/basics/jtb _biodiesel.pdf.

Viaroli, P., Azzoni, R., Bartoli, M., Giordani, G., and Taje, L. 2001. Evolution of the trophic conditions and dystrophic outbreaks in the Sacca di Goro lagoon (Northern Adriatic Sea). In *Mediterranean Ecosystems: Structure and Processes*, edited by F.M. Faranda, L. Guglielmo, and G. Spezie, 467–475. Milano: Springer Verlag.

Vijayaraghavan, K., and Hemanathan, K. 2009. Biodiesel production from freshwater algae. *Energy Fuels* 23:5448–5453.

Wen, Z., and Johnson, M.B. 2009. Microalgae as a feedstock for biofuel production. Publication 442–886. Blacksburg, VA: Virginia Polytechnic Institute and State University.

Xu, H., Miao, X., and Wu, Q. 2006. High quality biodiesel production from a microalga Chlorella protothecoides by heterotrophic growth in fermenters. *J. Biotechnology* 126:499–507.

Zhang, Y., Dube, M.A., McLean, D.D., and Kates, M. 2003. Biodiesel production from waste cooking oil: 1. Process design and technological assessment. *Bioresour. Technol.* 89:1–16.

Zheng, S., Kates, M., Dube, M.A., and McLean, D.D. 2006. Acid-catalyzed production of biodiesel from waste frying oil. *Biomass Bioenerg.* 30:267–272.

Ziejewski, M., Kaufman, K.R., Schwab, A.W., and Pryde, E.H. 1984. Diesel engine evaluation of a nonionic sunflower oil-aqueous ethanol microemulsion. *J. Am. Chem. Soc.* 61:1620–1626.

CHAPTER 15

Pretreatment Techniques for Enhancement of Biogas Production from Solid Wastes

Velmurugan Sivasubramanian,
Balakrishnan Vinothraj, Veerapadaran
Velumani, and Krishnan Manigandan

CONTENTS

Abstract

The anaerobic digestion process is used to produce biogas, which is a clean and renewable form of energy. Biogas is a substitute for conventional energy sources (coal, oil, etc.) that are causing more environmental problems and at the same time depleting at a faster rate. Despite its several advantages, this anaerobic digestion technology could not be fully harnessed or tapped, as many constraints are associated with it. A large hydraulic retention time and low biogas production in winter are the constraints of the process. Therefore, efforts are needed to remove its limitations so as to promote this technology in the rural areas. This chapter presents a review of different pretreatment techniques that are used to enhance biogas production in the anaerobic digestion process. Pretreatment techniques are used widely to increase the degradability and hydrolysis rate of the substrate being fed into digesters. The pretreatment facilitates the substrates for full biodegradation during anaerobic digestion in spite of their complex physical and chemical structure. Various pretreatment techniques such as mechanical, thermal, chemical, and biological methods are used to change the structure of the biomass. These techniques are used to reduce the hydrolysis time by improving the process rate, which in turn increases the anaerobic digestion performance. In this review, the importance of pretreatment and some important results from literature are covered.

15.1 INTRODUCTION

As the world's population continues to grow above the 7 billion mark, energy demand is becoming an ever more critical challenge for the world's energy leaders. Governments are searching for sustainable remedies that give the most competitive energy supplies from secure sources, and trying to balance the short-term and long-term energy needs of the environment (Frei, 2012). Conversion of biomass (organic waste) into biogas by anaerobic digestion is one solution. Anaerobic digestion of organic waste not only gives energy, but it also reduces the quantity of sludge by 40% to 50%.

Anaerobic digestion is carried out in four stages: hydrolysis, acidogenesis, acetogenesis, and methanogenesis. Among these, hydrolysis is the initial and very important stage that breaks the complex structure into simpler molecules. Improving hydrolysis by pretreatments can increase biogas yield by increasing the performance of anaerobic digestion. Various pretreatments can be used to improve biogas yield such as mechanical, thermal, chemical, biological, and combined pretreatments (Bougrier et al., 2006; Muller et al., 1998; Weemaes et al., 1998). These treatments help to solubilize and/or reduce the size of organic compounds in order to make them more easily biodegradable (Lehne et al., 1998; Weemaes et al., 2000). This chapter presents a comprehensive review of various pretreatment methods used for improving biogas production through the anaerobic digestion process.

15.2 MECHANICAL PRETREATMENT

The mechanical treatment of biomass is an important method used for intensifying the biomethanation process. The most commonly used mechanical pretreatment methods include milling/grinding, high-pressure homogenizers, ultrasonication, and mechanical jets (Mudhoo, 2012).

15.2.1 Milling/Grinding

Particle size of the substrate plays a major role in the biomethanation process. A smaller particle size will increase substrate utilization because a smaller particle size can provide increased microbial activity (Nayono, 2009; Palmowski and Muller, 2000). By breaking the large structures into shorter chains, the speed and the efficiency of the hydrolysis increases, which makes biodegradability easy. Milling/grinding is used to reduce the particle size of the substrate. For materials like hardwoods, instead of milling/grinding, chipping can be applied (Kumar et al., 2009). The size of the substrate will be 0.2 to 2 mm after milling/grinding and 10 to 30 mm after chipping (Sun and Cheng, 2002). Sharma et al. (1988) evaluated agricultural and forest residues with five particle sizes (0.088, 0.40, 1.0, 6.0, and 30.0 mm) and found that the maximum quantity of biogas was produced from 0.088 and 0.40 mm particle size. Nah et al. (2000) investigated the mechanical pretreatment of waste activated sludge by jetting and colliding with a collision-plate at 30 bar, and found solid retention time was decreased from 13 days to 6 days without affecting process performance. In most cases, after the pretreatment the total hydrolysis yield increased by 5%–25% and the digestion time reduced by 23%–59% (Fang et al., 2011).

15.2.2 Ultrasonication

Ultrasonication is a useful and effective mechanical pretreatment method to enhance the biodegradability of a material (Khanal et al., 2007; Pilli et al., 2011). Application of high-intensity ultrasound modifies the structure of the material, which makes biodegradability easier compared to an untreated sample (Dewil et al., 2006). Available literature reported that the application of ultrasonic pretreatment in anaerobic digestion of waste activated sludge increases biogas production and reduces volatile solids (Kim et al., 2003; Mudhoo, 2012). Wang et al. (1999) investigated the effect of ultrasonication on waste activated sludge and reported that methane generation was increased by 64% due to ultrasonic pretreatment. Braguglia et al. (2011) tested the waste activated sludge in a laboratory-scale digester with ultrasonic pretreatment and found that biogas production of sonicated sludge was up to 30% higher than untreated sludge. Some of the results available in the literature for ultrasonication are given in Table 15.1.

15.3 THERMAL PRETREATMENT

Thermal pretreatment can be used to treat the biomass before it is taken into the anaerobic digestion process. This promotes splitting of complex organic wastes into simpler and more biodegradable constituents (Fang et al., 2011). During the thermal pretreatment, the organic and inorganic compounds in the feedstock are partially solubilized before hydrolysis, which reduces digester volume and increases biogas production (Mudhoo, 2012; Tanaka et al., 1997). From literature, it is found that the optimum temperature is 160°C to 200°C and the treatment time is 30 to 60 minutes (Table 15.2). Temperatures above 250°C should be avoided during pretreatment, because unwanted pyrolysis reactions start to take place (Brownell et al., 1986). Some of the results available in the literature for thermal pretreatment are given in Table 15.2. Thermal treatment can be done by conventional method, steam explosion or by autoclave/microwave method.

15.3.1 Steam Explosion

In steam explosion, biomass is out into a closed vessel or container, then steamed with high temperature, and pressure is passed directly in it

TABLE 15.1 Details of Experiments and Results with Ultrasonication

Substrate	Digester Condition	Treatment Conditions	Results	Reference
Waste activated sludge (WAS)	Batch reactor (37°C)	42 kHz, 10–120 minutes	Increase in biogas production from 3657 l/m³ WAS to 4413 l/m³ WAS and increase in methane production from 2507 l/m³ WAS to 3007 l/m³ WAS.	Kim et al., 2003
Sewage sludge	Semicontinuous reactor (37°C)	31 kHz, 64 seconds	Improvement in volatile solid removal from 45.8% to 50.3%.	Tiehm et al., 1997
Biological sludge	Batch reactor (36°C)	20 kHz, (9690 kJ kg⁻¹ TS)	Increase of methane production by 44%. The soluble chemical oxygen demand (sCOD), dissolved organic carbon (DOC), total nitrogen (TN), and total phosphorus (TP) in the sludge increased by 340%, 860%, 716%, and 207.5%, respectively.	Erden and Filibeli, 2009
Sewage sludge	Batch reactor (35°C)	20 kHz, 60 seconds	Increase in the biogas production from 20% to 24%.	Bien et al., 2004
Aquaculture effluent	Psychrophilic, semicontinuous reactor	50 kHz, 120 W, 50 minutes	Chemical oxygen demand (COD) was increased by 10% and methane production increased from 0.06 m³/kg of COD removed to 0.08 m³/kg of COD removed.	Dermott et al., 2001
Activated sludge	Continuous reactor	20 kWhm⁻³	Increase of volatile solid removal and biogas production by 25% and 37%, respectively.	Perez-Elvira et al., 2009

(Continued)

TABLE 15.1 (CONTINUED) Details of Experiments and Results with Ultrasonication

Substrate	Digester Condition	Treatment Conditions	Results	Reference
Sewage sludge	Batch reactor, with both mesophilic and thermophilic	20 kHz	In mesophilic condition, increase in biogas production from 0.76 l/g VS to 0.883 l/g VS and COD removal efficiency from 43.89% to 51.93%. In thermophilic condition, increase in biogas production from 0.821 l/g VS to 0.898 l/g VS and COD removal efficiency from 49.21% to 52.53%.	Benabdallah et al., 2007
Palm oil mill effluent	Batch reactor (32°C–37°C)	20 kHz	Increase in methane production by 16%.	Saifuddin and Fazlili, 2008
Dairy cattle slurry	Batch reactor (35°C)	30 kHz	Increase in methane production from 210 N m³/ton VS to 250 N m³/ton VS.	Luste and Luostarinen, 2011
Activated sludge	Batch reactor	20 kHz, 200 W, 1–30 minutes	Increase in soluble chemical oxygen demand by 35.5% and reduction in the volatile suspended solids concentration by 25%.	Yagci and Akpinar, 2011
Activated sludge	Mesophilic batch reactor (18°C–22°C)	25 kHz, 0–250 minutes	Increase in biogas production up to 40%.	Appels et al., 2008

TABLE 15.2 Details of Experiments and Results Obtained with Thermal Pretreatments

Substrate	Digester Type	Treatment Conditions	Results	Reference
Meat processing wastewater	Batch reactor	140°C, 30 minutes	Increase in biogas production by 23.6%.	Erden, 2013
Waste activated sludge	Semicontinuous reactor	170°C, 30 minutes	Increase in methane production from 145 ml/g VS to 256 ml/g VS.	Bougrier et al., 2006
Waste activated sludge	Semicontinuous reactor	190°C	Increase in methane production by 25%.	Bougrier et al., 2007
Waste activated sludge	UASB reactor	75°C, 7 hours	Increase in biogas production by 50%.	Borges and Chernicharo, 2009
Waste activated sludge	Semicontinuous mesophilic reactor	175°C (microwave irradiation)	Increase in biogas production from 636 l/kg VS to 839.6 l/kg VS.	Toreci et al., 2009
Sewage sludge	Mesophilic batch and continuous flow reactors	120°C, 150°C, and 175°C (microwave irradiation)	The ratio of soluble to total chemical oxygen demand increased from 9% to 24%, 28%, and 35% at 120°C, 150°C, and 175°C, respectively.	Eskicioglu et al., 2009
Waste activated sludge	Continuous reactor	60°C–70°C, 900 W (microwave irradiation)	Increase in methane production by 35%.	Kuglarz et al., 2013
Sewage sludge	Batch reactor (35°C)	170°C–180°C	Increase in methane production by 25%.	Bien et al., 2004

(Continued)

TABLE 15.2 (CONTINUED) Details of Experiments and Results Obtained with Thermal Pretreatments

Substrate	Digester Type	Treatment Conditions	Results	Reference
Primary sludge	Batch reactor (35°C)	35°C, 65°C, and 90°C, (microwave irradiation)	Biogas production increased up to 37%, and the time required for digestion of pretreated sludge is 27% less than untreated sludge.	Zheng et al., 2009
Primary and secondary sludge	Batch and semicontinuous reactors	70°C with preheating time 9, 24, 48, and 72 hours	Increase in biogas production up to 30% both in batch and semicontinuous experiments.	Ferrer et al., 2008
Primary sludge	Continuous reactor (35°C)	50°C–70°C, 2 days	Volatile substance destruction increases from 44% to 54% and methane production increased by 25%.	Ge et al., 2010
Activated sludge	Batch reactor	170°C and 190°C (autoclave)	Ratio of VSS/TSS decreased from 78% to 73%. Increased methane production from 221 ml/g COD added to 333 ml/g COD added (at 170°C) and 328 ml/g COD added (190°C).	Bougrier et al., 2006

(Hendriks and Zeeman, 2009). Hemicellulose degradation and lignin transformation are caused by this high temperature, which results in the increase of hydrolysis potential of the process. During steam explosion, the pretreatment time depends on the moisture content of the biomass. When the moisture content is high, the optimum steam pretreatment will be long (Brownell et al., 1986). For better solubilization and hydrolysis, either the temperature should be high with low pretreatment time or the pretreatment time should be high with lower temperature (Duff and Murray, 1996). Compared to conventional treatments, steam explosion provides better result in treating certain wastes such as paper (Fang et al., 2011).

15.3.2 Liquid Hot Water (LHW)

In pretreatment using liquid hot water (LHW), steam is replaced by hot water. The hot water easily solubilizes the hemicellulose material and prevents inhibitor formations. The pH level should be maintained between 4 and 7, which helps to minimize monosaccharides formation and also acts as a catalyst for the hydrolysis of the cellulosic material in the pretreatment process (Mosier et al., 2005; Weil et al., 1997). The difference between LHW and steam pretreatment is the amount and concentration of solubilized products. Compared to steam pretreatment, the amount of solubilized products in the liquid hot water treatment is higher and has lower concentration products (Bobleter, 1994). This is because of the usage of a high amount of water in liquid hot water treatment.

15.3.3 Autoclave/Microwave Method

Instead of steam explosion and liquid hot water pretreatment techniques, the biomass material can be thermally pretreated by microwave irradiation and autoclave methods. These methods ensure better performance through effective anaerobic digestibility and less holding time when compared to steam explosion and hot water treatments (Fang et al., 2011; Qiao et al., 2007). Carrere et al. (2008) investigated six different waste activated sludge samples using autoclave type pretreatment technique with a temperature range of 60°C to 210°C, and reported that the biodegradability of the substrate and the COD solubilization increases with an increase in temperature due to pretreatment.

15.4 CHEMICAL PRETREATMENT

Chemical pretreatments are used to solubilize the main components of lignocellulosic biomass such as cellulose, hemicellulose, and lignin. Alkaline pretreatment, acid pretreatment, oxidation pretreatment, and ozonolysis pretreatment are the techniques available for chemical pretreatment. Some of the results available in the literature for chemical pretreatment are given in Table 15.3.

15.4.1 Alkaline Pretreatment

Alkali pretreatment is the process in which alkaline solutions are used to remove lignin and a part of hemicellulose. This pretreatment results in increasing the accessibility of enzyme to the cellulose (Taherzadeh and Karimi, 2008). During alkaline pretreatment, solvation, saponification, peeling, and hydrolytic reactions take place (Fang et al., 2011). These reactions result in a swollen state for the biomass, which enhances the accessibility of bacteria and enzymes (Hendriks and Zeeman, 2009). Various alkali agents like KOH, NaOH, $Ca(OH)_2$, and $Mg(OH)_2$ are used in alkaline pretreatments (Mudhoo, 2012). Alkaline pretreatment causes swelling of organic/biomass particles, making them more susceptible to enzymatic attack by improving the biodegradability (Baccay and Hashimoto, 1984). Addition of 2% of NaOH with paper waste increases the methane yield from 235 Nml/g VS to 403 Nml/g VS (Teghammar et al., 2010).

15.4.2 Acid Pretreatment

In acid pretreatment, dilute or strong acids such as acetic acid and nitric acid are used as agents to treat the biomass chemically. The purpose of acid treatment is to create better accessibility to cellulose for bacteria/enzymes by solubilizing the hemicellulose and removing the lignin (Hendriks and Zeeman, 2009). During strong acid pretreatment, the solubilization of hemicellulose and precipitation of solubilized lignin are more pronounced compared to dilute acid pretreatment.

15.4.3 Oxidation Pretreatment

Oxidation pretreatment involves the addition of oxidizing compounds such as hydrogen peroxide or peracetic peroxide with the biomass. The purpose of oxidation treatment is to increase cellulose accessibility and to remove the hemicelluloses and lignin (Hendriks and Zeeman, 2009). Care should be taken in the addition of oxidation compounds, because this can also form aromatic compounds, which may inhibit the digestion process.

TABLE 15.3 Details of Experiments and Results Obtained with Chemical Pretreatments

Substrate	Digester Condition	Treatment Conditions	Results	Reference
Waste activated sludge	Mesophilic, single stage batch reactor	Peracetic acid	Increase in biogas production by 21%.	Appels et al., 2011
Pig slurry (codigestion of 10%, 20%, 40%, and 60% of acidified slurry with raw pig slurry)	CSTR (50°C)	H_2SO_4 (7–8 kg per ton of slurry)	Inclusion of acidified slurry above 30% with raw slurry causes reduction in methane yield and below 10% acidified slurry codigestion increases methane yield by 20%.	Moset et al., 2012
Sunflower oil cake	Mesophilic reactor	H_2SO_4 (1% by weight)	Increase in methane production from 195 mL CH_4/g VS to 302 mL CH_4/g VS.	Monlau et al., 2013
Padauk Angsana leave	Batch reactor	NaOH (2% by weight)	NaOH-treated substrates gave higher biogas production than untreated substrate.	Juntarasiri et al., 2011
Plant residues and cattle dung	—	NaOH (1% for 7 days)	The digestion efficiency of the NaOH-treated plant residues is almost 31%–42% higher than untreated cattle dung.	Dar and Tandon, 1987
Pulp and paper sludge	Batch reactor	NaOH (8 g per 100 g of TS)	0.32 m^3 CH_4/kg $VS_{removal}$ was produced, which is 183.5% of the untreated sludge. Soluble chemical oxygen demand increased up to 83%.	Lin et al., 2009

(Continued)

TABLE 15.3 (CONTINUED) Details of Experiments and Results Obtained with Chemical Pretreatments

Substrate	Digester Condition	Treatment Conditions	Results	Reference
Oil palm empty fruit bunches	Batch reactor	NaOH (8% for 60 min)	Methane production increased by 85% after NaOH pretreatment.	Nieves et al., 2011
Water hyacinth	CSTR (25°C–36°C)	KOH	Methane composition of biogas increased from 60% to 71%.	Ofoefule et al., 2009
Activated sludge	Batch reactor	Ozone (0.10 and 0.16 g of O_3 per g of TS)	Ratio of VSS/TSS decreased from 78% to 66%. Increased methane production from 221 ml/g COD added to 246 ml/g COD added (0.10 g of O_3), and 272 ml/g COD added (0.16 g of O_3).	Bougrier et al., 2006
Wastewater sludge	Batch reactor	Ozone (0.1 g of O_3 per g of SS)	Increased biodegradation from 36% to 77% after ozone pretreatment.	Yeom et al., 2002
Biological sludge	Batch reactor (35°C–37°C)	Ozone (0.15 g of O_3 per g of TS)	Soluble COD increased from 4% to 37% and the biogas production increased by 2.4 times than untreated sludge.	Bougrier et al., 2007
Textile wastewater	Continuous reactor	Ozone	Increase of COD removal up to 65%.	Perkowski et al., 1996

15.4.4 Ozonolysis Pretreatment

Ozonolysis pretreatment uses ozone, which can effectively remove lignin and part of hemicellulose (Taherzadeh and Karimi, 2008). Ozone pretreatment is normally carried out at room temperature, and the main parameters involved in the pretreatment are moisture content of the sample, particle size, and ozone concentration in the gas flow. Among these, moisture content has the most significant effect on the solubilization. This method is the best of the chemical pretreatments, because it does not leave acidic, basic, or toxic residues (Neely, 1984).

15.5 BIOLOGICAL PRETREATMENT

Biological pretreatment is carried out by using biological agents (microorganisms) with the biomass materials (Carrere et al., 2010). The microorganisms usually degrade most of the lignin and hemicellulose, but a very small part of cellulose. This is because cellulose is more resistant to biological attack (Taherzadeh and Karimi, 2008). Various microorganisms such as brown-rot fungi, white-rot fungi, and soft-rot fungi are commonly used for the degradation of lignin and hemicellulose in the substrate. Brown-rot fungi are mainly used to attack cellulose and the white-rot and soft-rot fungi are used to attack both cellulose and lignin (Kumar et al., 2009). Some of the results available in the literature for biological pretreatment are given in Table 15.4.

15.6 COMBINED PRETREATMENT

In some processes, the combined treatment of any two of the aforementioned pretreatments is used to increase the biogas yield. For example, thermochemical (combination of thermal and chemical pretreatments), combined ultrasonication and chemical, combined ultrasonication and thermal, and so forth may be used for better overall pretreatment effect. Table 15.5 shows the results of some combined pretreatment technique applications.

15.7 CONCLUSION

Mechanical, thermal, chemical, and biological techniques have been developed for the pretreatment of biomass in order to enhance the biogas yield in anaerobic digestion processes. Many researchers have investigated those pretreatments with different feedstock and the results showed significant improvement in the biogas production. Combining treatments

TABLE 15.4 Details of Experiments and Results Obtained with Biological Pretreatments

Substrate	Digester Condition	Treatment Conditions	Results	Reference
Corn straw	UASB mesophilic anaerobic digester	Yeast and cellulolytic bacteria	Biogas production increased by 33.07% and methane yield increased by 75.57%. Digestion time reduced by 34.6%.	Zhong et al., 2011
Pulp and paper sludge	Batch reactor	Mushroom compost extracts	Methane yield under pretreatment condition was 134.2% of the control.	Yunqin et al., 2010
Olive mill wastewater	Continuous plug flow reactor	White-rot fungi	COD removal was increased from 34% to 65%.	Dhouib et al., 2006
Rice hull	Only enzymatic hydrolysis (not biogas generation)	*Pleurotus ostreatus*	71% of hemicellulose and 27% of cellulose were degraded after enzymatic pretreatment.	Yu et al., 2009
Orange processing waste	1500 liter KVIC plant	*Sporotrichum, Aspergillus, Fusarium,* and *Penicillium*	Allowed to use organic loading rate of 8%–10% of dry weight compared with untreated of 4%. Increase in biogas production was achieved.	Srilatha et al., 1995
Wheat straw	Batch reactor (37°C)	White-rot fungi	Biogas yield from pretreated straw was doubled than the untreated straw.	Muller and Trosch, 1986

TABLE 15.5 Details of Experiments and Results Obtained with Combined Pretreatments

Substrate	Digester Condition	Treatment Conditions	Results	Reference
Waste activated sludge	Semicontinuous reactor (37°C)	Thermochemical Chemical: NaOH Temperature: 160°C by microwave irradiation	43.5% and 55% improvement in biogas and methane production, respectively.	Dogan and Sanin, 2009
Dewatered pig manure	Batch reactor	Thermochemical Chemical: Ca(OH)$_2$ Temperature: up to 150°C	Compared to untreated substrate, the cumulative biogas production increased 139%, 97%, and 86% after 7, 19, and 29 days, respectively.	Rafique et al., 2010
Waste activated sludge	Thermophilic batch reactor	Thermochemical Chemical: NaOH Temperature: 110°C–210°C	Increase in methane production by 27%.	Chi et al, 2011
Spent microbial biomass	Batch reactor	Thermochemical Chemical: NaOH Temperature: 140°C by autoclave	75%–80% COD solubilization was achieved when heating at 140°C with 5 g of NaOH per liter.	Penaud et al., 1999

(Continued)

TABLE 15.5 (CONTINUED) Details of Experiments and Results Obtained with Combined Pretreatments

Substrate	Digester Condition	Treatment Conditions	Results	Reference
Waste activated sludge	Batch reactor	Combined ultrasonic and thermal Ultrasonication: 1000, 5000, and 10,000 kJ/kg TSS Temperature: 50°C, 70°C, and 90°C	Reduced volatile suspended solids by 29%–38% and increase in methane production by 30%.	Dhar et al., 2012
Rice stalk	One liter batch reactor	Combined chemical and ultrasound Chemical: 2% NaOH Ultrasonication: 30 kHz for 60 min	Increase in biogas production by 67%–76% than untreated stalk. Almost 41% of lignin content in the stalk was degraded after combined pretreatment.	Wang et al., 2012
Waste activated sludge	CSTR	Combined chemical and ultrasound Chemical: 10 mg of NaOH/g of TS Ultrasonication: 3.8 kJ g⁻¹ TS	Highest volatile solids removal was achieved with combined chemical-ultrasonic pretreatment compared to individual chemical, ultrasonic, and untreated sludge.	Seng et al., 2010
Waste activated sludge	Semicontinuous reactor (30°C)	Combined chemical and ultrasound Chemical: 100 g of NaOH/kg of dry solid Ultrasonication: 7500 kJ/kg of dry solid	Increase in degradation efficiency of organic matter from 38.0% to 50.7%, which is higher than individual pretreatments.	Jin et al., 2009

was also analyzed by various researchers. The pretreatment techniques and results presented are from some of the most important literature available in this area of research.

REFERENCES

Appels, L., Dewil, R., Baeyens, J., Degreve, J. 2008. Ultrasonically enhanced anaerobic digestion of waste activated sludge. *Int. J. Sust. Eng.* 1:94–104.

Appels, L., Assche, A.V., Willems, K., Degreve, J., Impe, J.V., Dewil, R. 2011. Peracetic acid oxidation as an alternative pre-treatment for the anaerobic digestion of waste activated sludge. *Bioresour. Technol.* 102:4124–4130.

Baccay, R.A., Hashimoto, A.G. 1984. Acidogenic and methanogenic fermentation of causticized straw. *Biotechnol. Bioeng.* 26:885–891.

Benabdallah El-Hadj, T., Dosta, J., Serrano, R.M., Alvarez, J.M. 2007. Effect of ultrasound pretreatment in mesophilic and thermophilic anaerobic digestion with emphasis on naphthalene and pyrene removal. *Water Res.* 41:87–94.

Bien, J.B., Malina, G., Bien, J.D., Wolny, L. 2004. Enhancing anaerobic fermentation of sewage sludge for increasing biogas generation. *J. Environ. Sci. Health, Part A–Toxic/Hazard. Subst. Environ. Eng.* 39:939–949.

Bobleter, B. 1994. Hydrothermal degradation of polymers derived from plants. *Prog. Polym. Sci.* 19:797–841.

Borges, E.S.M., Chernicharo, C.A.S. 2009. Effect of thermal treatment of anaerobic sludge on the bioavailability and biodegradability characteristics of the organic fraction. *Braz. J. Chem. Eng.* 26:469–480.

Bougrier, C., Albasi, C., Delgenes, J.P., Carrere, H. 2006. Effect of ultrasonic, thermal and ozone pre-treatments on waste activated sludge solubilisation and anaerobic biodegradability. *Chem. Eng. Process* 45:711–718.

Bougrier, C., Battimelli, A., Delgenes, J.P., Carrere, H. 2007. Combined ozone pretreatment and anaerobic digestion for the reduction of biological sludge production in wastewater treatment. *Ozone Sci. Eng.* 29:201–206.

Bougrier, C., Delgenes, J.P., Carrere, M. 2006. Combination of thermal treatments and anaerobic digestion to reduce sewage sludge quantity and improve biogas yield. *Process Saf. Environ. Protect.* 84:280–284.

Bougrier, C., Delgenes, P., Carrere, H. 2007. Impacts of thermal pre-treatments on the semi-continuous anaerobic digestion of waste activated sludge. *Biochem. Eng. J.* 34:20–27.

Braguglia, C.M., Gianico, A., Mininni, G. 2011. Laboratory-scale ultrasound pretreated digestion of sludge: Heat and energy balance. *Bioresour. Technol.* 102:7567–7573.

Brownell, H., Yu, E.K.C.C., Saddler, J.N. 1986. Steam-explosion pretreatment of wood: Effect of chip size, acid, moisture content and pressure drop. *Biotechnol. Bioeng.* 28:792–801.

Carrere, H., Bougrier, C., Castets, D., Delgenes, D. 2008. Impact of initial biodegradability on sludge anaerobic digestion enhancement by thermal pretreatment. *J. Environ. Sci. Health, Part A–Toxic/Hazard. Subst. Environ. Eng.* 43:1551–1555.

Carrere, H., Dumas, C., Battimelli, A., Batstone, D.J., Delgenes, J.P., Steyer, J.P., Ferrer, I. 2010. Pretreatment methods to improve sludge anaerobic degradability: A review. *J. Hazard. Mater.* 183:1–15.

Chi, Y., Li, Y., Fei, X., Wang, S., Yuan, H. 2011. Enhancement of thermophilic anaerobic digestion of thickened waste activated sludge by combined microwave and alkaline pretreatment. *J. Environ. Sci.* 23:1257–1265.

Dar, G.H., Tandon, S.M. 1987. Biogas production from pretreated wheat straw, lantana residue, apple and peach leaf litter with cattle dung. *Biological Wastes* 21:75–83.

Dermott, B.L.M., Chalmers, A.D., Goodwin, J.A.S. 2001. Ultrasonication as a pre-treatment method for the enhancement of the psychrophilic anaerobic digestion of aquaculture effluents. *Environ. Technol.* 22:823–830.

Dewil, R., Baeyens, J., Goutvrind, R. 2006. Ultrasonic treatment of waste activated sludge. *Environ. Prog.* 25:121–128.

Dhar, B.R., Nakhla, G., Ray, M.B. 2012. Techno-economic evaluation of ultrasound and thermal pretreatments for enhanced anaerobic digestion of municipal waste activated sludge. *Waste Manage.* 32:542–549.

Dhouib, A., Ellouz, M., Aloui, F., Sayadi, S. 2006. Effect of bioaugmentation of activated sludge with white-rot fungi on olive mill wastewater detoxification. *Lett. Appl. Microbiol.* 42:405–411.

Dogan, I., Sanin, F.D. 2009. Alkaline solubilization and microwave irradiation as a combined sludge disintegration and minimization method. *Water Res.* 43:2139–2148.

Duff, S.J.B., Murray, W.D. 1996. Bioconversion of forest products industry waste cellulosics to fuel ethanol: A review. *Bioresour. Technol.* 55:1–33.

Erden, G. 2013. Combination of alkaline and microwave pretreatment for disintegration of meat processing wastewater sludge. *Environ. Technol.* 34:711–718.

Erden, G., Filibeli, A. 2009. Ultrasonic pre-treatment of biological sludge: Consequences for disintegration, anaerobic biodegradability, and filterability. *J. Chem. Technol. Biotechnol.* 85:145–150.

Eskicioglu, C., Kennedy, C.J., Droste, R.L. 2009. Enhanced disinfection and methane production from sewage sludge by microwave irradiation. *Desalination* 248:279–285.

Fang, S., Ping, L., Yang, Z., Mao, J. 2011. A review of different pretreatment techniques for enhancing biogas production. In *Proceedings of International Conference on Materials for Renewable Energy & Environment (ICMREE)*, Shanghai.

Ferrer, I., Ponsa, S., Vazquez, F., Font, X. 2008. Increasing biogas production by thermal (70°C) sludge pre-treatment prior to thermophilic anaerobic digestion. *Biochem. Eng. J.* 42:186–192.

Frei, C. 2012. Global and regional issues: The energy challenges for the future. In *World Energy Insight*, World Energy Council, 7–10.

Ge, H., Jensen, P.D., Batstone, D.J. 2010. Pre-treatment mechanisms during thermophilic–mesophilic temperature phased anaerobic digestion of primary sludge. *Water Res.* 44:123–130.

Hendriks, A.T.W.M., Zeeman, G. 2009. Pretreatments to enhance the digestibility of lignocellulosic biomass. *Bioresour. Technol.* 100:10–18.

Jin, Y., Li, H., Mahar, R.B., Wang, Z.Y., Nie, Y.F. 2009. Combined alkaline and ultrasonic pretreatment of sludge before aerobic digestion. *J. Environ. Sci.* 21:279–284.

Juntarasiri, P., Nijsunkij, S., Buatick, T., Jamkrajang, E., Wacharawichanant, S., Seadan, M., Wasantakorn, A., Suttiruengwong, S. 2011. Enhancing biogas production from padaukangsana leave and wastewater feedstock through alkaline and enzyme pretreatment. *Energy Procedia* 9:207–215.

Khanal, S.K., Grewell, D., Sung, S., Leeuwen, J.V. 2007. Ultrasound applications in wastewater sludge pretreatment: A review. *Crit. Rev. Environ. Sci. Technol.* 37:277–313.

Kim, J., Park, C., Kin, T.K., Lee, M., Kim, S., Kim, S.W., Lee, J. 2003. Effects of various pretreatments for enhanced anaerobic digestion with waste activated sludge. *J Biosci. Bioeng.* 95:271–275.

Kuglarz, M., Karakashev, D., Angelidaki, I. 2013. Microwave and thermal pretreatment as methods for increasing the biogas potential of secondary sludge from municipal wastewater treatment plants. *Bioresour. Technol.* 134: 290–297.

Kumar, P., Barrett, D.M., Delwiche, M.J., Stroeve, P. 2009. Methods for pretreatments of lignocellulosic biomass for efficient hydrolysis and biofuel production. *Ind. Eng. Chem. Res.* 48:3713–3729.

Lehne, G., Muller, A., Schwedes, J. 2001. Mechanical disintegration of sewage sludge. *Water Sci. Technol.* 43:9–26.

Lin, Y., Wang, D., Wu, S., Wang, C. 2009. Alkali pretreatment enhances biogas production in the anaerobic digestion of pulp and paper sludge. *J. Hazard. Mater.* 170:366–373.

Luste, S., Luostarinen, S. 2011. Enhanced methane production from ultrasound pre-treated and hygienized dairy cattle slurry. *Waste Manage.* 31:2174–2179.

Monlau, F., Latrille, E., Costa, A.C.D., Steyer, J.P., Carrere, H. 2013. Enhancement of methane production from sunflower oil cakes by dilute acid pretreatment. *Applied Energy* 102:1105–1113.

Moset, V., Cerisuelo, A., Sutaryo, S., Moller, H.B. 2012. Process performance of anaerobic co-digestion of raw and acidified slurry. *Water Res.* 46:5019–5027.

Mosier, N., Hendrickson, R., Ho, N., Sedlak, M., Ladisch, M.R. 2005. Optimization of pH controlled liquid hot water pretreatment of corn stover. *Bioresour. Technol.* 96:1986–1993.

Mudhoo, A. 2012. *Biogas Production: Pretreatment Methods in Anaerobic Digestion.* Hoboken, NJ: John Wiley & Sons.

Muller, H.W., Trosch, W. 1986. Screening of white-rot fungi for biological pretreatment of wheat straw for biogas production. *Appl. Microbiol. Biotechnol.* 24:180–185.

Muller, J., Lehne, G., Schwedes, J., Battenberg, S., Naeveke, R., Kopp, J., Dichtl, N., Scheminski, A., Krull, R., Hempel, D.C. 1998. Disintegration of sewage sludges and influence on anaerobic digestion. *Water Sci. Technol.* 18:425–433.

Nah, I.W., Kang, Y.W., Hwang, K.W., Song, W.K. 2000. Mechanical pretreatment of waste activated sludge for anaerobic digestion process. *Water Res.* 34:2362–2368.

Nayono, S.E. 2009. *Anaerobic Digestion of Organic Solid Waste for Energy Production.* Karlsruhe: KIT Scientific Publishing.

Neely, W.C. 1984. Factors affecting the pretreatment of biomass with gaseous ozone. *Biotechnol. Bioeng.* 26:59–65.

Nieves, D.C., Karimi, K., Horvath, I.S. 2011. Improvement of biogas production from oil palm empty fruit bunches (OPEFB). *Ind. Crops Prod.* 34:1097–1101.

Ofoefule, A.U., Uzodinm, E.O., Onukwuli, O.D. 2009. Comparative study of the effect of different pretreatment methods on biogas yield from water Hyacinth (*Eichhorniacrassipes*). *Int. J. Phys. Sci.* 4:535–539.

Palmowski, L.M., Muller, J.A. 2000. Influence of size reduction of organic waste on their anaerobic digestion. *Water Sci. Technol.* 41:155–162.

Penaud, V., Delgenes, J.P., Moletta, R. 1999. Thermo-chemical pretreatment of a microbial biomass: Influence of sodium hydroxide addition on solubilization and anaerobic biodegradability. *Enzyme Microb. Technol.* 25:258–263.

Perez-Elvira, S., Fdz-Polanco, M., Plaza, F.I., Garralon, G., Fdz-Polanco, F. 2009. Ultrasound pre-treatment for anaerobic digestion improvement. *Water Sci. Technol.* 60:1525–1532.

Perkowski, J., Kos, L., Ledakowicz, S. 1996. Application of ozone in textile waste-water treatment. *Ozone Sci. Eng.* 18:73–85.

Pilli, S., Bhunia, P., Yan, S., Blanc, R.J.L., Tyagi, R.D., Surampalli, R.Y. 2011. Ultrasonic pretreatment of sludge: A review. *Ultrason. Sonochem.* 18:1–18.

Qiao, W., Wang, W., Xun, R. 2007. Microwave thermal hydrolysis of sewage sludge as a pretreatment stage for anaerobic digestion. Paper presented at 5th International Workshop on Water Dynamics, Japan, September 25–27.

Rafique, R., Poulsen, T.J., Nizami, A.S., Asam, Z.Z., Murphy, J.D., Kiely, G. 2010. Effect of thermal, chemical and thermo-chemical pre-treatments to enhance methane production. *Energy* 35:4556–4561.

Saifuddin, N., Fazlili, S.A. 2008. Effect of microwave and ultrasonic pretreatments on biogas production from anaerobic digestion of palm oil mill effluent. *Am. J. Eng. Applied Sci.* 2:139–146.

Seng, B., Khanal, S.K., Visvanathan, C. 2010. Anaerobic digestion of waste activated sludge pretreated by a combined ultrasound and chemical process. *Environ. Technol.* 31:257–265.

Sharma, S.K., Mishra, I.M., Sharma, M.P., Saini, J.S. 1988. Effect of particle size on biogas generation from biomass residues. *Biomass* 17:251–263.

Srilatha, H.R., Krishna, N., Sudhakar Babu, K. 1995. Fungal pretreatment of orange processing waste by solid-state fermentation for improved production of methane. *Process Biochem.* 30:327–331.

Sun, Y., Cheng, J. 2002. Hydrolysis of lignocelluosic materials for ethanol production—A review. *Bioresour. Technol.* 83:1–11.

Taherzadeh, M.J., Karimi, K. 2008. Pretreatment of lignocellulosic wastes to improve ethanol and biogas production: A review. *Int. J. Mol. Sci.* 9:1621–1651.

Tanaka, S., Kobayashi, T., Kamiyama, K., Lolita, M., SigneyBildan, N. 1997. Effects of thermochemical pretreatment on the anaerobic digestion of waste activated sludge. *Water Sci. Technol.* 35:209–215.

Teghammar, A., Yngvesson, J., Lundin, M., Taherzadeh, M.J., Horvath, I.S. 2010. Pretreatment of paper tube residuals for improved biogas production. *Bioresour. Technol.* 101:1206–1212.

Tiehm, A., Nickel, K., Neis, U. 1997. The use of ultrasound to accelerate the anaerobic digestion of sewage sludge. *Water Sci. Technol.* 36:121–128.

Toreci, I., Kennedy, K.J., Droste, R.L. 2009. Evaluation of continuous mesophilic anaerobic sludge digestion after high temperature microwave pretreatment. *Water Res.* 43:1273–1284.

Wang, Q., Kuninobu, M., Kakimoto, K., Ogawa, H.I., Kato, Y. 1999. Upgrading of anaerobic digestion of waste activated sludge by ultrasonic pretreatment. *Bioresour. Technol.* 68:309–313.

Wang, Y.Z., Chen, X., Wang, Z., Zhao, J.F., Fan, T.F., Li, D.S., Wang, J.H. 2012. Effect of low concentration alkali and ultrasound combination pretreatment on biogas production by stalk. *Adv. Mater. Res.* 383:3434–3437.

Weemaes, M., Grootaerd, H., Simoens, F., Verstraete, W. 2000. Anaerobic digestion of ozonized biosolids. *Water Res.* 34:2330–2336.

Weemaes, M., Verstraete, W. 1998. Evaluation of current wet sludge disintegration techniques. *J. Chem. Technol. Biotechnol.* 73:83–92.

Weil, J.R., Brewer, M., Hendrickson, R., Sarikaya, A., Ladisch, M.R. 1997. Continuous pH monitoring during pretreatment of yellow poplar wood sawdust by pressure cooking in water. *Appl. Biochem. Biotechnol.* 68:21–40.

Yagci, N., Akpinar, I. 2011. The investigation and assessment of characteristics of waste activated sludge after ultrasound pretreatment. *Environ. Technol.* 32:221–230.

Yeom, I.T., Lee, K.R., Ahn, R.H., Lee, S.H. 2002. Effects of ozone treatment on the biodegradability of sludge from municipal wastewater treatment plants. *Water Sci.Technol.* 46:421–425.

Yu, J., Zhang, J., He, J., Liu, Z., Yu, Z. 2009. Combinations of mild physical or chemical pretreatment with biological pretreatment for enzymatic hydrolysis of rice hull. *Bioresour. Technol.* 100:903–908.

Yunqin, L., Dehan, W., Lishang, W. 2010. Biological pretreatment enhances biogas production in the anaerobic digestion of pulp and paper sludge. *Waste Manag. Res.* 28:800–810.

Zheng, J., Kennedy, K., Eskicioglu, C. 2009. Effect of low temperature microwave pretreatment on characteristics and mesophilic digestion of primary sludge. *Environ. Technol.* 30:319–327.

Zhong, W., Zhang, Z., Luo, Y., Sun, S., Qiao, W., Xiao, M. 2011. Effect of biological pretreatments in enhancing corn straw biogas production. *Bioresour. Technol.* 102:11177–11182.

A Comprehensive Review on Source Apportionment of Pollutants in the Atmosphere by Receptor Models

N. Anu and N. Selvaraju

CONTENTS

Abstract

Particulate matter (PM_{10} and $PM_{2.5}$) emission from different sources can transport and deposit to distant locations due to meteorological conditions such as wind speed, wind direction, temperature, and humidity. Emission criteria and limitations are implemented by various countries and autonomous bodies for protecting and maintaining an aesthetic environment through continuous monitoring and analysis of atmospheric pollutants. Seasonal variations in the ambient particulate as well as gaseous concentrations are also a major concern. The current review focuses on the source apportionment studies conducted in various parts of the world using receptor models such as factor analysis, chemical mass balance (CMB), positive matrix factorization (PMF), principal component analysis (PCA), and Unmix to reveal the source contribution and source profile estimation to characterize different sources of air pollution. Modifications in the basic equations (robotic chemical mass balance, absolute principal component scores, etc.) were also observed in various studies to obtain more accurate source contribution in comparison with the conventional receptor model approach. The chapter discusses various receptor models, and their strengths and weaknesses to apportion air pollution sources.

16.1 INTRODUCTION

Environmental aesthetics is a major problem faced by the world in the current century due to population explosion and natural calamities. Environmental pollution due to anthropogenic activities as well as natural processes results in damage to the atmosphere due to various categories of pollutants such as particulate matter (PM), gaseous compounds, and secondary organic and inorganic species.

Particulate matter is one of the major carriers of hazardous pollutants that cause adverse effects on human health (Schauer et al., 2006; Viana et al., 2007), such as asthma induction and exacerbations, respiratory symptoms, and impaired lung function (Seagrave et al., 2006; Wichmann et al., 2009; Stanek et al., 2011); degradation on structures (Brimblecombe and Grossi, 2010); and visibility problems (Y. Lee and Sequeira, 2001; Cheung et al., 2005). In order to design active control strategies, we must identify the contributions of air pollutants from different sources and evaluate the source profile health impacts (Lall et al., 2011; Heal et al., 2012). Hence, implementation of protocols and regulations has to be made for emission of $PM_{2.5}$ and PM_{10} (particulate matter with aerodynamic size less than 2.5 μm and 10 μm) at industrial, urban, and rural locations with increasing focus on $PM_{2.5}$ control. However, in order to design effective programs and strategies for reduction of PM concentration in the ambient air, it is necessary to have information about the sources and their respective contributions (B. Kim and Henry, 2000).

The method used to quantify contribution of different sources to atmospheric PM concentrations is known as source apportionment (SA) study. There is a wide range of published literature on source apportionment using dispersion models and monitoring data (Colvile et al., 2003; Fushimi et al., 2005; Evangelista et al., 2007; Laupsa et al., 2009). The method may or may not require any emission inventory or source profile data to apportion the sources of pollutants in the atmosphere. To quantify the contribution of sources, the source compositions for typical sources are available (Chow et al., 1992; Watson et al., 2001). Specific source identification, such as differences in wood smoke compositions that arise from differences in the type of wood burned in various regions, must be understood (McDonald et al., 2000; Fine et al., 2002). However, emission rate estimation from industries using industrial source complexes (ISC) is difficult due to the locations being surrounded with many emission sources. Receptor-model-based source apportionment studies are underway to assess the source contribution of particulate matter (Thurston and Spengler, 1985; Harrison et al., 1997; Kumar et al., 2001; Larsen and Baker, 2003; Lewis et al., 2003; Begum et al., 2004; E. Kim et al., 2004; Lai et al., 2005; Qin et al., 2006; Y. Song et al., 2006; Tsai and Chen, 2006; Chowdhury et al., 2007; Chow et al., 2008; Guo et al., 2009; Kong et al., 2010; Stone et al., 2010; Gu et al., 2011, 2013; Mansha et al., 2012). The nature of the sources and species concentrations based on the receptor concentration data has been used as the model input parameters to get the source contribution (Henry et al., 1984; Kim and Henry, 2000). The percentage source contributions (univariate and multivariate method) and source profiles (multivariate method) of different sources of particulate matter are the key output of a receptor model. Hence it is not necessary for a complete emissions inventory data of pollutants to apportion the sources (Hopke, 1991).

Detailed analyses and cost estimates are needed to produce quantitative results from large-scale data assessments (Cooper and Watson, 1980). Chemical methods, such as enrichment factor analysis, times series analysis, chemical mass balance (CMB) analysis, multivariate factor analysis (including principal component analysis [PCA] and positive matrix factorization [PMF]), Unmix, species series analysis, and multi-linear regression analysis (MLRA) were applied when the source information were not available (Cooper and Watson, 1980; Henry et al., 1984; Hopke, 1991; Ramadan et al., 2003; Amato et al., 2009). PMF integrates the nonnegativity constraints into the optimization process, which uses a least squares approach to solve the factor analysis problem. Unmix will reduce the complexity to identify the sources of pollution, but trace elements have been used for the assumption that they can serve as tracers for some specific sources. Multiple source identifications based on correlation between the sources and elemental data will be difficult due to collinear (similar emission character) sources. Thus, molecular markers (MM) have been shown to be good tracers of certain sources for exact source identification (Schauer et al., 1996; Simoneit et al., 1999; Schauer and Cass, 2000).

Recent studies about source apportionment through receptor models use different source markers, since both inorganic (trace metals, cations and anions) and organic species (organic carbon [OC] and elemental carbon [EC]) are composed of particulate matter. To reduce the source ambiguity and sources that are difficult to apportion solely on the basis of inorganic tracers (e.g., levoglucosan for biomass burning; Simoneit et al., 1999), identification and development of organic molecular markers are necessary (Harrison et al., 1996, 2003; Schauer et al., 1996; Robinson et al., 2006; Wang et al., 2012). Hopke (2008) developed effective and efficient air quality management plans and refined emission inventories for input into deterministic models to predict changes in air quality following implementation of various management plans. Apportionment studies also provide exposure estimates for health effects models to identify those components of the PM that are most closely related to observed adverse health effects.

16.2 RECEPTOR MODELS (RMs)

The basic receptor model equation is given by

$$X = G \cdot F \tag{16.1}$$

$$x_{ij} = \sum_{k=1}^{p} g_{ik} f_{kj} \tag{16.2}$$

where x_{ij} is the concentration of species j measured on sample i, p is the number of factors contributing to the samples, f_{kj} is the concentration of species j in factor profile k, and g_{ik} is the relative contribution of factor k to sample i. A chemical mass balance (CMB) model can be solve by using one day or average receptor concentration and source composition data. Hence $i = 1$ for solving the CMB receptor model. Multivariate data interpretation can be possible through the following equation:

$$X = G \cdot F + E \tag{16.3}$$

or in parameter expression,

$$x_{ij} = \sum_{k=1}^{p} g_{ik} f_{kj} + e_{ij} \tag{16.4}$$

where e_{ij} is error of the multivariate model for the species j measured on sample i. In the literature, factors resolved by the models (PMF, Unmix, etc.) are often interpreted as sources, although they are not necessarily synonymous. The goal is to find values of g_{ik}, f_{kj}, and p that best reproduce x_{ij}. The values of g_{ik} and f_{kj} are adjusted until a minimum value of Q for a given p is found. Q is defined as

$$Q = \sum_{i=1}^{n} \sum_{j=1}^{m} \left(\frac{e_{ij}}{\sigma_{ij}} \right)^2 \tag{16.5}$$

The input for a fundamental receptor model are the receptor concentration data and source composition data for solving the matrices to obtain the source contribution matrix to understand the effect of various sources of atmospheric particulate emission in that locality. Based on the number of receptors used, as well as the period of sampling, the particulate matter will lead to a large quantum of data, which can be solved through multilinear regression such as principal component analysis, positive matrix factorization, or Unmix.

16.2.1 Classical Approach in Receptor Modeling

PM_{10} collected on a filter paper was subjected to speciation analysis to find information on the concentration of different species, i.e., elements, ions and carbon fraction distributed in the aerosol phase of the atmosphere. Sampling was carried out by replacing filter paper every 24 h and the data were collected. The input data consists of the concentrations of species (G) present in PM_{10} over a period of days (D). The data was represented in the form of $G \times D$ matrix R. The obtained data set contains the information about daily variation in the concentration of species measured.

16.2.2 Factor Analysis (FA)

Factor analysis looks for variations in the concentrations of different components and identifies a minimal set of "unobservable factors" that can explain the variations (X. Song et al., 2001). The entire data set were analyzed to determine if the variations in the data could be explained by a minimum number of unobservable factors. Since these factors are unobservable, linear regression techniques cannot be used for the analysis and for obtaining quantitative estimates of the contribution of the individual factors. However, one can identify the variance percentage attributed to various factors. These factors are characterized by factor loadings and can be viewed as sources containing a dominant set of species. This was used to match the factors with different sources identified in the emission inventory. This identification with sources is highly subjective.

16.2.3 Positive Matrix Factorization (PMF)

Factor analysis, discussed in the previous section, helps to determine the number of factors or possible sources in order to explain the variations in the data. FA cannot be used to determine the quantitative source contributions and source profiles, but FA contributes to explaining the trends in the receptor data set. The application of PMF over FA allows us to determine source profiles and source contributions that can generate a data set from X. Here the concentration of various species ($j = 1, 2, \ldots, J$) is measured in different batches of samples ($i = 1, 2, \ldots, I$) represented in the form of a matrix X ($i \times j$). The matrix X was written in the form of the product of two matrices $G \times F$, where G ($i \times k$) contains the contributions of the various sources and F ($k \times j$) contains the source profiles of the various sources. The main challenge in PMF is to choose p, which represents an effective number of sources. Since the source profiles and source contributions cannot be negative, the matrices F and G are done in such a way that all elements are positive or at least nonnegative (Chelani et al., 2010). The goal of multivariate receptor modeling, for example with PMF, is to identify a number of factors p, the species profile (F) of each source, and the amount of mass (G) contributed by each factor to each individual sample given in Equation 16.6:

$$X_{ij} = \sum_{k=1}^{p}(G_{ik} \cdot F_{kj}) + e_{ij} \qquad (16.6)$$

where

X_{ij} = Receptor concentration of the jth species in ith sample
G_{ik} = Source contribution of the kth source on the ith sample
F_{kj} = Source profile of the jth species in the kth source
e_{ij} = Residual of the jth species on the ith day

For the PMF approach based on minimizing a least square error Q,

$$Q = \sum_{i=1}^{I} \sum_{j=1}^{J} \left(\frac{X_{ij} - \sum_{k=1}^{p}(G_{ik} \cdot F_{kj})}{u_{ij}} \right)^{2} \qquad (16.7)$$

where u_{ij} represents the uncertainty of the concentration of the jth species in the ith sample at the receptor X.

The uncertainty measurement can found based on the method detection limit (MDL).

$$\text{Uncertainty} = 2 \times \text{MDL, if Concentration} \le \text{MDL}$$

$$\text{Uncertainty} = \sqrt{(percentage \times concentration)^2 + (MDL)^2},$$
$$\text{if Concentration} > \text{MDL}$$

A disadvantage of PMF is that multiple solutions of G and F may exist, which minimize the objective functions. Under this condition, it is challenging to select the physically relevant solution.

16.2.4 Chemical Mass Balance (CMB)

In chemical mass balance (CMB) the source contribution is determined by using the information from both the receptor concentration and source profiles on a daily basis. Here a solution to Equation 16.8 is obtained for $D = 1$. Similar to PMF, CMB uses the principle of conservation of mass in such a way to obtain the quantitative contribution of various sources to pollution levels at a point (E. Kim et al., 2003; Kothai et al., 2008; Gianini et al., 2013). Based on the mass balance of particulate matter species:

$$X_{(j \times 1)} = \sum F_{(j \times k)} \cdot G_{(k \times 1)} \tag{16.8}$$

where,

$X_{j \times 1}$ = Concentration of the jth species at the receptor site ($\mu g/m^3$)
$F_{j \times k}$ = Source profile composition of jth species in the kth source (mg/g)
$G_{j \times 1}$ = Contribution of the kth source ($\mu g/m^3$) to the receptor site

A successful application of the CMB model will be possible only when all the important sources are identified, accurate source profiles are known, and uncertainties are estimated for both ambient concentrations and source profiles or fingerprints. A major challenge in CMB modeling is addressing the collinearity issue in source profiles, since it makes CMB difficult to discriminate between various sources. This problem can be overcome by means of considering a subset of species or sources in the analysis. Then the exercise becomes purely mathematical and loses physical relevance.

16.2.5 Unmix

Unmix tries to resolve the general mixture problems where the data are assumed to be a linear combination of an unknown number of sources of unknown composition, which contribute an unknown amount to each sample. Unmix also assumes that the compositions and contributions of the sources are all positive. Unmix assumes that for each source, there are some samples that contain little or no contribution from that source. Unmix is a type of factor analysis but is geometrically constrained to generate source contributions and profiles with the physically meaningful attribute of non-negativity. Factor analysis and PMF face the problem of rotational ambiguities, which are eliminated through the multivariate analysis method of Unmix (Lewis et al., 2003). The model can be expressed as

$$X_{ij} = \sum_{j=1}^{J} \left(\sum_{k=1}^{p} G_{ik} \cdot F_{kj} \right) V_{jj} + \varepsilon_{ij} \tag{16.9}$$

where G, F, and V are $I \times p$, $p \times J$ diagonal, and $J \times J$ matrices, respectively; and ε_{ij} is the error term consisting of all the variability in X_{ij} not accounted for by the first p principal components. The edges represent the samples that characterize the source. Unmix uses data to find data points (edges) where one of the sources is missing or small. Such edges in point sets are then used to calculate the vertices, which are used with the matrices decomposed by singular value decomposition (SVD) and physical constrains (tracer compound) supplied by the user can also used (Hellen et al., 2003). The Unmix 6.0 tool, developed by the U.S. Environmental Protection Agency (USEPA), was used for the receptor data analysis. Uncertainty in the data is not considered by the model, which implicitly assumes a certain standard of accuracy in the data for a good model fit. There are an infinite number of solutions to Equation 16.9, so the model restricts solutions to those that are physically meaningful in that factor compositions and contributions are nonnegative. Also, the sum of the concentrations of an element from all factors in one day cannot be greater than the measured mass concentration of the element for that day. The model operates under the assumption that the data form well-defined "edges," meaning that each factor, when plotted against another, will have some dates during which the contribution is zero for one factor and nonzero for the other factor, creating an edge in a multidimensional space. The results are limited to a maximum of seven factors but the user must select which measured species to use in the model. The model does not use data from any date from which one or more species concentrations are missing (Lewis et al., 2003).

16.3 MODIFIED RECEPTOR MODELS

In order to improve the source contribution predictions using receptor models, several modifications in the basic receptor model have been implemented. The CMB model requires information of the composition of all sources contributing to airborne pollution. The measured air quality is assumed to be a linear sum of the contributions of the known sources, whose contributions are summed over each different sampling period to give the best match to the concentrations of the many chemical species measured in the atmosphere. In more recent studies, organic molecular markers, which may be only minor constituents of emissions, are measured, as these help to discriminate between similar sources (e.g., gasoline and diesel engines). Source profiles within the same source category or class (e.g., biomass burning) can have varying chemical compositions that influence apportionment results. Furthermore, source profiles that represent actual sources in every region are not always available, in which case the "best existing" profile must be selected as a surrogate even though it may not accurately represent the source emissions in the sampling area of interest (Bullock, 2008).

The USEPA developed a PMF software tool that has been used in various source apportionment studies (Heo et al., 2009, 2013; Gugamsetty et al., 2012). The method requires no a priori knowledge of source composition, but any information on source emissions characteristics is helpful in discriminating between similar sources. The method requires large data (at least 50) air samples and works best with a large data set in which the number of samples far exceeds the number of analytical variables. A minimum variable-to-case ratio of 1:3 should be maintained in order to obtain accurate results (Thurston and Spengler, 1985). For a clearer distinction, it is better to have short sampling times so that overlap of multiple point source contributions to a given sample is minimized. The samples are analyzed for the chemical constituents, and those constituents from the same source have the same temporal variation, and if unique to that source are perfectly correlated. Typically, however, a given chemical constituent will have multiple sources and the program is able to view correlations in multidimensional space and can generate chemical profiles of "factors" with a unique temporal profile characteristic of a source. Past knowledge of source chemical profiles is used to assign factors to sources, and typically identification of six or seven different sources is a good outcome. Before PMF became widely adopted, principal components analysis (PCA) was widely used for the same purpose, but is less refined than PMF. Input data plays an important role in the final results, and care has to be taken to ensure that this is of good quality and, where possible, uncertainties can be assigned to individual analytes.

The methods based upon receptor models for various source apportionment studies are summarized in Table 16.1. Studies have been conducted to compare results from different models (Chow et al., 1993; Larsen and Baker, 2003; Ramadan et al., 2003; Srivastava and Jain, 2007; Bullock et al., 2008; S. Lee et al., 2008; Viana et al., 2008; Yatkin and Bayram, 2008; Callen et al., 2009; Tauler et al., 2009). Multicollinearity can affect the model estimates, particularly in cases where different sources have similar signatures, although multivariate models help to reduce that problem substantially (Henry et al., 1984; Thurston and Lioy, 1987). It has been reported that in cases where two different sources have similar signatures,

TABLE 16.1 Methods Based upon Receptor Model for Various Source Apportionment Studies

Particulate Matter	Species Analyzed	Receptor Model Used	Sources Identified and Their Contributions	Reference
PM_{10}	*Elements*: Na, K, Al, Si, P, S, Cl, K, Ca, Ti, V, Cr, Mn, Fe, Ni, Cu, Zn, Se, Br, Rb, Sr, Zr, Ba, Pb *Ions*: NO_3^-, SO_4^{2-}, NH_4^+ *Carbon*: EC, OC	CMB	Geological material 54% NH_4NO_3 15% Motor vehicle 10% Construction 8% Others 4%	Chow et al., 1992
PM_{10} and $PM_{2.5}$	*Elements*: Na, K, Al, Si, P, S, Cl, K, Ca, Ti, V, Cr, Mn, Fe, Ni, Cu, Zn, Se, Br, Rb, Sr, Zr, Ba, Pb *Ions*: NO_3^-, SO_4^{2-}, NH_4^+ *Carbon*: EC, OC	CMB	Primary geological material 54% Secondary ammonium nitrate 15% Primary motor vehicle exhausts 10% Primary construction 8% Unexplained 4%	Chow et al., 1993
$PM_{2.5}$	*Elements*: Al, Si, Ca, Mg, Na, K, Cl, Ti, V, Cr, Mn, Fe, Ni, Cu, Zn, As, Se, Br, Pb *Ions*: SO_4^{2-}, NO_3^-, NH_4^+ *Carbon*: EC, OC	PMF, CMB	From PMF and CMB Biomass burning 11%, 6% Secondary sulfate 17%, 17% NO_3^- 14%, 10% Coal combustion 19%, 7% Industry 6% Motor vehicle 6%, 7% Road dust 9%, 20% Yellow dust 48%	Y. Song et al., 2006
$PM_{2.5}$	*Heavy metals*: Ag, As, Ba, Be, Ca, Cd, Ce, Co, Cr, Cs, Cu, Fe, Ga, K, Li, Mg, Mn, Mo, Ni, Pb, Rb, Sb, Se, Sr, Ti, Tl, V, Zn *Ions*: SO_4^{2-} *Carbon*: OC, EC	PMF, Unmix	Crustal material 13% Regional transport 68% Nitrate 14% Fe, Mn, and Zn 4% Specialty steel industry <1%	Pekney et al., 2006

(Continued)

TABLE 16.1 (CONTINUED) Methods Based upon Receptor Model for Various Source Apportionment Studies

Particulate Matter	Species Analyzed	Receptor Model Used	Sources Identified and Their Contributions	Reference
$PM_{2.5}$ Total No: 51	32 molecular markers, OC, EC, Al, Si, SO_4^{2-}, NO_3^-, NH_4^+	CMB-MM (32 molecular markers)	Secondary SO_4^{2-} 40% Secondary NH_4^+ 17% Secondary NO_3 11% Wood combustion 18% Gasoline exhaust 16% Other organic 18%	Ke et al., 2008
$PM_{2.5}$ Total: 932	*Heavy metals:* As, Ba, Br, Cu, Mn, Pb, Se, Ti, Zn, Al, Si, K, Ca, Fe *Secondary species:* SO_4^{2-}, NO_3^{3-}, NH_4^+ *Organic carbon:* OC1, OC2, OC3, OC4 Organic phosphorus *Elemental carbon:* EC1, EC2, and EC3 *Gaseous species:* CO, SO_2, NO, HNO_3, and NO_y	PMF	Diesel exhaust 10.26% Gasoline exhaust 1.12% Wood combustion 8.62% Road dust 4.5% Coal combustion 1.32% Sulfate factor 36.5% Nitrate factor 2.6% Cement industry 2.92% Zinc industry 0.36% Unapportioned 20.32%	
$PM_{2.5}$, PM_{10}, TSP	*Elements:* Mg, Al, Si, Ca, Mn, Fe, Co, Ni, Cu, Zn, Sc, V, Cr, Cd, Ba, Bi, Ge, Sn, and Pb *Ions:* Na^+, K^+, NH_4^+, Cl^-, NO_3^-, and SO_4^{2-} *Carbon:* EC and OC	CMB and PCA	Coal combustion 5%–31% Marine aerosol 1%–13% Vehicular emission 13%–44% Soil dust 3%–46%	Kong et al., 2010

(Continued)

TABLE 16.1 (CONTINUED) Methods Based upon Receptor Model for Various Source Apportionment Studies

Particulate Matter	Species Analyzed	Receptor Model Used	Sources Identified and Their Contributions	Reference
$PM_{2.5}$, $PM_{10-2.5}$	Black carbon (BC), Na, Mg, Al, Si, P, S, Cl, K, Ca, Ti, Cr, Mn, Fe, Ni, Cu, Zn, Br, Pb	PMF	Motor vehicle 40% Metal smelters, sea salt, two-stroke engine 10% Soil dust and road dust 50%	Begum et al., 2004
PM_{10} Total: 168 Winter: 123 μg/m³ Summer–fall: 93 μg/m³ Number of receptors: 8	*Heavy metals:* Na, Mg, Al, Si, K, Ca, Ti, Cr, Mn, Fe, Co, Ni, Cu, Zn, Pb *Ions:* SO_4^{2-}, NO_3^-, Cl^- *Carbon:* TC, OC, EC	Nested chemical mass balance (NCMB)	Resuspended dust 39.5% Coal combustion dust 14.6% Vehicle 9.4% Construction and cement 6.7% Secondary aerosol: Soil dust 4.3% and 7.9%	Han et al., 2011
PM_{10} Season: Cold and warm Urban 44.42 ± 33.6 μg/m³ Rural 22.7 ± 26.9μg/m³	Mg, Al, Si, S, Cl, Ca, K, Ti, V, Cr, Mn, Fe, Co, Ni, Cu, Zn, Se, Br, Sr, Sb, Ba, Pb	Robotic CMB	Geological (road/soil dust): Urban 32.6%; Rural 22% and 29% Secondary aerosol: Urban 22.1%; rural 30.6% and 28.7% Biomass burning: Urban 28.2%; rural 14.6% and 24.6% Vehicular exhaust: Urban 21.9%; rural 11.5% and 10.5% Fly ash: Urban 5.4% and 2.7%; rural 33.2% and 7.2% Mine field: Minor Lignite dust and/or wet ash deposit: 3.1% to 11% in summer	Argyropoulos et al., 2013
$PM_{2.5}$ and PM_{10}	*Heavy metals:* Al, Si, Fe, Zn, Cu *Carbon:* EC, OC, and their Radiocarbon (^{14}C)		OC (Modern carbon 70%; fossil fuel 30%) EC (Modern carbon 17%; fossil fuel 83%)	Keuken et al., 2013

it becomes difficult to distinguish between them and neither CMB nor multivariate models can distinguish between sources with similar signatures when additional information (for e.g., meteorology data) is missing (Henry et al., 1984).

Kong et al. (2010) studied atmospheric particulate matter ($PM_{2.5}$, PM_{10}, and total suspended particulate matter [TSP]) from June 2007 to February 2008 at a coastal site in Tianjin Economic–Technological Development Area (TEDA) in China. It was determined that the PM contains 19 elements, 6 water-soluble ions, and organic and elemental carbons. PCA and CMB modeling were applied to determine the PM sources and their contributions with the assistance of NSS SO_4^{2-}, and the mass ratios of NO_3^- to SO_4^{2-} and OC to EC. An air mass backward trajectory model was compared with source apportionment results to evaluate the origin of PM. Local emissions causes SO_4^{2-} in $PM_{2.5}$ during the summer, fall, and winter seasons. Most of it was below zero in summer for PM_{10} indicating the influence of sea salt. The ratios of NO^{3-} to SO_4^{2-} for $PM_{2.5}$, PM_{10}, and TSP in winter indicate high amounts of coal consumption. The secondary organic aerosol observed is due to the high OC/EC ratio. The major sources were construction activities, road dust, vehicle emissions, marine aerosol, metal manufacturing, secondary sulfate aerosols, soil dust, biomass burning, some pharmaceutical industries, and fuel–oil combustion according to PCA. Air parcels originating from the sea accounted for 39% in summer, while in autumn and winter the air parcels were mainly related to continental origin and were determined from backward trajectory analysis.

Stedman et al. (2007) developed a model to map PM_{10} and $PM_{2.5}$ concentrations across the United Kingdom at background and roadside locations. Separate models have been calibrated using gravimetric measurements and tapered element oscillating microbalance (TEOM) instruments using source apportionments appropriate to the size fractions and sampling methods. Maps were prepared for a base year of 2004 and predictions were calculated for 2010 and 2020 on the basis of current policies. Comparisons of the modeling results with air quality regulations suggest that exceedences of the EU Daughter Directive stage 1, 24 h limit value for PM_{10} at the roadside in 2004 will be largely eliminated by 2020. The concentration cap of 25 μgm^{-3} for $PM_{2.5}$ proposed within the Clean Air for Europe (CAFE) Directive is expected to be met at all locations. Projections for 2010 and 2020 suggest that the proposed exposure reduction (ER) target is likely to be considerably more stringent and require additional measures beyond current policies. Thus the model results suggest that the balance between the stringency of the concentration cap and the ER target in the proposed directive is appropriate. Measures to achieve greater reductions should therefore have the maximum public health benefit, and air quality policy is not driven by the need to reduce concentrations at isolated hotspots.

Source contribution results obtained from CMB and PMF models are compared by selecting the organic species as input. Both models identified the biomass burning and mobile sources (vehicular source) as the sources of PM. A combination of cooking and secondary organic aerosol was observed as a third source of particulate matter (Bullock, 2008). The emission inventory for fine particulate ($PM_{2.5}$) were assessed by source contributions from a receptor model (PMF), dispersion model (AirQUIS-EPISODE), and multiple linear regression (MLR) of the dispersion model source contribution together with the observed $PM_{2.5}$ concentrations. Although the measured and modeled total $PM_{2.5}$ concentrations were, on average, in good agreement at all sites in Oslo, the analysis showed large deviations for individual sources that compensate when combined. The largest deviations are revealed for wood burning and traffic induced suspension where receptor model calculations differ from the dispersion model by a factor of 0.54 and 7.1, respectively. This is also confirmed by the inverse modeling using MLR (Laupsa, 2009).

Samples of fine and coarse fractions of airborne PM were collected in a semiresidential (AECD) area from June 2001 to June 2002 of Dhaka, and in an urban area of Rajshahi, a city in the northwestern region of Bangladesh, from August 2001 to May 2002. The samples were collected using a "Gent" stacked filter sampler in two fractions of 2.5 mm fine and 2.5–10 mm coarse sizes. The samples were analyzed for elemental concentrations by particle-induced x-ray emission (PIXE). The data sets were then analyzed by the PMF technique to identify the possible sources of atmospheric aerosols in these areas. The best solutions were found to be six and seven factors for elemental compositions for coarse and fine PM fractions in semiresidential Dhaka, and five factors for elemental compositions of each of the coarse and fine PM in the urban area of Rajshahi. The sources are soil dust, road dust, cement, sea salt, motor vehicles, and biomass burning. The PMF results show that a large fraction of about more than 50% of the $PM_{2.5-10}$ mass at both sites comes from soil dust and road dust. The motor vehicle including two strokes contributes about 48% of the $PM_{2.5}$ mass in case of the semiresidential area Dhaka. On the other hand, the biomass-burning factor contributes about 50% of the $PM_{2.5}$ mass in Rajshahi (Begum et al., 2004).

Source contributions to primary airborne particulate matter calculated using the source-oriented UCD/CIT air quality model and the receptor-oriented CMB model were compared for two air quality episodes in different parts of California. The first episode occurred in the San Joaquin Valley on January 4–6, 1996, with peak 24 h average $PM_{2.5}$ concentrations exceeding 100 mg/m³. This episode was characterized by low photochemical activity and high particulate nitrate concentrations, with localized regions of high particulate carbon concentrations around urban centers. The second episode occurred in the South Coast Air Basin on September 7–9, 1993, with peak 4 h average $PM_{2.5}$ concentrations reaching 86 mg/m⁻³. This episode was characterized by high photochemical activity and high secondary organic aerosol concentrations. The results from the two independent source apportionment calculations show strong agreement for source contributions to primary $PM_{2.5}$ total organic mass at seven receptor sites across the two studies, with a correlation slope of 0.84 and a correlation coefficient (R^2) of 0.70. Agreement for source contributions to primary $PM_{2.5}$ total mass was similarly strong, with a correlation slope of 0.83 and a correlation coefficient (R^2) of 0.55. Wood smoke was identified as the dominant source of primary $PM_{2.5}$ at urban locations in the San Joaquin Valley by both source apportionment techniques. Transportation sources, including paved road dust, gasoline engines, and diesel engines, were identified as the dominant source of primary $PM_{2.5}$ at all locations in the South Coast Air Basin by both models. The amount of secondary particulate matter (organic and inorganic) was in good agreement with the measured values minus the primary material identified by the CMB calculation (Held, 2005).

Polycyclic aromatic hydrocarbons (PAHs) were observed as severe pollutants in urban atmosphere. Several PAHs are known carcinogens or are the precursors to carcinogenic daughter compounds. Understanding the contributions of the various emission sources is critical to appropriately managing PAH levels in the environment. The sources of PAHs to ambient air in Baltimore, Maryland, were determined by using three source apportionment methods: principal component analysis with multiple linear regression, Unmix, and positive matrix factorization. Determining the source apportionment through multiple techniques mitigates weaknesses in individual methods and strengthens the overlapping conclusions. Overall, source contributions compared well among the methods.

Vehicles, both diesel and gasoline, contribute on average 16%–26%, coal 28%–36%, oil 15%–23%, and wood/other having the greatest disparity of 23%–35% of the total (gas plus particle phase) PAHs. Seasonal trends were found for both coal and oil. Coal was the dominant PAH source during the summer, whereas oil dominated during the winter. Positive matrix factorization was the only method to segregate diesel from gasoline sources. These methods indicate the number and relative strength of PAH sources to the ambient urban atmosphere. As with all source apportionment techniques, these methods require the user to objectively interpret the resulting source profiles.

Samples of fine particulate matter were collected in a roadway tunnel near Houston, Texas, over a period of 4 days during two separate sampling periods: one sampling period from 12.00 to 14.00 local time and another sampling period from 16.00 to 18.00 local time. During the two sampling periods, the tunnel traffic contained roughly equivalent numbers of heavy-duty diesel trucks. However, during the late afternoon sampling period, the tunnel contained twice as many light-duty gasoline-powered vehicles. The effect of this shift in the vehicle fleet affects the overall emission index (grams pollutant emitted per kilogram carbon in fuel) for fine particles and fine particulate elemental carbon. Additionally, this shift in the fraction of diesel vehicles in the tunnel is used to determine if the chemical mass balancing techniques used to track emissions from gasoline-powered and diesel-powered emissions accurately separates these two emission categories. The results show that the chemical mass balancing calculations apportion roughly equal amounts of the particulate matter measured to diesel vehicles between the two periods and attribute almost twice as much particulate matter in the late afternoon sampling period to gasoline vehicles. Both of these results are consistent with the traffic volume of gasoline and diesel vehicles in the tunnel in the two separate periods and validate the ability for chemical mass balancing techniques to separate these two primary sources of fine particles (Fraser et al., 2003).

Investigation on the effect of uncertainties in the estimated source contributions from CMB through Monte Carlo analysis with Latin hypercube sampling (MC-LHS) was carried out in Atlanta to evaluate the source impact uncertainties and to quantify how uncertainties in ambient measurement and source profile data affect source contribution. In general, uncertainties in the source profile data contribute more to the final uncertainties in source apportionment results than do those in ambient measurement data. Uncertainty contribution estimates suggest that nonlinear interactions among source profiles also affect the final uncertainties, although their influence is typically less than uncertainties in source profile data (S. Lee and Russell, 2007).

In order to reduce the influence of collinearity on the CMB model, the nested chemical mass balance (NCMB) method was used (Feng et al., 2002). The NCMB approach involves dividing the traditional CMB model into three stages to estimate the percent contribution of (1) each individual primary source to ambient PM; (2) each individual primary source to resuspended dust (RD); and (3) RD to ambient PM_{10}. By incorporating the results from this three-stage modeling approach, the collinearity-controlled net contribution from each source can be obtained (Feng et al., 2002; Bi et al. 2007). NCMB identified six sources of pollution and its contributions with less collinearity between the sources using PM_{10}, carbon (TC, OC), and secondary aerosols (SO_4^{2-}, NO_3^-, Cl^-) samples collected during winter and summer–fall seasons at Wuxi, China (Han et al., 2011).

Apart from sources of local air pollution, urban activities are significant contributors to transboundary pollution and to the rising global concentrations of greenhouse gases. Attempts to solve urban problems by introducing cleaner, more energy-efficient technologies will generally have a beneficial impact on these large-scale problems. Attempts based on city planning with a spreading of the activities, on the other hand, may generate more traffic and may thus have the opposite effect (Fenger, 1999).

In Europe, source apportionment of PM and its organic fractions over the past two decades with a variety of receptor models have shifted from principal component analysis techniques, enrichment factors, and classical factor analysis toward models able to handle uncertainties on the input and output such as, e.g., positive matrix factorization. A wider use of advanced factor analysis techniques able to deal with heterogeneous and complex data and to provide improved uncertainty estimations should be promoted. On the other hand, the PCA technique should preferably be used for qualitative or preliminary estimations. The CMB model is still topical and has gained new impetus from the application of molecular markers for the apportionment of the carbonaceous fraction. Nevertheless, the scarcity of measured source profiles for European sources and the lack of long-term, speciated PM series, especially in urban areas, should be dealt with early to remove the limiting factors that hamper the development of receptor model studies in Europe. The definition and documentation of the source categories in Europe has improved swiftly, but there is still a need for harmonization of the different approaches in order to facilitate the interpretation and comparability of the results and their application in the design of abatement measures. In particular, more efforts and methods are needed to estimate and constrain the uncertainties of the resulting source contribution estimates, and more studies are required with a widespread geographical distribution to improve the estimations of biomass burning contributions to PM (Belis, 2013).

The population living on streets with intense road traffic are exposed to elevated concentrations of EC from exhaust emissions in $PM_{2.5}$ and a factor 2 to 3 for heavy metals from brake and tire wear, and resuspended road dust in PM_{10}. It follows that local air quality management may focus on local measures to streets with intense road traffic (Keuken, 2013). A comprehensive comparison of PMF and molecular-marker-based chemical mass balance (CMB-MM) modeling on $PM_{2.5}$ source contributions was conducted for particulate matter measurements taken at Jefferson Street, in Atlanta, Georgia (Ke et al., 2008). Source apportionment of PM_{10} and associated components by were done using the modified robotic chemical mass balance model (RCMB) in urban and rural areas, using chemical compositions from power plant fly ash, flue gas desulfurization wet ash, feeding lignite, infertile material from the opencast mines, paved and unpaved road dusts, soil, vehicular traffic, residential oil combustion, biomass burning, uncontrolled waste burning, marine aerosol, and secondary aerosol formation (Argyropoulos et al., 2013).

16.4 CONCLUSION

A source apportionment study is an important step in order to practice air pollutant emission management programs for urban as well as rural locations. Receptor models are the most powerful tools to identify the available sources and their contributions at a receptor based on the input data of speciate concentration of particulate matters $PM_{2.5}$ and PM_{10}. In this chapter reviewed various receptor model studies to understand the sources and its contributions. The basic approach in a receptor model is mass balance analysis, and human interpretation is required in order to identify the sources based on the obtained source profile data from the model. There is research underway to improve the source contribution predictions using the available input data by the implementation of various modifications in the basic models. Molecular markers and nested chemical mass balance models are the best examples for this approach. Also optimization using fmincon and genetic algorithms, for example, are other approaches to obtain the source contribution in a precise manner. Without the dispersion model approach, air pollution prediction and estimation of source contribution cannot be fulfilled. There are several approaches such as backward trajectory evaluation and the coupled rector-dispersion model that were developed to estimate the source contribution.

REFERENCES

Amato, F., Pandolfi, M., Escrig, A., Querol, X., Alastuey, A., Pey, J., Perez, N., and Hopke, P.K. 2009. Quantifying road dust resuspension in urban environment by multilinear engine: A comparison with PMF2. *Atmospheric Environment* 43, no. 17:2770–2780.

Argyropoulos, G., Grigoratos, T., Voutsinas, M., and Samara, C. 2013. Concentrations and source apportionment of PM_{10} and associated elemental and ionic species in a lignite-burning power generation area of southern Greece. *Environmental Science and Pollution Research*. doi: 10.1007 /s11356-013-1721-y.

Begum, B.A., Kim, E., Biswasa, S.K., and Hopke, P.K. 2004. Investigation of sources of atmospheric aerosol at urban and semi-urban areas in Bangladesh. *Atmospheric Environment* 38:3025–3038.

Belis, C.A., Karagulian, F., Larsen, B.R., and Hopke, P.K. 2013. Critical review and meta-analysis of ambient particulate matter source apportionment using receptor models in Europe. *Atmospheric Environment* 69:94–108.

Bi, X.H., Feng, Y.C., Wu, J.H., Wang, Y.Q., and Zhu, T. 2007. Source apportionment of PM_{10} in six cities of northern China. *Atmospheric Environment* 41, no. 5:903–912.

Brimblecombe, P., and Grossi, C.M. 2010. Potential damage to modern building materials from 21st century air pollution. *The Scientific World Journal* 10:16–125.

Bullock, K.R., Duvall, R.M., Norris, G.A., McDow, S.R., and Hays, M.D. 2008. Evaluation of the CMB and PMF models using organic molecular markers in fine particulate matter collected during the Pittsburgh Air Quality Study. *Atmospheric Environment* 42:6897–6904.

Callen, M.S., de la Cruz, M.T., Lopez, J.M., Navarro, M.V., and Mastral, A.M. 2009. Comparison of receptor models for source apportionment of the PM_{10} in Zaragoza (Spain). *Chemosphere* 76:1120–1129.

Chelani, A.B., Gajghate, D.G., Chalapati Rao, C.V., and Devotta, S. 2010. Particle size distribution in ambient air of Delhi and its statistical analysis. *Bulletin of Environmental Contamination and Toxicology* 85, no. 1:22–27.

Cheung, H.C., W. Tao, B. Karstern, and Hai, G. 2005. Influence of regional pollution outflow on the concentrations of fine particulate matter and visibility in the coastal area of southern China. *Atmospheric Environment* 39, no. 34:6463–6474.

Chow, J.C., Watson, J.G., Lowenthal, D.H., Solomon, P.A., Magliano, K.L., Ziman, S.D., and Richard, L.W. 1993. PM_{10} and $PM_{2.5}$ compositions in California's San Joaquin Valley. *Aerosol Science and Technology* 18:105–128.

Chow, J.C., Watson, J.G., Lowenthal, D.H., Solomon, P. A., Magliano, K.L., Ziman, S.D., and Richards, L.W. 1992. PM_{10} source apportionment in California's San Joaquin Valley. *Atmospheric Environment* 26A, no. 18:3335–3354.

Chow, J.C., Watson, J.G., Lowenthal, D.H., and Magliano, K.L. 2008. Size-resolved aerosol chemical concentrations at rural and urban sites in Central California, USA. *Atmospheric Research* 90, no. 2–4:243–252.

Chowdhury, Z., Zheng, M., Schauer, J.J., Sheesley, R.J., Salmon, L.G., Cass, G.R., and Russell, A.G. 2007. Speciation of ambient fine organic carbon particles and source apportionment of $PM_{2.5}$ in Indian cities. *Journal of Geophysical Research* 112, no. D15. doi: 10.1029/2007JD008386.

Colvile, R.N., Gómez-Perales, J.E., and Nieuwenhuijsen, M.J. 2003. Use of dispersion modelling to assess road-user exposure to PM2.5 and its source apportionment. *Atmospheric Environment* 37, no. 20:2773–2782.

Cooper, J.A., and Watson Jr., J.G. 1980. Receptor oriented methods of air particulate source apportionment. *Journal of the Air Pollution Control Association* 30, no. 10:1116–1125.

Evangelista, H., Maldonado, J., Godoi, R.H.M., Pereira, E.B., Koch, D., Tanizaki-Fonseca, K., Van Grieken, R. et al. 2007. Sources and transport of urban and biomass burning aerosol black carbon at the South–West Atlantic Coast. *Journal of Atmospheric Chemistry* 56:225–238.

Feng, Y.C., Bai, Z.P., and Zhu, T. 2002. The principle and application of improved-source-apportionment technique of atmospheric particulate matter. *Environmental Sciences* 23:106–108.

Fenger, J. 1999. Urban air quality, *Atmospheric Environment* 33:4877–4900.

Fine, P.M., Glen R. Cass, and Simoneit, B.R.T. 2002. Organic compounds in biomass smoke from residential wood combustion: Emissions characterization at a continental scale. *Journal of Geophysical Research* 107, no. D21. doi: 10.1029/2001JD000661.

Fraser, M.P., Buzcu, B., Yue, Z.W., Mcgaughey, G.R., Desai, N.R., Allen, D.T., Seila, R.L., Lonneman, W.L., and Harley, R.A. 2003. Separation of fine particulate matter emitted from gasoline and diesel vehicles using chemical mass balancing techniques. *Environmental Science & Technology* 37:3904–3909.

Fushimi, A., Kawashima, H., and Kajihara, H. 2005. Source apportionment based on an atmospheric dispersion model and multiple linear regression analysis. *Atmospheric Environment* 39:1323–1334.

Gianini, M.F.D., Piot, C., Herich, H., Besombes, J.L., Jaffrezo, J.L., and Hueglin, C. 2013. Source apportionment of PM_{10}, organic carbon and elemental carbon at Swiss sites: An intercomparison of different approaches. *Science of the Total Environment* 454–455:99–108.

Gu, J. Kreis, J.S., Pitz, M., Diemer, J., Reller, A., Zimmermann, R., Soentgen, J., Peters, A., and Cyrys, J. 2013. Spatial and temporal variability of PM_{10} sources in Augsburg, Germany. *Atmospheric Environment* 71:131–139.

Gu, J., Pitz, M., Kreis, J.S., Diemer, J., Reller, A., Zimmermann, R., Soentgen, J. et al. 2011. Source apportionment of ambient particles: Comparison of positive matrix factorization analysis applied to particle size distribution and chemical composition data. *Atmospheric Environment* 45, no. 10:1849–1857.

Gugamsetty, B., Wei, H., Liu, C.N., Awasthi, A., Hsu, S.C., Tsai, C.J., Roam, G.D., Wu, Y.C., and Chen, C.F. 2012. Source characterization and apportionment of PM_{10}, $PM_{2.5}$ and $PM_{0.1}$ by using positive matrix factorization. *Aerosol and Air Quality Research* 12:476–491.

Guo, H., Ding, A.J., So, K.L., Ayoko, G., Li, Y.S., and Hung, W.T. 2009. Receptor modeling of source apportionment of Hong Kong aerosols and the implication of urban and regional contribution. *Atmospheric Environment* 43, no. 6:1159–1169.

Han, B., Bi, X., Xue, Y., Wu, J., Zhu, T., Zhang, B., Ding, J., and Du, Y. 2011. Source apportionment of ambient PM_{10} in urban areas of Wuxi, China. *Frontiers of Environmental Science & Engineering* 5, no. 4:552–563.

Harrison, R.M., Smith, D.I.T., and Luhana, L. 1996. Source apportionment of atmospheric polycyclic aromatic hydrocarbons collected from an urban location in Birmingham, U.K. *Environmental Science and Technology* 30, no. 3:825–832.

Harrison, R.M., Smith, D.J.T., Pio, C.A., and Castro, L.M. 1997. Comparative receptor modelling study of airborne particulate pollutants in Birmingham (United Kingdom), Coimbra (Portugal) and Lahore (Pakistan). *Atmospheric Environment* 31:3309–3321.

Harrison, R.M., Tilling, R., Callen Romero, M.S., Harrad, S., and Jarvis, K. 2003. A study of trace metals and polycyclic aromatic hydrocarbons in the roadside environment. *Atmospheric Environment* 37, no. 17:2391–2402.

Heal, M.R., Kumar, P., and Harrison, R.M. 2012. Particles, air quality, policy and health. *Chemical Society Reviews* 41, no. 19:6606–6630.

Held, T., Yinga, Q., Kleeman, M.J., Schauer, J.J., and Fraser, M.P. 2005. A comparison of the UCD/CIT air quality model and the CMB source-receptor model for primary airborne particulate matter. *Atmospheric Environment* 39:2281–2297.

Hellen, H., Hakola, H., and Laurila, T. 2003. Determination of source contributions of NMHCs in Helsinki (60°N, 25°E) using chemical mass balance and the Unmix multivariate receptor models. *Atmospheric Environment* 37:1413–1424.

Henry, R.C., Lewis, C.W., Hopke, P.K., and Williamson, H.J. 1984. Review of receptor model fundamentals. *Atmospheric Environment* 18:1507–1515.

Heo, J.B., Dulger, M., Olson, M.R., McGinnis, J.E., Shelton, B.R., Matsunaga, A., Sioutas, C., and Schauer, J.J. 2013. Source apportionments of $PM_{2.5}$ organic carbon using molecular marker positive matrix factorization and comparison of results from different receptor models. *Atmospheric Environment* 73:51–61.

Heo, J.B., Hopke, P.K., and Yi, S.M. 2009. Source apportionment of $PM_{2.5}$ in Seoul, Korea. *Atmospheric Chemistry and Physics* 9:4957–4971.

Hopke, P.K. 1991. An introduction to receptor modelling. *Chemometrics and Intelligent Laboratory Systems* 10:21–43.

Hopke, P.K. 2008. The use of source apportionment for air quality management and health assessments. *Journal of Toxicology and Environmental Health, Part A: Current Issues* 71, no. 9–10:555–563.

Ke, L., Liu, W., Wang, Y., Russell, A.G., Edgerton, E.S., and Zheng, M. 2008. Comparison of $PM_{2.5}$ source apportionment using positive matrix factorization and molecular marker-based chemical mass balance. *Science of the Total Environment* 394, no. 2–3:290–302.

Keuken, M.P., Moerman, M., Voogt, M., Blom, M., Weijers, E.P., Rockmann, T., and Dusek, U. 2013. Source contributions to $PM_{2.5}$ and PM_{10} at an urban background and a street location. *Atmospheric Environment* 71:26–35.

Kim, B.M., and Henry, R.C. 2000. Application of SAFER model to the Los Angeles PM_{10} data. *Atmospheric Environment* 34:1747–1759.

Kim, E., Hopke, P.K., and Edgerton, E.S. 2003. Source identification of Atlanta aerosol by positive matrix factorization. *Journal of the Air and Waste Management Association* 53, no. 6:731–739.

Kim, E., Hopke, P.K., Larson, T.V., Maykut, N.N., and Lewtas, J. 2004. Factor analysis of Seattle fine particles. *Aerosol Science and Technology* 38, no. 7:724–738.

Kong, S., Han, B., Bai, Z., Chen, L., Shi, J., and Z. Xu. 2010. Receptor modelling of $PM_{2.5}$, PM_{10} and TSP in different seasons and long-range transport analysis at a coastal site of Tianjin, China. *Science of the Total Environment* 408, no. 20:4681–4694.

Kothai, P., Saradhi, I.V., Prathibha, P., Pandit G.G., and Puranik, V.D. 2008. Source apportionment of coarse and fine particulate matter at Navi Mumbai, India. *Aerosol and Air Quality Research* 8:423–436.

Kumar, A.V., Patil, R.S., and Nambi, K.S.V. 2001. Source apportionment of suspended particulate matter at two traffic junctions in Mumbai, India. *Atmospheric Environment* 35:4245–4251.

Lai, C.H., Chen, K.S., Ho, Y.T., Peng, Y.P., and Chou, Y.M. 2005. Receptor modeling of source contributions to atmospheric hydrocarbons in urban Kaohsiung, Taiwan. *Atmospheric Environment* 39:4543–4559.

Lall, R., Ito, K., and Thurston, G.D. 2011. Distributed lag analyses of daily hospital admissions and source-apportioned fine particle air pollution. *Environmental Health Perspectives* 119:455–460.

Larsen, R.K., III, and Baker, D.J. 2003. Source apportionment of polycyclic aromatic hydrocarbons in the urban atmosphere: A comparison of three methods. *Environmental Science & Technology* 37:1873–1881.

Laupsa, H., Denby, B., Larssen, S., and Schaug, J. 2009. Source apportionment of particulate matter ($PM_{2.5}$) in an urban area using dispersion, receptor and inverse modelling. *Atmospheric Environment* 43, no. 31:4733–4744.

Lee, S., and Russell, A.G. 2007. Estimating uncertainties and uncertainty contributors of CMB $PM_{2.5}$ source apportionment results. *Atmospheric Environment* 41:9616–9624.

Lee, S., Liu, W., Wang, Y., Russell, A.G., and Edgerton, E.S. 2008. Source apportionment of $PM_{2.5}$: Comparing PMF and CMB results for four ambient monitoring sites in the southeastern United States. *Atmospheric Environment* 42:4126–4137.

Lee, Y.L., and Sequeira, R. 2001. Visibility degradation across Hong Kong: Its components and their relative contributions. *Atmospheric Environment* 35, no. 34:5861–5872.

Lewis, C.W., Norris, G.A., Conner ,T.L., and Henry, R.C. 2003. Source apportionment of Phoenix PM$_{2.5}$ aerosol with the Unmix receptor model. *Journal of Air and Waste Management Association* 53, no. 3:325–338.

Mansha, M., Ghauri, B., Rahman, S., and Amman, A. 2012. Characterization and source apportionment of ambient air particulate matter (PM$_{2.5}$) in Karachi. *Science of the Total Environment* 425:176–183.

McDonald, J.D., B. Zielinska, E.M. Fujita, J.C. Sagebiel, J.C. Chow, and Watson, J.G. 2000. Fine particle and gaseous emission rates from residential wood combustion. *Environmental Science and Technology* 34:2080–2091.

Pekney, N.J., Davidson, C.I., Robinson, A., Zhou, L., Hopke, P., Eatough, D., and Rogge, W.F. 2006. Major source categories for PM2.5 in Pittsburgh using PMF and UNMIX. *Aerosol Science and Technology* 40, no. 10:910–924.

Qin, Y., Kim, E., and Hopke, P.K. 2006. The concentrations and sources of PM$_{2.5}$ in metropolitan New York City. *Atmospheric Environment* 40:S312–S332.

Ramadan, Z., Eickhout, B., Song, X.H., Buydens, L.M.C., and Hopke, P.K. 2003. Comparison of positive matrix factorization and multilinear engine for the source apportionment of particulate pollutants. *Chemometrics and Intelligent Laboratory Systems* 66, no. 1:15–28.

Robinson, A.L., Subramanian, R., Donahue, N.M., Bricker, A.B., and Rogge, W.F. 2006. Source apportionment of molecular markers and organic aerosol—1. Polycyclic aromatic hydrocarbons and methodology for data visualization. *Environmental Science and Technology* 40, no. 24:7803–7810.

Schauer, J.J., and Cass, G.R. 2000. Source apportionment of wintertime gas-phase and particle phase air pollutants using organic compounds as tracers. *Environmental Science and Technology* 34:1821–1832.

Schauer, J.J., Lough, G.C., Shafer, M.M., Christensen, W.F., Arndt, M.F., DeMinter, J.T., and Park, J.S. 2006. Characterization of metals emitted from motor vehicles. Health Effects Institute.

Schauer, J.J., Rogge, W.F., Hildemann, L.M., Mazurek, M.A., Cass, G.R., and Simoneit, B.R.T. 1996. Source apportionment of airborne particulate matter using organic compounds as tracers. *Atmospheric Environment* 30, no. 22:3837–3855.

Seagrave, J.C., McDonald, J.D., Bedrick, E., Edgerton, E.S., Gigliotti, A.P., Jansen, J.J., Ke, L. et al. 2006. Lung toxicity of ambient particulate matter from southeastern U.S. sites with different contributing sources: Relationships between composition and effects. *Environmental Health Perspectives* 114, no. 9:1387–1393.

Simoneit, B.R.T., Schauer, J.J., Nolte, C.G., Oros, D.R., Elias, V.O., Fraser, M.P., Rogge, W.F., and Cass, G.R. 1999. Levoglucosan, a tracer for cellulose in biomass burning and atmospheric particles. *Atmospheric Environment* 33, no. 2:173–182.

Song, X.H., Polissar, A.V., and Hopke, P.K. 2001. Sources of fine particle composition in the northeastern US. *Atmospheric Environment* 35:5277–5286.

Song, Y., Zhang, Y., Xie, S., Zeng, L., Zhengc, M., Salmon, L.G., Shao, M., and S. Slanina. 2006. Source apportionment of PM$_{2.5}$ in Beijing by positive matrix factorization. *Atmospheric Environment* 40:1526–1537.

Srivastava, A., and Jain, V.K. 2007. Seasonal trends in coarse and fine particle sources in Delhi by the chemical mass balance receptor model. *Journal of Hazardous Materials* 144:283–291.

Stanek, L.W., Brown, J.S., Stanek, J., Gift, J., and Costa, D.L. 2011. Air pollution toxicology—A brief review of the role of the science in shaping the current understanding of air pollution health risks. *Toxicological Sciences* 120:S8–S27.

Stedman, J.R., Kenta, A.J., Gricea, S., Busha, T.J., and Derwent, R.G. 2007. A consistent method for modelling PM_{10} and $PM_{2.5}$ concentrations across the United Kingdom in 2004 for air quality assessment. *Atmospheric Environment* 41:161–172.

Stone, E., Schauer, J., Quraishi, T.A., and Mahmood, A. 2010. Chemical characterization and source apportionment of fine and coarse particulate matter in Lahore, Pakistan. *Atmospheric Environment* 44:1062–1070.

Tauler, R., Viana, M., Querol, X., Alastuey, A., Flight, R.M., Wentzell, P.D., and Hopke, P.K. 2009. Comparison of the results obtained by four receptor modelling methods in aerosol source apportionment studies. *Atmospheric Environment* 43:3989–3997.

Thurston, G.D., and Spengler, J.D. 1985. A quantitative assessment of source contributions to inhalable particulate matter in metropolitan Boston, Massachusetts. *Atmospheric Environment* 19:9–25.

Thurston, G.D., and Lioy, P.J. 1987. Receptor modelling and aerosol transport. *Atmospheric Environment* 21:687–698.

Tsai, Y.I., and Chen, C.L. 2006. Atmospheric aerosol composition and source apportionments to aerosol in southern Taiwan. *Atmospheric Environment* 40:4751–4763.

Viana, M., Pandolfi, M., Minguillon, M.C., Querol, X., Alastuey, A., Monfort, E., and Celades, I. 2008. Inter-comparison of receptor models for PM source apportionment: Case study in an industrial area. *Atmospheric Environment* 42:3820–3832.

Viana, M., Querol, X., Gotschi, T., Alastuey, A., Sunyer, J., Forsberg, B., Heinrich, J. et al. 2007. Source apportionment of ambient $PM_{2.5}$ at five Spanish centres of the European community respiratory health survey (ECRHS II). *Atmospheric Environment* 41:1395–1406.

Wang, Y., Hopke, P.K., Xia, X., Rattigan, O.V., Chalupa, D.C., and Utell, M.J. 2012. Source apportionment of airborne particulate matter using inorganic and organic species as tracers. *Atmospheric Environment* 55:525–532.

Watson, J.G., Chow, J.C., and Houck, J.E. 2001. $PM_{2.5}$ chemical source profile for vehicle exhaust, vegetative burning, geological material and coal burning in Northwestern Colorado during 1995. *Chemosphere* 43:1141–1151.

Wichmann, F.A., Muller, A., Busi, L.É., Cianni, N., Massolo, L., Schlink, U., Porta, A., and Sly, P.D. 2009. Increased asthma and respiratory symptoms in children exposed to petrochemical pollution. *Journal of Allergy and Clinical Immunology* 123, no. 3:632–638.

Yatkin, S., and Bayram, A. 2008. Source apportionment of PM_{10} and $PM_{2.5}$ using positive matrix factorization and chemical mass balance in Izmir, Turkey. *Science of the Total Environment* 390:109–123.

Index

Page numbers followed by f and t indicate figures and tables, respectively.

Milton Keynes UK
Ingram Content Group UK Ltd.
UKHW021900071024
449327UK00021B/1596